Electrical Machinery, Transformers, and Control

Electrical Machinery, Transformers, and Control

HAROLD W. GINGRICH

Westchester Community College

PRENTICE-HALL, INC. ENGLEWOOD CLIFFS, NEW JERSEY 07632

Library of Congress Cataloging in Publication Data

GINGRICH, HAROLD W
Electrical machinery, transformers, and control.

Includes index.
1. Electric machinery. 2. Electric transformers.
3. Electric controllers. 4. Power electronics.
I. Title.
TK2000.G56 621.31'04'2 78-16498
ISBN 0-13-247320-8

Printed in the United States of America

10 9 8 7 6 5 4 3 2 1

PRENTICE-HALL INTERNATIONAL, INC., *London*
PRENTICE-HALL OF AUSTRALIA PTY. LIMITED, *Sydney*
PRENTICE-HALL OF CANADA, LTD., *Toronto*
PRENTICE-HALL OF INDIA PRIVATE LIMITED, *New Delhi*
PRENTICE-HALL OF JAPAN, INC., *Tokyo*
PRENTICE-HALL OF SOUTHEAST ASIA PTE. LTD., *Singapore*
WHITEHALL BOOKS LIMITED, *Wellington, New Zealand*

Contents

PREFACE xi

1

BASIC CONCEPTS 1

1-1 Introduction *1*

Section 1-1. Units
1-2 Systems of Units *2*

Section 1-2. Magnetism
1-3 Magnetism and the Magnetic Field *4*
1-4 Electromagnetism *6*
1-5 Magnetic Categorization of Materials *8*
1-6 Magnetic Circuits *10*
1-7 Induced Electromotive Force *16*
1-8 Hysteresis *18*
1-9 Eddy Currents *20*
1-10 Electromagnetic Force and Torque *21*
1-11 Illustrated Application of Electromagnetic Phenomena *24*

Section 1-3. Analytical Methods
1-12 Electromagnetic Analysis *25*

1-13 Mechanical Analysis 26
1-14 Energy 27

Section 1-4. Standards
1-15 Standards 28

2 **DC MACHINERY 34**

Section 2-1. DC Generator
2-1 Generator Principles 34
2-2 Forms of the DC Generator 41
2-3 Effect of Speed on E_g 49
2-4 Armature Reaction 50
2-5 Line-Voltage Characteristics 55
2-6 Voltage Regulation 57
2-7 Torques in the Generator 60
2-8 Paralleling Generators 61

Section 2-2. Armature Windings and Commutation
2-9 Armature Winding 67
2-10 Commutation 76

Section 2-3. DC Motor
2-11 Motor Fundamentals 78
2-12 Speed–Torque Characteristics 81
2-13 Starting DC Motors 85
2-14 Speed Control 89
2-15 Electrical Braking 92
2-16 Starters and Speed Controllers 94

Section 2-4. Special DC Generators
2-17 Special DC Generators 103

Section 2-5. Testing
2-18 Electrical Measurements 108
2-19 Tests Common to the Motor and Generator 109
2-20 Efficiency 110
2-21 Generator Voltage Regulation 111
2-22 Motor Speed Regulation 111

Section 2-6. Ratings
2-23 Ratings 111

3 **INTRODUCTION TO FEEDBACK REGULATION AND CONTROL 120**

3-1 Block Diagrams 120
3-2 Feedback Regulation 123
3-3 Position Control 128

4 TRANSFORMERS 135

4-1 Introduction *135*
4-2 The Ideal Transformer *138*
4-3 The Realistic Transformer *139*
4-4 Voltage Regulation and Efficiency *143*
4-5 Transformer Tests *146*
4-6 Tapped Transformers *148*
4-7 Transformer Polarity *149*
4-8 Autotransformer *150*
4-9 Polyphase Transformation *154*
4-10 Paralleling Transformers *161*
4-11 Construction and Ratings *162*
4-12 Power Transmission and Distribution *164*
4-13 Transformer-Related Devices *164*

5 POLYPHASE SYNCHRONOUS MACHINES 176

Section 5-1. Synchronous Generator (Alternator)
5-1 Generated Voltage *176*
5-2 Advantages of Polyphase Systems *182*
5-3 Revolving Flux *184*
5-4 Equivalent Circuit *186*
5-5 Excitation Systems *188*
5-6 Voltage Regulation *190*
5-7 Power and Torque *194*
5-8 Efficiency *196*

Section 5-2. Synchronous Motor
5-9 Operating Principles *199*
5-10 Motor Circuit Analysis *200*
5-11 Starting a Synchronous Motor *203*
5-12 Torque *206*
5-13 Motor Power Factor *207*
5-14 Hunting *208*

Section 5-3. Testing
5-15 Resistance Measurement *209*
5-16 Open-Circuit Test *209*
5-17 Short-Circuit Test *211*

Section 5-4. Ratings and Application
5-18 Synchronous Machine Ratings *212*
5-19 Parallel Operation of Generators *214*
5-20 Power Factor Correction *223*

6 POLYPHASE INDUCTION MOTOR 230

6-1 Operational Principles *230*
6-2 Equivalent Circuit *232*
6-3 Performance Analysis *237*
6-4 Effect of Rotor Resistance on Performance *241*
6-5 Starting *244*
6-6 Speed Control *248*
6-7 Auxiliary Speed-Control Methods *252*
6-8 Braking and Generation *253*
6-9 Control Circuits *255*
6-10 Testing *260*
6-11 Ratings and Application *261*

7 SINGLE-PHASE MOTORS AND SERVO COMPONENTS 265

Section 7-1. Single-Phase Motors
7-1 Construction and Operating Principles—Induction Motors *265*
7-2 Phase-Splitting Induction Motors *268*
7-3 Shaded-Pole Induction Motor *274*
7-4 Reluctance Motor (Synchronous) *276*
7-5 Hysteresis Motor (Synchronous) *278*
7-6 Repulsion Motors *279*
7-7 Universal Motor *280*

Section 7-2. Servo Components
7-8 Synchros *282*
7-9 Resolver *289*
7-10 AC Control Motors *292*
7-11 AC Tachometer *293*

8 INTRODUCTION TO SOLID-STATE POWER AND CONTROL 298

8-1 Switching Function *298*
8-2 Circuit Analysis *299*
8-3 Diode *306*
8-4 Thyristor *311*
8-5 Power Transistor *317*
8-6 Motor Applications *319*

APPENDICES

A POLYPHASE CIRCUITS 323

A-1 Introduction *323*
A-2 Three-Wire Single-Phase and Polyphase Systems *323*

A-3 CBA System *325*
A-4 Three-Phase Currents and Power *327*
A-5 Measurement of Three-Phase Power *331*
A-6 Unbalanced Loading *332*
A-7 Measurement of Power in Unbalanced Loads *335*

B **CONVERSION FACTORS 337**

C **SI PREFIXES 338**

D **MATHEMATICAL SUMMARY 339**

E **ANSWERS TO SELECTED PROBLEMS 343**

INDEX 347

Preface

This text is designed to support a two-semester course within an electrical technology curriculum on the fundamentals of machinery and power transformers and basic concepts of control systems and power electronic circuits.

The level of the text assumes that the reader is schooled in basic electric circuit theory and in algebra and trigonometry. Since introductory circuits courses often do not include a study of three-phase power, this topic is discussed in an appendix. Further, circuits courses treat magnetism primarily from the viewpoint of inductance; to provide an adequate basis for understanding electromagnetic devices, magnetism and magnetic circuits are covered in the introductory chapter.

Despite temptation, there is no calculus in this text. Consequently, mathematical analysis generally is limited to steady-state operating conditions, but transients are discussed where they are a significant consideration in machine or system performance. The mathematical limitation is no millstone; ANSI and IEEE Test Codes applicable to equipment discussed in the text contain remarkably little calculus. A mathematical summary is included as an appendix.

The unified approach to machinery, common in engineering texts, is inappropriate at the technology level. It is highly unlikely that the idea of an induction *generator* ever entered Tesla's mind while he was developing the induction *motor*. Herein, the focus is on unique machines, but similarities are not ignored and generally are used in the transition from one to another. The stress is on deducing performance by considering the physical interactions of electricity and magnetism rather than by abstract interpretation of mathematical equations. One objective is to develop an ability to rationalize the performance of a device by analysis of the electrical and mechanical structure.

Systems are discussed to permit an appreciation of the role and performance of component equipment. The depth of the treatment necessarily is limited—whole textbooks are devoted to control systems, for example—but the presentation should adequately meet the needs of most students.

The rapid development and future potential of solid-state power electronics have made this topic a *must* in a contemporary discussion of power systems. The physics of semiconductors and the application of devices in low-power circuits are amply covered in electronics texts. In a typical electrical/electronics curriculum, the novel aspect of power electronics is the application of semiconductor devices in controlled converters—it is this that is discussed in the text. Logic circuits are important elements both in power electronics and as a replacement for many electromagnetic components in machine controls; however, because of practical limitations and a recognition that courses in logic are not uncommon in technology curriculums, this subject is generally excluded. Finally, to permit evaluation of output characteristics, particularly with delayed firing in controlled semiconductors, some discussion of transients in sinusoidal circuits is included.

Although intended for a two-semester course, the organization of the text permits selective use in courses of lesser duration, tailored to the desired local emphasis.

A conscious effort has been made to adhere to ANSI, IEEE, and NEMA standards in terminology, definitions, symbols, testing, and so on. Similarly, problems largely reflect standardized equipment and normal machine parameters and performance.

The contributions of my wife, Helen, to the production of this text are enormous. Without her encouragement, tolerance, critical comment, and Herculean performance in producing the final manuscript, the book never would have progressed beyond a tentative project in the mind of a confirmed procrastinator.

Also in order are thanks to my colleagues at Westchester Community College for making it possible for me to concentrate upon this task.

HAROLD W. GINGRICH

1

Basic Concepts

1-1. INTRODUCTION This book is a study of practical devices and systems all of which are functionally dependent upon interactions between electricity and magnetism. Properly exploited, electrical and magnetic phenomena allow transformation of electrical power between differing voltage–current combinations, conversion between electrical and mechanical energy, and regulation and control of electrical and mechanical system characteristics.

The explanation of the functioning and analysis of the devices in the following chapters relies upon a basic understanding of the fundamental principles of electromagnetic induction, force, and torque, and upon analytical methods appropriate to the development. This chapter highlights principles that are recurrent throughout the text.

The text assumes that the reader has a sound preparation in algebra, trigonometry, and electrical circuit theory. Regarding the latter, the modern interest in electronics has influenced circuit texts and courses to the detriment of the study of polyphase systems; since this topic is fundamental to a study of commercial and industrial power applications, the essentials are presented in Appendix A.

SECTION 1-1. UNITS

1-2. SYSTEMS OF UNITS The movement toward metrication in the United States must be recognized, but **U.S. customary units** (US) will remain for many years and cannot be ignored. Consequently, both systems are employed in this book, with the attendant requirement for interconversion.

In the fields of engineering and science, metrication has broader connotations than direct substitution of meters for yards or kilograms for pounds. An important aspect is standardization of physical units through adoption of the **International System of Units** (SI). SI defines seven base units: meter (m), kilogram (kg), second (s), ampere (A), kelvin (K), mole (mol), and candela (cd). All but the latter two are pertinent to the study of machinery.* Two supplementary units, the radian (rad) and steradian (sr) for plane and solid angles, respectively, also are specified.

All other units are derived and may or may not have special names. For example, the unit of angular velocity is radians per second (rad/s), but the force unit, the product of mass (kg) and acceleration (m/s^2) resulting in $kg{\cdot}m/s^2$, is termed newton (N). Derived units may be defined by combinations of base, supplementary, and previously derived units but always are traceable to base and supplementary units; for example, work or energy is the product of force and distance (N·m) and has the special designation joule (J), which, in base units, is $kg{\cdot}m^2/s^2$.

The base units of fundamental electrical properties may be derived from formula relationships. In the following development, the distinction between the terms *quantity* and *unit* must be recognized. A unit is a standard, finite amount of a property in terms of which any amount of the property may be measured.

Quantity appears in equations as an indefinite amount of a property; the equation, when mathematically satisfied, interrelates specific amounts (expressed in units) of the elements appearing in it. The symbols for unit and quantity of a specific property may or may not be identical: for example, V represents both the quantity (voltage) and the unit (volts), but the quantity for power is P while the unit is W (watts). W also symbolizes the quantity, work (or energy), whose unit is joule (J); duplications of this type are common, but no confusion arises if one is alert to the distinction and notes the sense in which the symbol is used.

CURRENT. SI unit: ampere (A)

This is a base unit.

VOLTAGE. SI unit: volt (V)

$$\text{(Quantity)} \quad V = \frac{W}{Q} = \frac{W}{It}$$

*The kelvin is a temperature unit. In lieu of it, the degree Celsius (C), formerly termed centigrade, is permitted in SI. 1°C = 1 K.

$$\text{(Unit)} \qquad \mathrm{V} = \frac{\mathrm{J}}{\mathrm{A \cdot s}} = \frac{\mathrm{kg \cdot m^2/s^2}}{\mathrm{A \cdot s}} = \frac{\mathrm{kg \cdot m^2}}{\mathrm{A \cdot s^3}}$$

POWER. SI unit: watt (W)

$$\text{(Quantity)} \qquad P = VI$$

$$\text{(Unit)} \qquad \mathrm{W} = \frac{\mathrm{kg \cdot m^2}}{\mathrm{A \cdot s^3}} \cdot \mathrm{A} = \frac{\mathrm{kg \cdot m^2}}{\mathrm{s^3}}$$

RESISTANCE. SI unit: ohm (Ω)

$$\text{(Quantity)} \qquad R = \frac{V}{I}$$

$$\text{(Unit)} \qquad \Omega = \frac{\mathrm{kg \cdot m^2/A \cdot s^3}}{\mathrm{A}} = \frac{\mathrm{kg \cdot m^2}}{\mathrm{A^2 \cdot s^3}}$$

Initially, some US units will seem strange when converted to SI. The kilowatt as an electrical power unit is commonplace, but a 74.6-kilowatt (not 100-horsepower) motor is not. US units will not disappear overnight but gradually will yield to SI. In technical professions, transition will be more rapid than in the public sector—pending common acceptance of SI, the technician has a continuing requirement for dual-system competency.

Fortunately, interchange between SI and US units usually requires only the application of a conversion factor; thus, 50 miles per hour is readily converted to ($50 \times 1.6 =$) 80 kilometers per hour. SI and US units and conversion factors appearing in this book are listed in Appendix B. The tabulation of US units is limited to those most commonly used in this area of study.

One problem in the US system is multiplicity of units for a single characteristic. For example, volume may be expressed in cubic yards, cubic feet, cubic inches, quarts, gallons, bushels, etc. There is no uniform numerical relationship between these units: e.g., 1 $\mathrm{yd^3} = 27\ \mathrm{ft^3}$, 1 $\mathrm{ft^3} = 1728\ \mathrm{in^3}$. In SI, volume is the cubic meter. But a cubic meter is relatively large—one cubic inch is only $1.639 \times 10^{-5}\ \mathrm{m^3}$. Commonly, working quantities are significantly larger or smaller than the SI unit. Large or small quantities are specified by standardized multiples representing 10^{3n}, where n is a positive or negative integer—e.g., 10^{+6}, 10^{-9}. Each multiple is assigned a specific name and symbol that may prefix any SI unit. For example, the prefix kilo (10^3), symbol k, appears in kilometers (km = 1000 m), kilovolts (kV = 1000 V), kilowatts (kW), and so on. The SI prefixes are tabulated in Appendix C.

A special caution applies to computations with prefixed units. Formulas are expressed in fundamental units. Thus, Ohm's law, $E = IR$, relates the fundamental unit volt to the product of the fundamental units ampere and ohm. If the current quantity were in milliamperes (mA $= 10^{-3}$ A) and R in ohms, the product would be not volts but millivolts (mV)—that is, $E = (10^{-3}\ \mathrm{A})(\Omega) = 10^{-3}\ (\mathrm{A \cdot \Omega}) = 10^{-3}$ V. Similarly, the product of milliamperes and megohms is $(10^{-3}\ \mathrm{A})(10^6 \Omega) = 10^{+3}\ (\mathrm{A \cdot \Omega})$ = kilovolts (kV).

Yet another caution is in order when prefixed units are exponential. The exponent applies not only to the unit but also to the prefix. For example, $(0.001\ \text{m})^2 = (10^{-3}\ \text{m})^2 = 10^{-6}\ \text{m}^2$. Prefixed, the form is mm^2—that is, $(10^{-3}\ \text{m})^2$—*not* μm^2, which would be $(10^{-6}\ \text{m})^2$.

In converting between systems of units, rounding of the arithmetic answer is in order. How do you determine the proper rounding? There is no simple rule for universal application, but common sense and experience usually suggest the answer.

The converted unit can be no more accurate than the original quantity. Assume that an object is measured with a ruler graduated in sixteenths of an inch, and the length is $7\frac{1}{4}$ inches to the nearest graduation. Actual length may be $\frac{1}{32}$ inch greater or lesser ($\pm\frac{1}{32}$ in). One inch equals exactly 25.4 millimeters. Converting, the length is $(7\frac{1}{4} \times 25.4)$, or 184.15 mm. However, $\frac{1}{32}$ in $= (\frac{1}{32})(25.4) = 0.79375$ mm; based on the original measurement accuracy, the length is 184.15 ± 0.79375 mm. Under these circumstances, a converted length implying an accuracy greater than 184 mm is unjustifiable. However, since the actual length lies somewhere between 183 and 185 mm, rounding the answer to 180 mm sacrifices accuracy; whether or not this sacrifice is acceptable depends upon the purpose of the measurement. Four millimeters are approximately $\frac{5}{32}$ of an inch; if the purpose of the measurement were to cut a pane of glass to fit inside a molding with a $\frac{1}{4}$-in lip, 180 mm would be satisfactory, but if it were for the purpose of cutting one replacement leg for a four-legged stool, a 4-mm discrepancy would cause the stool to wobble.

In making a conversion, the converted unit—*not* the original quantity and the conversion factor—should be rounded. For the example, if the length were rounded to 7 in and the conversion factor to 25, the converted length would be 175 mm, a difference of 9 mm, or nearly $\frac{3}{8}$ in.

In analytical computations within this text, some unpredictable factors normally are involved. Further, relationships frequently are based upon simplified or idealized assumptions. There is no justification for computed answers with the 8 to 10 significant digits that commonly are provided by electronic pocket calculators. In general, slide-rule precision is appropriate. Therefore, unless there is reason for deviating, the examples and problem answers retain three significant digits or, if the first digit is one, four digits.

SECTION 1-2. MAGNETISM

1-3. MAGNETISM AND THE MAGNETIC FIELD

Michael Faraday (1791–1867) originated a concept of magnetic lines of force (also termed lines of induction) as a means of analyzing and explaining magnetic phenomena. In Fig. 1-1(a), visible lines, each representing thousands of force lines, portray typical paths around a permanent magnet. Some fundamental characteristics are apparent on the sketch.

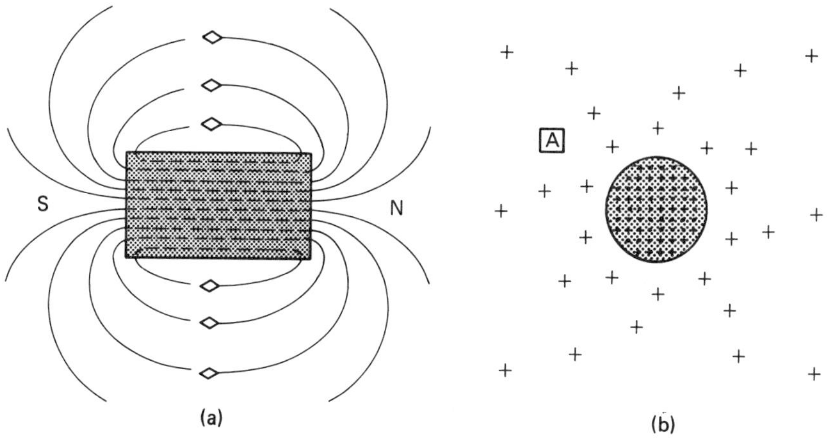

Figure 1-1 Force Lines Surrounding a Magnet

Magnetic Lines of Force Form Closed Loops. The lines exist within the body of the magnet and continue externally in order to close the loop.

Lines Are Directional. All lines emerge from one general area of the magnet, the **north** pole, and reenter at a different area, the **south** pole. Internally, the line direction is south to north; externally, north to south.

Lines Do Not Intersect. No two lines can occupy the same space or pass through a common point.

Although not graphically apparent, the following additional characteristics help explain those which can be represented.

Lines Are Mutually Repellent. Just as two like electrical charges exert mutually repellent forces, two proximate lines of induction experience forces which tend to separate them. The separating force is countered by yet another characteristic, as follows.

Lines Are Tensile. Blowing into a rubber balloon raises the internal pressure above atmospheric and forces the shape outward; this stretches the rubber, causing an increase in its elastic tensile forces. The size of the inflated balloon results from a balance between the forces of tension and pressure. Similarly, the path of a force line is fixed by a balance between the repellent forces of other lines and an inherent tensile force that causes the line to contract to the shortest permissible path.

Lines Seek Low-Reluctance Paths. Free space and most materials are difficult mediums for occupancy by force lines. However, one category of material, *ferromagnetic*, permits easy passage. Like resistance, the opposition to electrical current flow, *reluctance* is opposition to the creation of lines of force. Ferromagnetic materials have low reluctances. Thus, despite the contracting tendency, many force lines will deviate from a short, high-reluctance path and follow a longer, lower-reluctance route through a ferromagnetic body. Once the force lines are in the body, the tensile forces of the lines act to draw the body to the magnet; if the tensile forces are greater than opposing forces, the body moves. The result (Fig. 1-2) is the commonly known attraction of a magnet for certain materials.

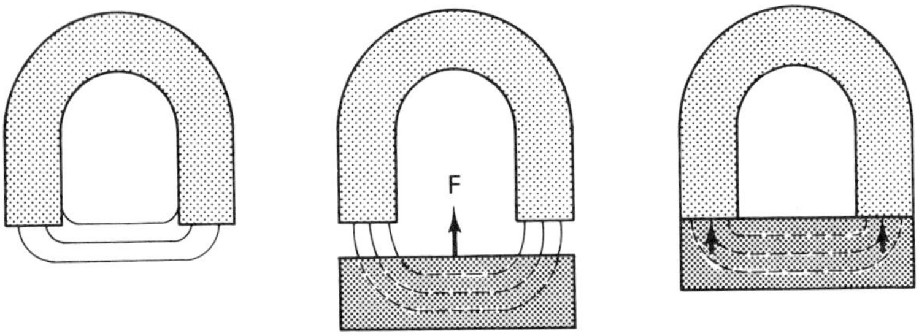

Figure 1-2 Magnetic Attraction of Iron

The general term for lines of force is *magnetic flux*, symbol Φ. The US unit is *line* or, interchangeably, *maxwell*. The SI unit is *weber* (Wb): 1 Wb $=10^8$ lines or maxwells.

Figure 1-1(b) shows the flux through an area perpendicular to the axis of the magnet, viewed along the traverse direction of the external lines. The spacing of the crosses indicates variable density; near the body, the flux per unit cross section is large in comparison to more remote areas. Area A represents a very small area (approaching a point); let Φ represent the flux through this area. The *flux density* (B) at this site is given by the equation $\mathbf{B} = \mathbf{\Phi}/\mathbf{A}$. A common US unit of flux density is *lines/in²* (or maxwells/in²); the SI unit is the *tesla* (T)—one tesla equals one weber per square meter (Wb/m²).

Laboratory-created densities may approach 10 T. The magnetic field of the earth is on the order of 50 μT, and practical densities in machines are on the order of 1 to 2 T.

1-4. ELECTROMAGNETISM

In 1820, Hans Christian Oersted (1777–1851) proved a direct connection between electric current and magnetic effects, thereby introducing the common study of electricity and magnetism. The evolutionary development in large measure resulted from the work of Faraday and James Clark Maxwell (1831–1879).

Oersted established the presence and direction of magnetic lines of force around any current-carrying conductor. In Fig. 1-3(a) the *right* hand grasps the conductor, with the thumb pointing in the conventional direction of current flow. The natural curl of the fingers around the conductor is in the direction of the force lines. These lines occur throughout the length, and form concentric circles both

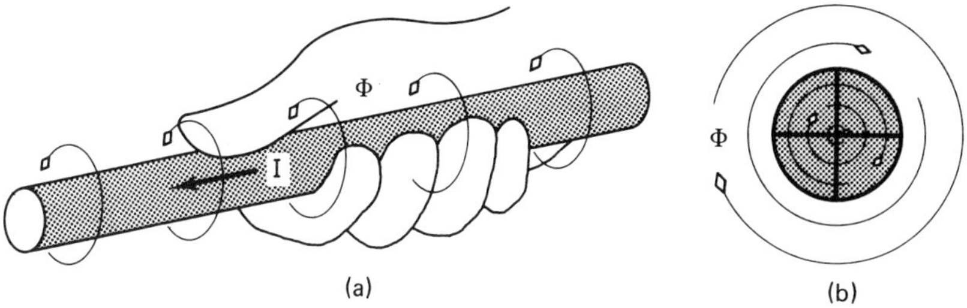

Figure 1-3 Right-Hand Rule

within and outside the conductor. Flux density is relatively large in the area close to the axis but decreases as the distance from the center of the conductor increases. Viewed in cross section, the spacing of the representative lines in Fig. 1-3(b) signifies varying external density.

Assume that a long, current-carrying conductor is formed into a coil with a number of closely packed loops [Fig. 1-4(c)]. The right-hand rule, applied to the adjacent conductors of Fig. 1-4(a), defines flux paths represented by the dashed lines. The space between the conductors would seem to have two vertical flux components, one up and the other down. But lines of force cannot occupy the same space and are mutually repellent; consequently, much of the flux will be forced into a path around both conductors as in Fig. 1-4(b). Thus, the flux created by each turn [Fig. 1-4(c)] follows a common path through the center of the coil, emerging at one end (north pole), traveling externally, and reentering at the south pole. Within the coil, the flux density is essentially constant throughout the cross section. Externally, except in the polar regions, the flux is distributed across broad

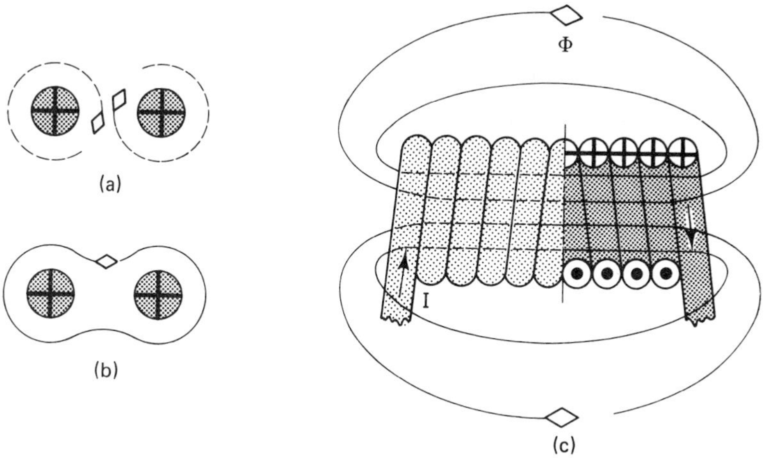

Figure 1-4 Electromagnet

areas and usually is considered to be of negligible density, even in regions close to the coil. From the magnetic viewpoint, although the field results from current flow through the coil, this electromagnet has characteristics identical to those of a comparable permanent magnet.

Both the permanent magnet and the electromagnet are *dipoles* (two poles, one north and one south). If either magnet were cut into segments, each section also would be a dipole. Considerable research effort has been expended in seeking a monopole in nature and trying to create one in the laboratory. These efforts have been unsuccessful. In the early 1970s, research on particles entering the earth's atmosphere detected one hitherto unidentified body; subsequent study suggests that it is a monopole. Interest in the monopole from the viewpoint of electromagnetic devices stems from the current unavoidable requirement for comparatively long ferromagnetic paths between relatively remote poles of a magnet. If it were possible to create monopoles, much of the iron structure could be eliminated and modern electrical machinery would be remarkably smaller, lighter, and cheaper.

1-5. MAGNETIC CATEGORIZATION OF MATERIALS

The magnetic classification of a material is determined by inserting a sample of the material into a magnetic field; the strengths of the field with and without the sample are compared.

A *paramagnetic* material causes an increase in field strength, and a *diamagnetic* material causes a decrease; however, these changes are slight and seldom consequential. In functional equipment, paramagnetic and diamagnetic materials are assumed to have the same magnetic characteristics as free space.

A small group of elements (notably iron, nickel, and cobalt) and alloys containing these elements (e.g., alnico—aluminum, nickel, and cobalt) will, when inserted into the field, cause tremendous increases in the field strength, by factors in the thousands. These materials are classified *ferromagnetic*.

The magnetic field surrounding a current is the basis for explanation of the magnetic characteristics of materials. Atoms contain a nucleus and one or more orbiting electrons, the number being unique to each element. Recalling that electrical current is a directed movement of electrons, the orbital path of each electron is the path of a minute electrical current around which there is a small magnetic field (Fig. 1-5). Each electron contributes an *orbital magnetic component* within the atom.

While orbiting the nucleus, each electron also spins around its axis. The electron is not a solid but rather a surface cloud of negative charge, analogous not to a solid rubber ball but to the rubber in an inflated balloon. Axial rotation of the electron is charge in motion (i.e., electric current), which produces a *spin magnetic component.*

The combined spin and orbital effects establish the magnetic characteristics of the atom. (The nucleus may also contribute, but the relative magnitude of the nuclear component is small and generally inconsequential.) The result is either a magnetically neutral atom or one with a net magnetic field.

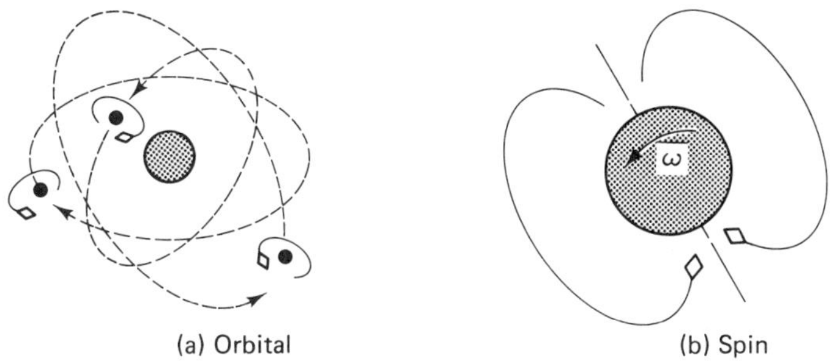

Figure 1-5 Magnetic Components in the Atom

A visible sample of an element contains millions of atoms. The collective effects of these atoms in ferromagnetic materials are conveniently analyzed by magnetic domain theory.

Domain theory recognizes that the sample consists of a number of domains (Fig. 1-6), each containing a large number of atoms. Within each domain, atomic contributions combine to produce a *domain magnetic vector*, unique in magnitude and direction. The characteristic of the overall sample results from the sum of the domain vectors; it may be neutral (i.e., no external magnetic field is detectable), or it may have a net effect, as in a natural or manufactured magnet.

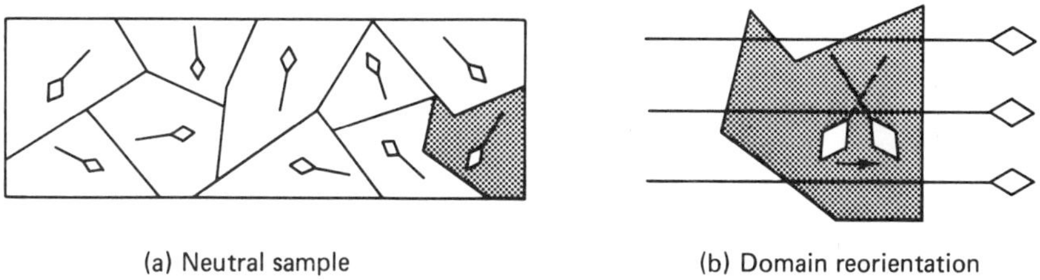

Figure 1-6 Magnetic Domains

When a ferromagnetic material is inserted into an external magnetic field, there are two reactions: (1) domains favorably oriented with respect to the external field grow in size at the expense of their neighbors, and (2) unfavorably oriented domains rotate toward alignment with the field. If the material was magnetically neutral before insertion into the field (sum of the domain vectors is zero), reorientation of domains in the direction of the field will produce a resultant that adds to and greatly increases the original field strength.

The amount of displacement of the domain vectors is governed by the strength of the external field. Increasing the strength of the field causes increasing domain alignment until, ultimately, all domain vectors are parallel with the field. The material now is *saturated*; it can contribute nothing more, regardless of further increases in the external field strength.

When the sample is removed from the external field, the domain vectors tend to return to their original orientation, but recovery is not complete. Consequently, although originally neutral, the sample now has *residual magnetism*. The direction of the residual is that of the field to which the sample was exposed, and the amount of the residual is governed both by the strength of the external field and by the unique material characteristics.

A manufactured permanent magnet is formed from magnetically neutral material, placed in a strong magnetic field, then removed; the material used in its construction has high *retentivity*—that is, it retains a strong residual after removal from the field.

1-6. MAGNETIC CIRCUITS Figure 1-7(a) shows a series electrical circuit whose characteristics are summarized here. The driving force is the electromotive force (E) which creates the current (I): the current is identical at all points in the circuit and is determined by the total resistance (R) of the circuit. A potential drop (V) occurs across each resistor. From Ohm's law, $I = E/(R_1 + R_2) = V_1/R_1 = V_2/R_2$. Kirchhoff's voltage law states that $E = V_1 + V_2$. Ignoring temperature effects, the ohmic value of each resistor is determined by ρ, the resistivity of the material from which the resistor is constructed, and by the physical dimensions of length (l) and the cross-sectional area (A). Mathematically, $R = \rho l/A$.

Parallelism exists between the electrical circuit and the series magnetic circuit of Fig. 1-7(b). Analogous relationships are E (and V) to $\mathfrak{F}$, I to Φ, and R to $\mathfrak{R}$.

The driving force in the magnetic circuit is the *magnetomotive force* (mmf),

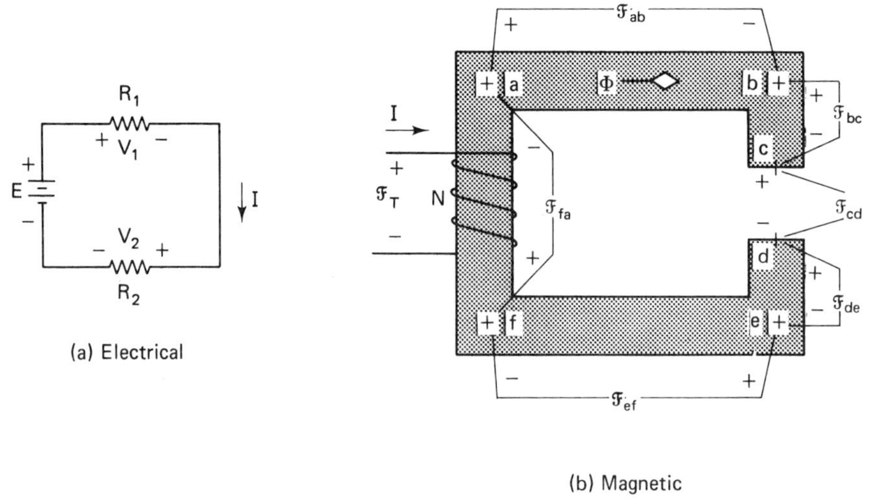

Figure 1-7 Series Circuits

symbol $\mathfrak{F}$; it creates the flux in the circuit. This flux results from current flow in an electrical conductor wound around a portion of the magnetic circuit, forming N turns (or loops). A single turn carrying current I produces a flux whose magnitude is determined by the reluctance $\mathcal{R}$ of the circuit; multiple turns, all carrying the same current, create proportionately larger fluxes. Therefore, $\mathfrak{F}$ is a function of both the number of turns and the current; $\mathfrak{F} = NI$. Note that N turns carrying I amperes is the same mmf as one turn carrying $(N \times I)$ amperes. The SI unit for mmf is the ampere (past practice commonly used amp-turn as the unit).

The reluctance of the circuit, like the resistance, is governed by characteristics of the component material and by the physical dimensions of the circuit. Electrically, the reciprocal of resistivity (ρ) is conductivity (σ); the formula for resistance can be written $R = l/\sigma A$. Magnetically, $\mathcal{R} = l/\mu A$, where l is the mean length of the circuit element, A the effective cross-sectional area, and μ the *permeability* of the material. Earlier, it was noted that nonferromagnetic materials are considered to be identical to free space, whose permeability is uniquely identified as μ_0 with an SI numerical value of $4\pi \times 10^{-7}$. In ferromagnetic materials, μ is *not* single-valued, for reasons discussed in following paragraphs.

Isolating a segment of the total magnetic circuit and identifying the flux and reluctance in that segment, one can identify a drop in mmf across the segment by the Ohm's law relationship $\mathfrak{F}_{\text{drop}} = \Phi\mathcal{R}$. Viewing the circuit as an entity, the applied mmf equals the sum of the mmf drops (Kirchhoff voltage-law analogy): in Fig. 1-7(b), for example,

$$\mathfrak{F}_T = \mathfrak{F}_{ab} + \mathfrak{F}_{bc} + \mathfrak{F}_{cd} + \mathfrak{F}_{de} + \mathfrak{F}_{ef} + \mathfrak{F}_{fa}$$

The Ohm's law relationship $E = IR$ is easily applied in electrical circuits if the resistance is constant, resulting in a linear relationship between E and I regardless of magnitudes. However, circuit elements may be nonlinear. Figure 1-8(a) shows the relationship between E and I in a semiconductor diode. At any point on the curve, Ohm's law could be used to determine a unique R at that point, only; elsewhere, it would have a different value. Similarly, the reluctance in a ferromagnetic segment of a magnetic circuit is nonlinear.

The relationships $\Phi = \mathfrak{F}/\mathcal{R}$, $\mathfrak{F} = NI$, and $\mathcal{R} = l/\mu A$ may be combined to pro-

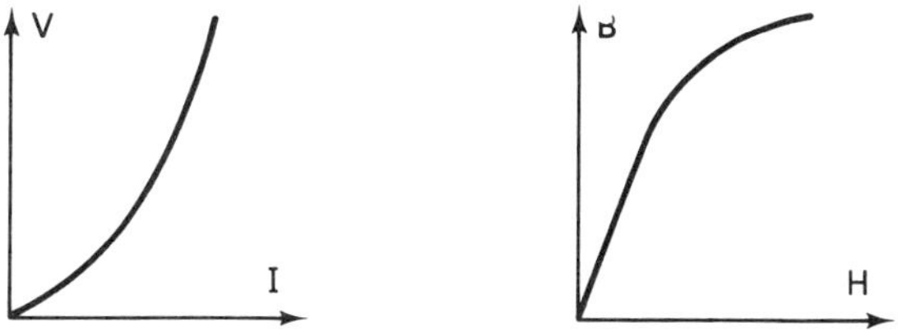

(a) Diode (b) Ferromagnetic element

Figure 1-8 Nonlinear Elements

duce $\Phi = NI\mu A/l$. Rearranged, $\Phi/A = \mu(NI/l)$. This can be written as $\boldsymbol{B} = \boldsymbol{\mu H}$, where B has been defined as flux density and $H(= NI/l)$ is the magnetic field *intensity* in amperes per meter (A/m). $B = \mu H$ is applicable to all matter; μ is constant in nonferromagnetic materials but not in a ferromagnetic substance. Like the graph of V and I for the diode, the relationship between B and H is graphically presented by *saturation curves* [Figs. 1-8(b) and 1-10]. The curves of Fig. 1-10 are used in example and end-of-chapter problems.

In Fig. 1-7(b), the flux is identical in all portions of the circuit. If all legs of the circuit are constructed from ferromagnetic (commonly termed iron) materials, only the air gap has a linear reluctance; in this gap the mmf drop is $\mathfrak{F}_{cd} = Hl = (B/\mu_0)l$. For the iron legs of the circuit, the relationship between $\mathfrak{F}$ and Φ is determined by use of the magnetization curve. Entering the graph at $B(= \Phi/A)$, the corresponding value of H is determined, and $\mathfrak{F} = Hl$.

The circuit analogy extends to parallel magnetic circuits. In Fig. 1-9(a) Kirchhoff's laws identify the relationships $E = V_1 + V_{23}$ and $I_1 = I_2 + I_3$. Particularly, recall that a single circuit voltage, V_{23}, occurs across the parallel resistors R_2 and R_3—*not* two separate voltages. In Fig. 1-9(b), $\mathfrak{F}_T = \mathfrak{F}_{abc} + \mathfrak{F}_{ca}$ and $\Phi_1 = \Phi_2 + \Phi_3$. Again note that a single circuit mmf, $\mathfrak{F}_{ca}$, exists across the parallel paths *cda* and *cea*.

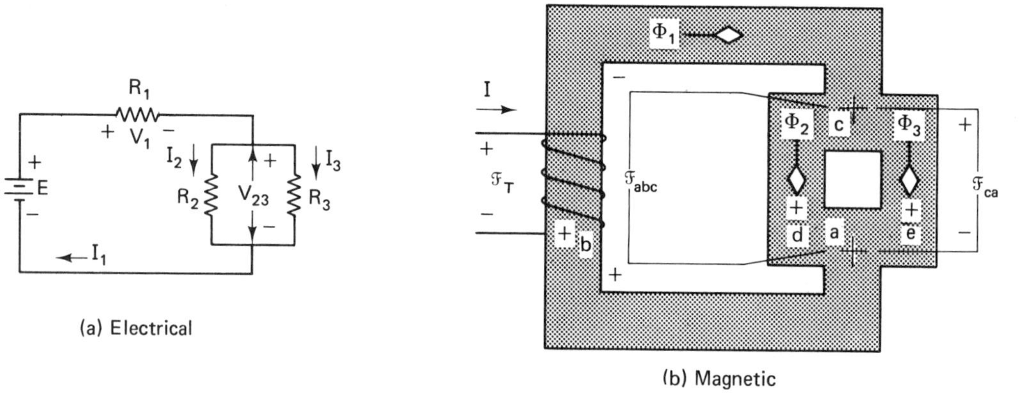

Figure 1-9 Parallel Circuits

The following examples illustrate the solution of magnetic circuits.

Example 1-1: Find the mmf that will establish an air-gap flux of 7×10^{-4} Wb in the circuit of Fig. 1-11.

Solution: Although this circuit is simple, problems of greater complexity are best solved by using a tabular organization. The tabular solution is presented here. Parenthetical entries are computed; all others are given data.

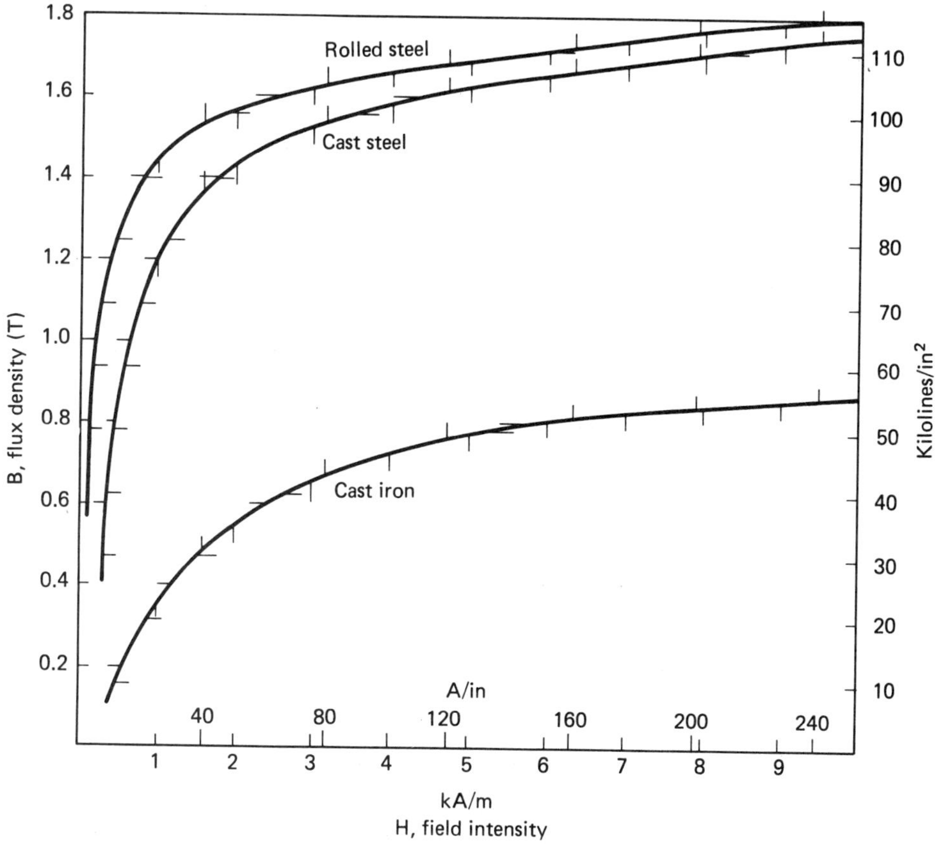

Figure 1-10 Saturation Curves

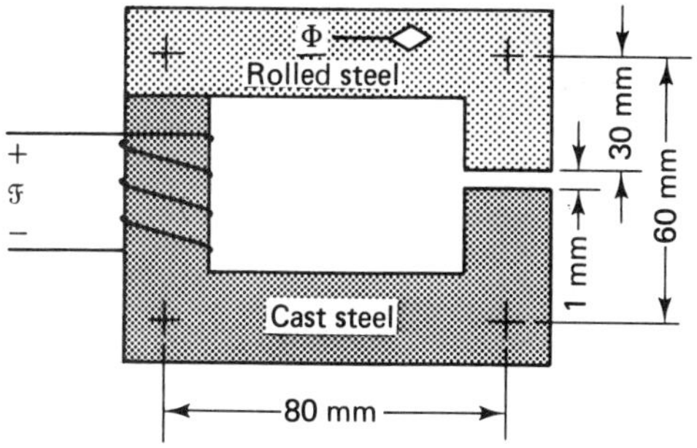

Figure 1-11 Example 1-1

Path	Type of Material	l (m)	A (m^2)	Φ (Wb)	B (T)	H (A/m)	$\mathfrak{F}$ (A)
A	Rolled steel	0.11	5×10^{-4}	7×10^{-4}	(1.4)	(800)	(88)
B	Cast steel	0.17	5×10^{-4}	7×10^{-4}	(1.4)	(1800)	(306)
ag	Air	0.001	5×10^{-4}	7×10^{-4}	(1.4)	(1.114×10^6)	(1114)
						$\mathfrak{F}_T$:	1508

$$B = \frac{\Phi}{A} = \frac{7 \times 10^{-4}}{5 \times 10^{-4}} = 1.4 \text{ T}$$

H for rolled and cast steel is extracted from the magnetization curves (Fig. 1-10).

$$H_{\text{air}} = \frac{B}{\mu_0} = \frac{1.4}{4\pi \times 10^{-7}} = 1.114 \times 10^6$$

$$\mathfrak{F}_A = Hl = (800)(0.11) = 88 \text{ A} \qquad \mathfrak{F}_B = (1800)(0.17) = 306 \text{ A}$$

$$\mathfrak{F}_{ag} = (1.114 \times 10^6)(10^{-3}) = 1114 \text{ A}$$

$$\mathfrak{F}_T = \sum \mathfrak{F} = 1508 \text{ A}$$

Example 1-2: Find the current that will create a flux of 1.44×10^{-3} Wb in the air gap of the circuit of Fig. 1-12. $N = 500$ turns on each leg of the yoke.

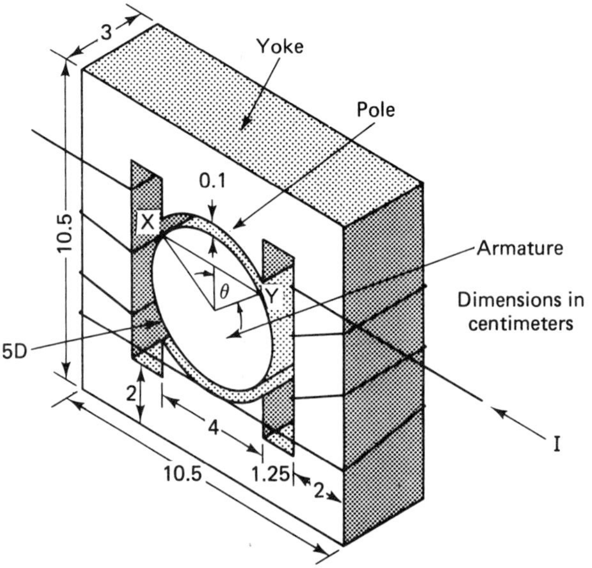

Figure 1-12 Example 1-2

SOLUTION: *Analysis of the problem*: Physically and electromagnetically, the circuit is symmetrical with respect to the center leg. Therefore, $\Phi_1 = \Phi_2 = \Phi_T/2$, and $\mathfrak{F}_1 = \mathfrak{F}_2$.

This is a series–parallel circuit. The series branch is the center leg, comprised of two poles, two air gaps, and the armature. The parallel branches are the identical right and left legs, each consisting of a core and windings.

The effective lengths and areas of the segments of the magnetic circuit must be determined. Several considerations apply:

(a) For most practical circuits, the mmf drop in an air gap is substantially larger than the drops in the iron portions of the circuit. Computations involving the length of the air gap require realistic dimensions, but reasonable approximations of the lengths of the ferromagnetic paths are adequate and produce computed results within the tolerances expected in reading the magnetization curve and due to irregularities (small air pockets, impurities, etc.) in the material.

(b) (1) Lengths of the magnetic paths may be determined by assuming that all the flux in any leg is concentrated in the center of that leg. The mean path length is used.

(2) The length of the air gap is the radial distance from the surface of the armature to the pole face.

(c) Cross-section areas

(1) For ferromagnetic paths other than the armature, the effective cross section is the physical cross section in a plane perpendicular to the axis of the flux.

(2) In the armature, flux enters and leaves across the area of the air gap. This area is smaller than the cross section of the armature in a plane through the axis. However, it would be pointless to attempt a precise definition of the effective armature area. The mmf across the armature is very small in contrast to the mmf of the air gaps (less than 1% in this problem). Seldom is precision possible in magnetic circuit calculations—in fact, it is not uncommon to find circuits in which the relative magnitudes of iron and air mmfs justify neglecting the iron and assuming that all the applied mmf is for the purpose of creating flux in the air gap. For this problem it is valid to assume that the effective armature cross section is the same as the air-gap area.

(3) The air-gap-area determination is based on two considerations: (a) flux follows the shortest (radial) path, but (b) fringing occurs at the edges of the area. The first consideration requires finding the cylindrical surface area of the armature directly opposite a pole face, and the second requires application of a corrective factor.

Determination of the cylindrical surface requires the length of the arc XY in Fig. 1-12. The chord XY is the width dimension of the pole, and the radius of the circle is the armature radius. The angle θ shown on the sketch is $\sin^{-1}(2.0/2.5) = 0.927$ rad. The angle subtending the arc XY is 2θ, and the length of the arc is $2\theta r = (2)(0.927)(0.025) = 0.0464$ m.

In correcting for fringing, a common practice is to add the length of the air gap to both the depth and width dimensions when computing the effective area. Therefore, the air-gap area is $(0.0464 + 0.001)(0.03 + 0.001) = 1.47 \times 10^{-3}\ \text{m}^2$.

A complete solution will result from applying Kirchhoff's voltage-law analogy to the left and center leg, again noting that all parameters of the right leg will be identical to those of the left because of symmetry.

The path through the yoke has the same ferromagnetic material, flux, and cross-sectional area in each segment and can be treated as a single component. (Note that if any one of these elements differed in a segment, that segment would require separate treatment.) The center leg can be treated as three separate segments: the combined air gaps, the combined poles, and the armature.

Tabular solution:

Path	Type of Material	l (m)	A (m²)	Φ (Wb)	B (T)	H (A/m)	$\mathfrak{F}$ (A)
Yoke	Cast steel	(0.17)	(0.6×10^{-3})	(0.72×10^{-3})	(1.2)	(1000)	(170)
2 poles	Cast steel	(0.035)	(1.2×10^{-3})	(1.44×10^{-3})	(1.2)	(1000)	(35)
Armature	Rolled steel	(0.05)	(1.47×10^{-3})	(1.44×10^{-3})	(0.98)	(150)	(8)
2 air gaps	—	(0.002)	(1.47×10^{-3})	1.44×10^{-3}	(0.98)	(7.8×10^5)	(1560)
							$\sum \mathfrak{F} = 1773$ A

$$\mathfrak{F}_1 = \mathfrak{F}_2 = 1773\ \text{A} \qquad I = \frac{\mathfrak{F}_1}{N} = \frac{1773}{500} = 3.55\ \text{A}$$

(*Note:* Tabular entries are determined by the methods applied in Example 1-1—that is, $B = \Phi/A$, H_{steel} extracted from B–H curve, $H_{\text{air}} = B/\mu_0$, $\mathfrak{F} = Hl$. Again, parenthetical entries are computed.)

1-7. INDUCED ELECTROMOTIVE FORCE

Independent experiments of Michael Faraday and Joseph Henry (1797–1878) demonstrated that a voltage is induced in a looped wire if a magnetic flux, enclosed by the loop, is time-variant. Shortly thereafter, Heinrich Lenz (1804–1865) developed relationships that define the polarity of the voltage.

Faraday's law of induction states that the magnitude of an induced voltage in a single loop (turn) equals the *rate* at which the flux through the loop is changing. The rate of change of flux, itself, can be constant or time-variant. In Fig. 1-13(a), the rate is constant; applying this rate in Faraday's law results in a dc voltage per turn whose magnitude is $e_{\text{dc}} = -\Delta\phi/\Delta t$. With flux in webers and time in seconds, e is in volts.

Figure 1-13(b) shows the common circumstance of a sinusoidally varying flux with amplitude Φ_m, frequency f, and period $1/f$. The voltage induced in a loop

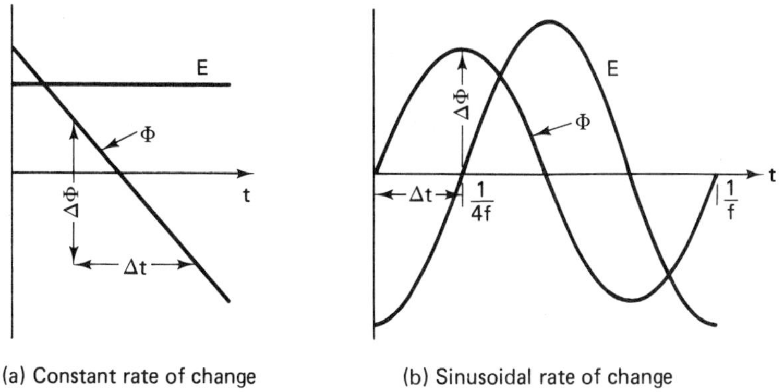

Figure 1-13 Induced Voltages

enclosing this flux will be a sinusoidal voltage lagging the flux by 90°. The average voltage induced in a loop can be determined by applying Faraday's law to the first quarter of the flux cycle.

$$e_{av} = \frac{\Phi_m}{\frac{1}{4f}}$$
$$= 4f\Phi_m \quad (1\text{-}1)$$

For a sine wave, e_{av} equals $0.636e_m$ (e_m is the peak value of the wave) and the effective (rms) value, e, equals $0.707e_m$. Relating these values, the rms induced voltage is

$$e = 1.11e_{av}$$
$$= 4.44f\Phi_m \quad (1\text{-}2)$$

If the loop is close-circuited, an induced voltage will cause a current to flow. This current, in turn, will produce a flux whose direction can be determined by the right-hand rule (Fig. 1-14). Thus, the total flux through the loop has two components: one, the inducing flux, establishes the loop voltage, while the second results from the induced current. **Lenz's law** states that the direction of the induced current will be such that the flux it creates opposes *changes* in the inducing flux.

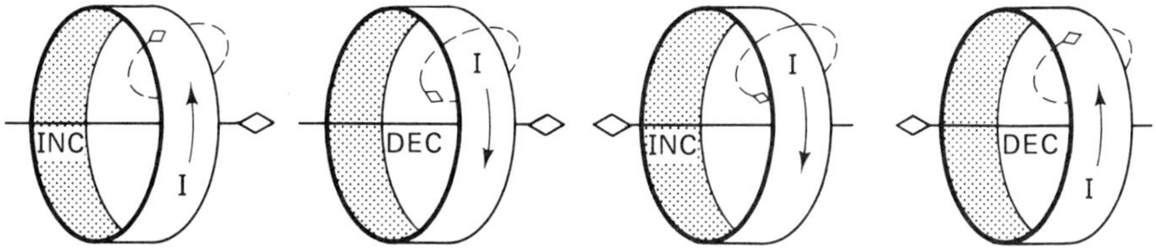

(INC, flux increasing; DEC, flux decreasing)

Figure 1-14 Direction of Induced Currents

The term "change" is significant in this law: specifically, if the inducing flux is increasing in strength, the flux resulting from the induced current will *reduce* (i.e., be in the opposite direction to) the inducing flux. But if the strength of the inducing flux is decreasing, the induced current flux will *add to* (i.e., be in the same direction as) the inducing flux. Note that the reference direction for each flux is the direction of its path through the area enclosed by the loop. The possibilities are illustrated in Fig. 1-14.

1-8. HYSTERESIS Figure 1-15(a) represents a hypothetical circuit for developing the characteristics of a ferromagnetic sample. By altering the position of the slide on the rheostat, both positive and negative currents of controlled magnitude can be produced in the coil. Assume that the rheostat is manipulated to produce currents that alternate in polarity, increasing in magnitude on each successive change. Further, assume that the strength and direction of both the current and the resulting flux is constantly measured and graphically recorded as a *B–H* plot. (*Note:* Since the circuit has a constant area, length, and number of

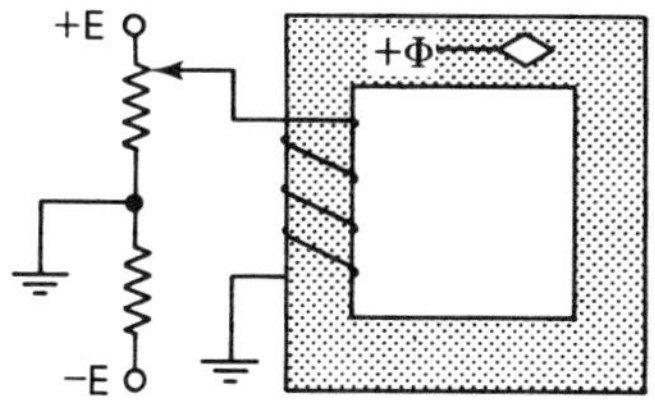

(a) Test circuit

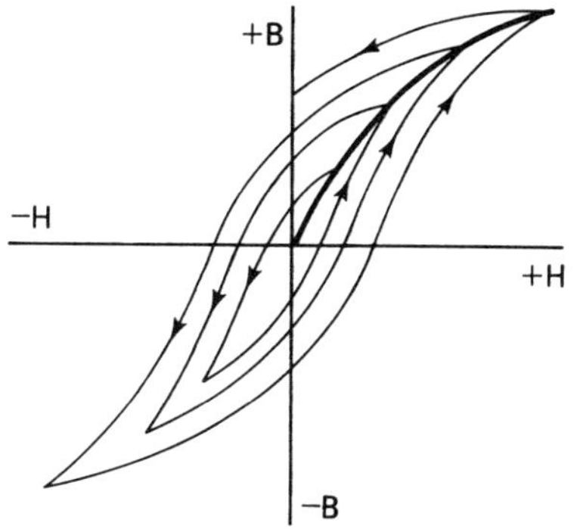

(b) Magnetizing characteristics

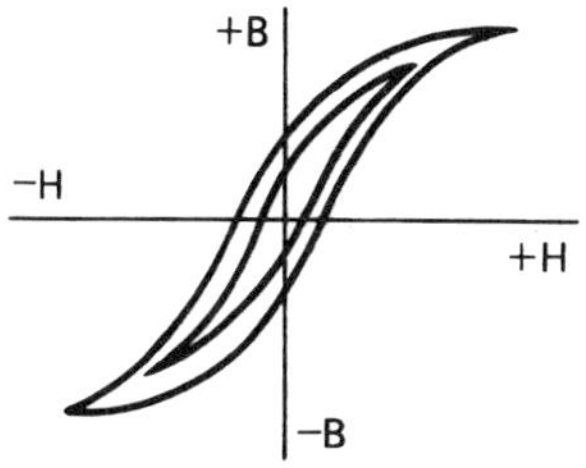

(1) Low retentivity

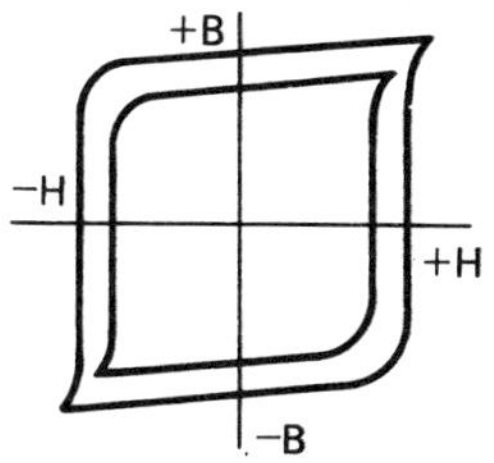

(2) High retentivity

(c) Hysteresis loops

Figure 1-15 B-H Curves

turns, $B \propto \Phi$ and $H \propto I$.) If the sample originally was unmagnetized, the resulting plot will be similar to Fig. 1-15(b).

Domain theory permits explanation of the results. Applying intensity H to the magnetic circuit causes domain vectors to rotate in the direction of H, with the amount of displacement governed by the strength of H. The rotation, in turn, alters the sum of the domain vectors, resulting in changes in flux strength. When H is reduced, the vectors rotate toward their original orientation, but they retain a displacement in the direction of the preceding H. As H increases, it ultimately reaches a strength where all domain vectors are aligned with it; this is saturation. Any additional increment of H will increase the flux only to the extent that the increment would produce flux in a nonferromagnetic substance.

The phenomena represented by the B–H curve is termed **hysteresis**. If the test circuit were modified by replacing the rheostat with an ac source of constant amplitude, the trace of the B–H curve would be repetitive, as shown in Fig. 1-15(c), a *hysteresis loop*. The specific trace results from both the magnitude of the applied voltage and the inherent characteristics of the ferromagnetic material. Figure 1-15(c, 1) and (c, 2) represent two different materials, with curves on each for two different amplitudes of the applied voltage. These hysteresis loops have several noteworthy features:

1. For a given value of H, there are two magnitudes of B, with the particular value dependent upon whether the intensity has been increasing or decreasing.
2. When $H = 0$, the magnetic circuit retains a residual magnetism whose polarity is determined by the immediate magnetization history. This residual is relatively large in a material with high retentivity and small if the retentivity is low.
3. The residual can be eliminated only by the application of a *coercive force*, an intensity of opposite polarity to the residual.
4. Since the hysteresis loop results from cyclic internal realignment of the atomic structure, energy is expended (i.e., converted to nonproductive heat). This expenditure is a *hysteresis loss*, represented by the area enclosed by the hysteresis loop. For the curves of Fig. 1-15(c), if the maximum B and H are identical, the hysteresis loss in the high-retentivity material is greater than that in the low-retentivity material.

Once established, a residual may decay with time or from heat or mechanical shock; normally, in order to demagnetize the material it is necessary to reverse the procedure used to develop Fig. 1-15(b). One feasible method is to apply a controlled ac voltage and gradually reduce it from maximum to zero.

Differing applications require differing characteristics of the hysteresis loop. In large computers, data may be stored in magnetic-core memories whose key elements are thousands of tiny ferromagnetic rings. The hysteresis loop of the material is "square" as in Fig. 1-15(c, 2). Each ring can represent one of two "intelligence" conditions by the polarity of the residual; thus, the ring stores either the number 0 or the number 1 of the binary number system. A grid of fine wires threaded through the rings permits selective application of a magnetizing force to a ring in order to store or retrieve information.

The square loop of Fig. 1-15(c, 2) would be totally inappropriate for a current transformer whose purpose is to step down very large currents in order that they may be read on a standard ammeter. To be of value, B must be directly proportional to H. The desired hysteresis loop is very narrow, essentially a straight line passing through the origin.

If a curve is drawn through the apexes of the hysteresis loops of Fig. 1-15(b), the resulting plot is the saturation curve of the material. It is in this form that material characteristics frequently are presented (Fig. 1-10).

1-9. EDDY CURRENTS

Consider a ferromagnetic circuit to which a sinusoidal voltage is applied (Fig. 1-16). A sinusoidal flux will be established in the circuit. Viewing the cross section A–A, flux with a positive direction out of the plane of the paper is shown in Fig. 1-16(b).

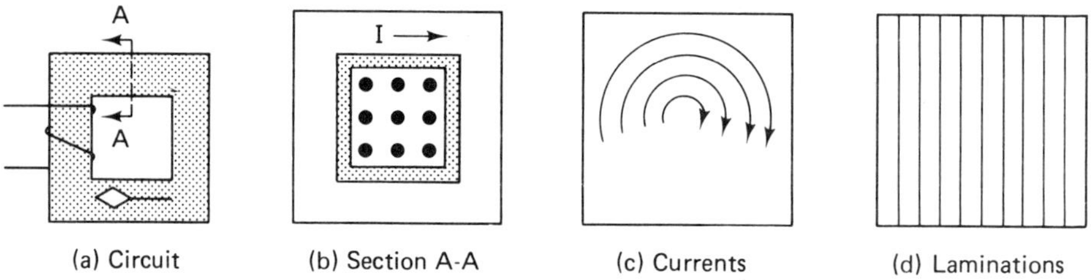

Figure 1-16 Eddy Currents

Ferromagnetic material is conductive; a conducting loop on the surface can be visualized as illustrated. This loop is enclosing a time-variant flux—a voltage will be induced and a current will flow. For the first quarter-cycle, the positive flux defined by Fig. 1-16(a) is increasing and the induced current will be in the direction shown on Fig. 1-16(b).

There are many additional conductive paths on this surface, resulting in a large number of currents, represented in Fig. 1-16(c). These are eddy currents. Eddy currents also cause a power loss in the magnetic circuit. Collectively, eddy-current and hysteresis losses are termed *core losses*, detectable by a rise in temperature of the core material. Neither component can be eliminated, but each can be controlled within limits.

Hysteresis loss can be expressed by an empirical formula, $P_H = K_H f B_m^n$. The hysteresis constant K_H is unique to each type of core material; f is the frequency and B_m the amplitude of the sinusoidal flux density. The exponent n also varies with types of core material, generally lying between 1.5 and 2.5. The hysteresis characteristics can be controlled by the choice of a particular core material and by manufacturing techniques during production.

Eddy-current losses are electrical power losses definable by the formula $P = E^2/R$, where E is the effective voltage induced in the core and R is the effective resistance of the core material. Any measures causing a reduction in the magnitude of E or an increase in R will reduce the power loss. Both E and R can be favorably altered by constructing the core from *insulated laminations* rather than a solid casting. Figure 1-16(d) illustrates section A–A when the core is constructed of laminations. Comparing the cross section of a single lamination to the core cross section of Fig. 1-16(b), it is apparent that the total flux, and therefore the induced voltage, is less in the lamination; also, the area through which eddy currents can flow is considerably smaller in the lamination, resulting in increased resistance. Finally, if the lamination is formed from a material in which the specific resistivity, ρ, is intentionally large, eddy-current losses will be reduced.

1-10. ELECTROMAGNETIC FORCE AND TORQUE

Figure 1-17(a) shows a flux field of uniform density B. From outside, a negative charge q is fired into the field with velocity v. The observed course of the charge while in the field will be curved as in the sketch; for this to occur a force must be operating on the charge. Performed under laboratory conditions, knowing B, q, v, and the mass of the charge, and with a means of observing and recording the trace of the path, the magnitude of the force and its relationship to the electromagnetic parameters may be determined. The result is $F_m = Bqv$, with the force in newtons, flux density in tesla, charge in coulombs, and velocity in meters per second.

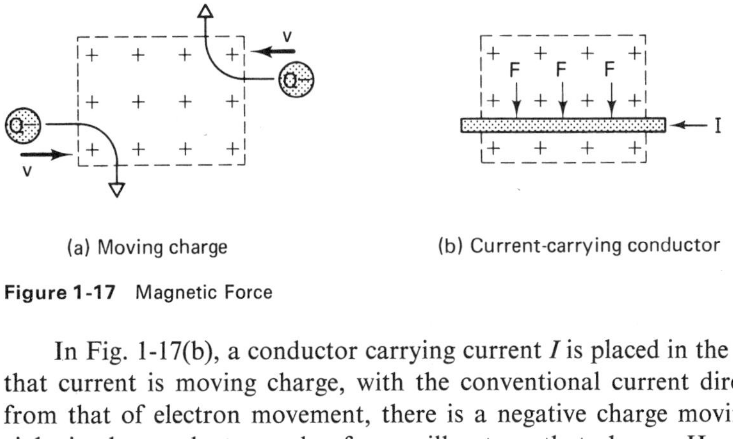

(a) Moving charge

(b) Current-carrying conductor

Figure 1-17 Magnetic Force

In Fig. 1-17(b), a conductor carrying current I is placed in the field. Recalling that current is moving charge, with the conventional current direction opposite from that of electron movement, there is a negative charge moving from left to right in the conductor and a force will act on that charge. However, since the electrons cannot exit the conductor, the net force is transmitted to the conductor in a direction to move it downward. If t is the time (seconds) for charge q to traverse length l (meters) of the conductor, the velocity of the charge is l/t and the force $F_m = Bql/t$; however, $q/t = I$, the current, and the force becomes $F_m = BlI$ newtons.

In Fig. 1-17, the direction of movement of the charge is in a plane perpendicular to the direction of B. If the movement is parallel to B as in Fig. 1-18(a), no force operates on the charge and its path is undeflected. With an initial velocity in a direction neither perpendicular nor parallel to B but at an angle θ as defined in Fig. 1-18(b), the force would be proportional to the quadrature component, $v \sin \theta$—that is, $F_m = Bqv \sin \theta$. Solving this equation for flux density, $B = F_m/qv \sin \theta$. This relationship commonly appears in physics texts; it is a practical basis for measuring B and, with an orientation such that $\theta = 90°$, is used to define flux density.

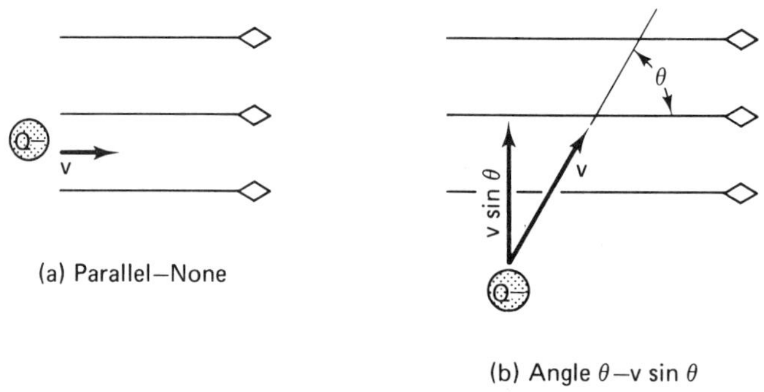

Figure 1-18 Force Component of Velocity

Example 1-3: The flux density B at a point in space located at a perpendicular distance r from a long, straight conductor carrying current I can be determined by the formula $B = \mu_0 I/2\pi r$. What will be the force on a conductor, length 1 m, that is parallel to and located at a distance of 1 m from a long, second conductor if both conductors are carrying a current of 1 A?

SOLUTION:

$$F = BlI = \frac{\mu_0 I_1}{2\pi r} l I_2 = \frac{4\pi \times 10^{-7} \cdot 1}{2\pi \cdot 1} \cdot 1 \cdot 1 = 2 \times 10^{-7}\ \text{N}$$

Compare this result with the following definition of the ampere, established by the 9th General Conference on Weights and Measures in 1948: "The ampere is that constant current which, if maintained in two straight parallel conductors of infinite length, of negligible cross section, and placed one metre apart in vacuum would produce between these conductors a force equal to 2×10^{-7} newton per meter of length."

Commonly, one needs to determine the direction of the force on a current-carrying conductor lying in a magnetic field. The concept of magnetic force lines provides a ready answer. Figure 1-19 shows the cross section of a conductor with current going into the plane of the paper. Applying the right-hand rule, the flux resulting from the current is clockwise as shown. Placed in the separate magnetic

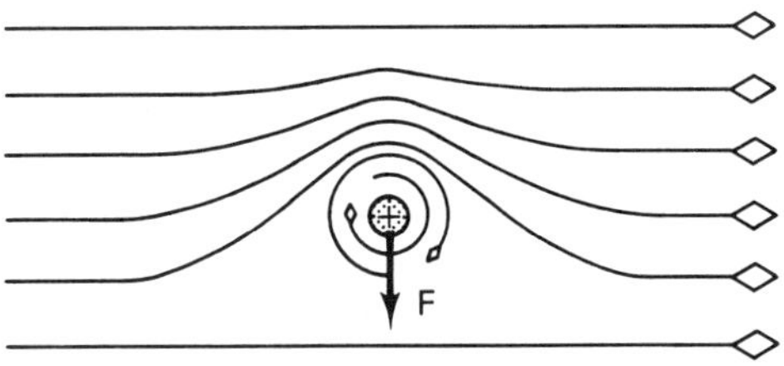

Figure 1-19 Direction of Electromagnetic Force

field, the conductor flux denies direct passage of the field flux at the bottom and forces it up and over the conductor. But this distorted flux is tensile and will contract, thereby exerting a downward force on the conductor. Note that the direction of the force in Fig. 1-19 is consistent with Fig. 1-17.

In Fig. 1-20(a) a conductor is formed into a loop with an axis around which it is free to rotate. The loop is placed in a magnetic field and a current, *I*, flows in the indicated direction. Each conductor experiences a vertical force as shown in Fig. 1-20(b), and therefore a clockwise *torque* occurs: $T_z = BlIr$, where T_z is the torque per conductor in newton-meters and r is the torque arm in meters. Since T_z is developed in each conductor, the loop torque is $2T_z$ or $BlId$, where d is the loop diameter.

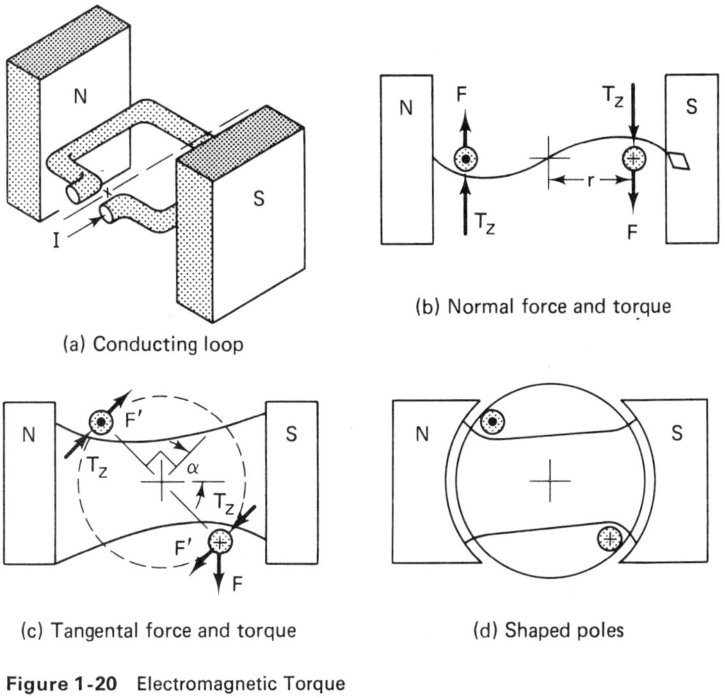

(a) Conducting loop

(b) Normal force and torque

(c) Tangental force and torque

(d) Shaped poles

Figure 1-20 Electromagnetic Torque

Repositioning the coil in Fig. 1-20(c), the forces on the conductors are the same as in Fig. 1-20(b); however, only the tangental component of this force, F', is effective in providing torque. Defining the angle α as shown, $F' = BlI \sin \alpha$ and $T_z = BlIr \sin \alpha$.

The torque production is greatly improved when the shape of the pole face is changed from the straight surface of Fig. 1-20(a) to the circular shape of Fig. 1-20(d) and the coil sides are placed in a ferromagnetic cylinder. If the length of the air gap between the pole faces and the surface of the cylinder is small, the flux will exit the pole and enter the cylinder at an angle perpendicular to the surfaces —that is, radially with respect to the center of the cylinder. Therefore, the magnetic force on the conductor always will be tangental and fully applied to torque production. Shaped poles are used in electrical machines.

While the force on a current-carrying conductor lying in a magnetic field and, by extension, the torque developed in a conducting loop are of predominant interest in a study of electrical machinery, the tractive force of a magnet is also useful. Figure 1-21 is a simplified representation of a relay. A current is introduced in the winding for the purpose of creating an air-gap flux, Φ_{ag}. If Φ_{ag} is sufficiently large, the tensile force of the flux will overcome the force of the spring and cause the movable portion of the magnetic circuit (the armature) to rotate around its pivot and close the air gap. The magnitude of the force exerted on the armature can be formulated in terms of B (the flux density in the air gap in teslas) and A (the cross-sectional area of the air gap in square meters), resulting in $F = B^2A/2\mu_0$.

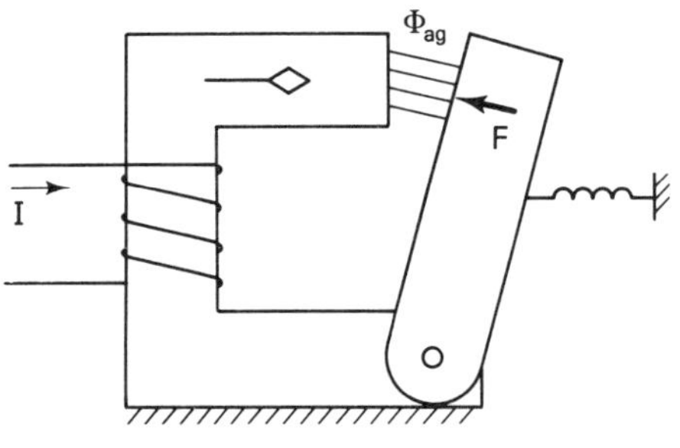

Figure 1-21 Tractive Forces in a Relay

1-11. ILLUSTRATED APPLICATION OF ELECTROMAGNETIC PHENOMENA

Figure 1-22 illustrates the exploitation of electromagnetic phenomena in a practical device. It shows an aluminum disc fastened to the end of a rotating shaft. A portion of the disc is positioned in the air gap between a north and a south pole. Viewed from the right with the magnet, but not the air-gap flux, removed, the disc is rotating in the direction shown. At time t_1, a portion of the disc, ΔA, enclosed by the dashed circle, is located at position 1. As the disc rotates, ΔA will successively arrive at positions 2 at t_2 and 3 at t_3.

At t_2, the flux through ΔA is increasing; applying Lenz's law and the right-hand rule, eddy currents will flow as shown (I_1). At position 3, the flux is decreasing,

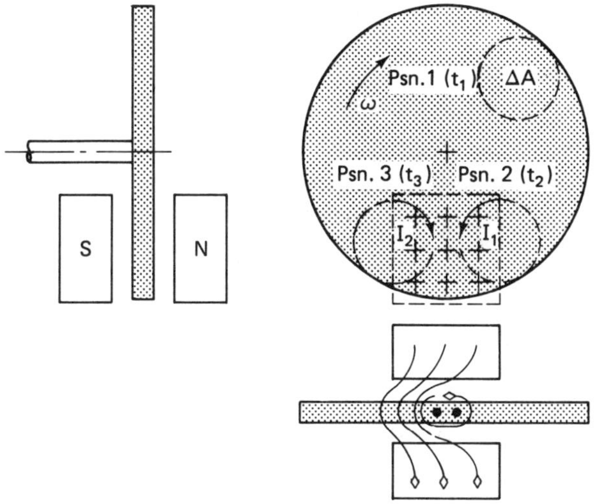

Figure 1-22 Eddy-Current Brake

and the current will be in the direction of I_2. A bottom view of the flux and the eddy currents shows that a magnetic force opposes the rotation and thus a countertorque is developed. This is an *eddy-current brake*, a practical braking device in which braking action may be controlled by adjusting the magnitude of the flux.

The same principles are exploited in eddy-current coupling, in which the disc may be replaced with the walls of a shaft-mounted, hollow cylinder concentrically located between rotating poles on a second, driven shaft. With no mechanical connection between the shafts, torque is transmitted through the coupling, adjustable by changing the strength of the flux. One application of the coupling is to control the speed of a driven load when the prime mover is a constant-speed motor.

SECTION 1-3. ANALYTICAL METHODS

The analysis of electromechanical devices and systems relies upon three basic tools: electrical and magnetic circuit analysis, mechanical system analysis, and the synthesizing element—the law of conservation of energy—which interrelates the electrical and mechanical.

1-12. ELECTROMAGNETIC ANALYSIS

A knowledge of electrical circuits has been assumed and the essentials of magnetic circuits have been presented. During early learning experiences many persons are slow to recognize the power of the circuit solution methods they have mastered. Regardless of the nature of a device, those portions that can be represented by equivalent circuit schematics are susceptible to analysis using any of the techniques for circuit solution. Particularly, Ohm's law, Kirchhoff's voltage and current laws, and power relationships are omnipotent. Whenever a new item of equipment is encountered, think "circuits."

1-13. MECHANICAL ANALYSIS

The mechanical principles applicable to the analysis of machines are covered in typical physics courses. Of particular interest are rotational *speed*, *torque*, and *power*.

The common US unit of rotational **speed** (S) is revolutions per minute (r/min). The SI unit is angular velocity, symbol ω, in radians per second (rad/s): $\omega = \theta/t$. Since there are 2π radians per revolution and the number of revolutions per second is $S/60$, it follows that $\omega = 2\pi S/60$.

Torque (T) is a turning force that produces rotation around an axis. It is the product of a linear force (F) times the perpendicular distance (r) between the line of action of the force and the axis of rotation: $T = Fr$ (Fig. 1-23). The US unit of predominant interest is pound-foot (lb-ft). The SI unit of torque is newton-meter (N·m).

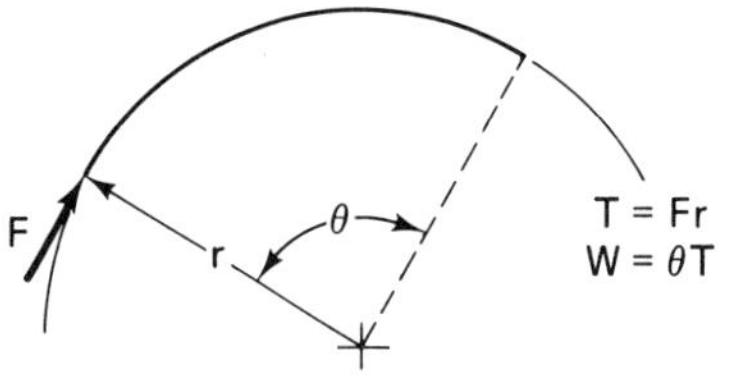

Figure 1-23 Force, Torque, and Work

Work or **energy** (W) is the product of a force and the distance (d) through which it operates: $W = Fd$. The US unit, with force in pounds and distance in feet, is foot-pound (ft-lb). In SI, with force in newtons and distance in meters, the unit is joule (J).

Particularly in the US system, the similarity between torque (lb-ft) and work (ft-lb) causes some confusion. In linear motion, an applied force greater than the sum of opposing forces causes a body to move and work is accomplished. But in rotary motion, a small applied force can cause motion of a body against larger opposing forces because of a mechanical advantage in the system; the classic case is the lever pivoting about a fulcrum. The true measure of the potential to do work involves not only the magnitude of the force but also the length of the moment arm; this measure is torque. When the *applied torque* exceeds *opposing torques* operating on a body, movement occurs and work is accomplished.

In Fig. 1-23 a tangental force is applied to a cylinder of radius r. If the cylinder rotates through the angle θ (in radians), the distance through which the force operates is θr, and the work accomplished is $F\theta r = \theta(Fr) = \theta T$ foot-pounds (US) or joules (SI).

Power (P) is the rate at which work is accomplished: $P = W/t$. With work expressed as θT, $P = \theta T/t = (\theta/t)T = \omega T$. If ω is radians per second and T is newton-meters, the power unit is in watts (SI). Substituting $2\pi S/60$ for ω and with

T in pound-feet, $P = 2\pi ST/60$ foot-pounds per second (US). The US unit, *horsepower*, is defined as 550 foot-pounds per second; thus, hp $= 2\pi ST/(60 \times 550) = 2\pi ST/33{,}000$.

Torque in newton-meters is related to torque in pound-feet by $T_{\text{N·m}} = 1.355818 T_{\text{lb-ft}}$. Using this relationship and writing the horsepower equation with torque in newton-meters,

$$\text{hp} = \frac{2\pi S T_{\text{N·m}}}{60 \times 550 \times 1.355818} = \frac{2\pi S}{60}\frac{T_{\text{N·m}}}{746} = \frac{\omega T}{746}$$

But ωT is in watts: therefore, *1 horsepower equals 746 watts.*

The watt is a common unit in both the mechanical and the electrical system. Thus, the characteristics of the two systems can be interrelated by expressing both in power terms.

1-14. ENERGY Among the more important laws of science is the principle of conservation of energy. This principle states that energy can be neither created nor destroyed, but may be converted from one form to another. In the process of transformation, some of the energy input to a system is lost in the sense that it is converted to nonuseful forms such as unwanted heat; consequently, the useful output energy is less than the energy input. For example, in an electric motor the electrical energy input is converted to mechanical energy output, but there are energy losses due to electrical resistance of the conductors (copper loss), hysteresis and eddy currents (core loss), friction, and windage (air resistance to the rotary motion).

Part of the energy input may be temporarily stored in the system in forms such as the charge on a capacitor, the magnetic field around an inductor, the compression of a spring, and the inertia of a rotating flywheel. Later, some stored energy may be recovered in useful form, but most of it will be nonproductively expended —an additional energy loss.

In analyzing a system, the energy balance is commonly used to relate input and output characteristics. For the motor, the friction, windage, and core losses (W_{fwc}) usually are lumped; the copper loss (W_{cu}) results from the familiar I^2R power loss of circuit theory, and the energy equation is

$$W_{\text{in}} = W_{\text{out}} + W_{\text{cu}} + W_{fwc} + W_{\text{stored}}$$

In general, analysis of a system by energy methods requires mathematical treatment at levels beyond algebra and trigonometry. Rigorous treatment must consider both *transient* and *steady-state* periods. For the motor, acceleration, deceleration, and momentary adjustments in response to load changes are transient states, but operation at a constant speed while driving a constant load is steady state. Transient-state periods often are of very short duration and inconsequential in assessing machine or system performance. In steady state, energy conversion in each component is at a constant rate and a power (W/t) equation can be applied. For the motor,

$$P_{\text{in}} = P_{\text{out}} + P_{fwc} + P_{\text{cu}} \qquad \text{(stored energy is neglected)}$$

The power equation is used in developing many of the relationships in the following chapters. Transient state is considered only when it is essential to an understanding of the performance of a device or system; where pertinent, the transient effects are described but little or no mathematical treatment is included.

SECTION 1-4. STANDARDS

1-15. STANDARDS With the evolution of technology there was early recognition of the need for standardization in such areas as technical vocabulary, product characteristics, and safety practices. In response to this need, numerous standards have been published in all areas of engineering by nongovernmental professional societies and associations. For electrical machinery, standards issued by the Institute of Electrical and Electronic Engineers (IEEE), National Electrical Manufacturers Association (NEMA), and National Fire Protection Association (NFPA) are particularly important.

IEEE publications include a dictionary of technical terms; symbols for schematics, quantities, and units; and practices for testing and using electrical equipment.

NEMA publications emphasize construction, performance, and guidance in application.

NFPA sponsors the National Electrical Code (NEC), which is directed toward safe installation of electrical wiring and equipment. The NEC is unique in that it has been widely adopted and written into law by local governments, whereas other standards, although very broadly observed, are not enacted laws.

The American National Standards Institute (ANSI) is an interdisciplinary association of professional societies. It does not write standards, but it reviews those developed by the represented organizations and accepts many for ANSI endorsement. An ANSI standard has at least two designating numbers: for example, *Graphic Symbols for Electrical and Electronics Diagrams* includes the markings *ANSI Y32.2–1975* and *IEEE* Std *315–1975*. Note that the promulgation year is shown; all standards are periodically reviewed and reconfirmed or reissued.

Domestic standardization agencies are active participants in international agencies such as the International Organization of Standardization (ISO) and the International Electrotechnical Commission (IEC). Where practical and advisable, national standards agree with international standards.

The material in this text is consistent with published standards. Where appropriate, specific references are included.

SUMMARY

The purpose of this chapter is to establish a common understanding of fundamentals that are broadly applied throughout the following chapters in analyzing the functioning and performance of electromagnetic and electromechanical devices. The coverage is not exhaustive—other common principles will be introduced at appropriate points in the text—but a basis has been established for initiating the study of direct-current machinery.

PROBLEMS AND QUESTIONS

1-1. The unit of capacitance is the farad (F). Using the relationships $C = Q/V$ and $Q = It$, express the farad in base units.

1-2. The unit of inductance is the henry (H). From the relationship $V = L(\Delta I/\Delta t)$, express the henry in base units.

1-3. Convert the following to SI units:
(a) 6 ft (b) 180 lb (force)
(c) 1 acre (4840 yd^2) (d) 1 ton (mass)
(e) 32 ft/sec^2

1-4. Research in nuclear fusion has developed a need for large flux densities. One superconducting air-core magnet produces a density of 8 T. What is the field intensity?

1-5. The SI value of μ_0 is $4\pi \times 10^{-7}$. What is the US value of μ_0 with flux density in kilolines/in^2 and H in A/in?

1-6. (Fig. 1-24) What current is required in order to establish 6.25 kilolines of flux in the toroid?

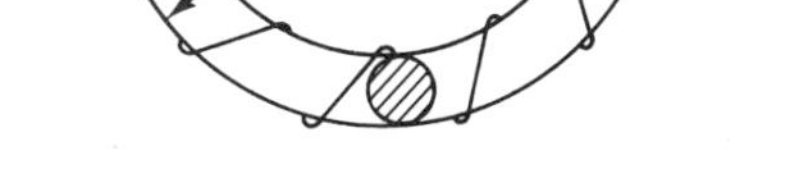

Figure 1-24 Problems 1-6 and 1-7

1-7. (Fig. 1-24) A 0.1-in segment of the iron is replaced with brass. What current is required for 6.25 kilolines of flux?

1-8. What mmf will establish a flux of 1.3×10^{-3} Wb in the air gap of Fig. 1-25? (Neglect fringing.)

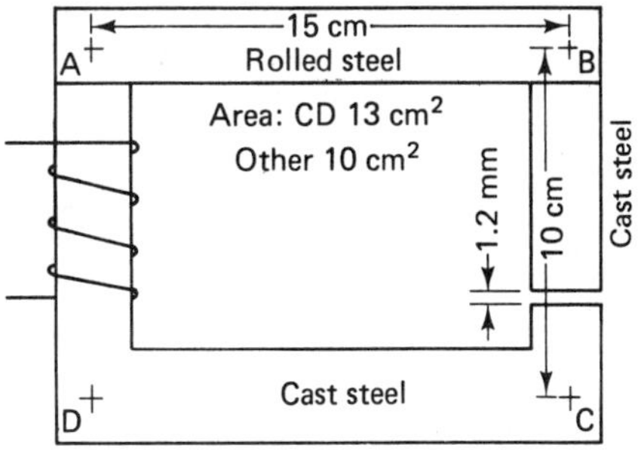

Figure 1-25 Problem 1-8

1-9. What current in the coil of Fig. 1-26 will produce an air-gap flux density of 1.2 T? (Neglect fringing.)

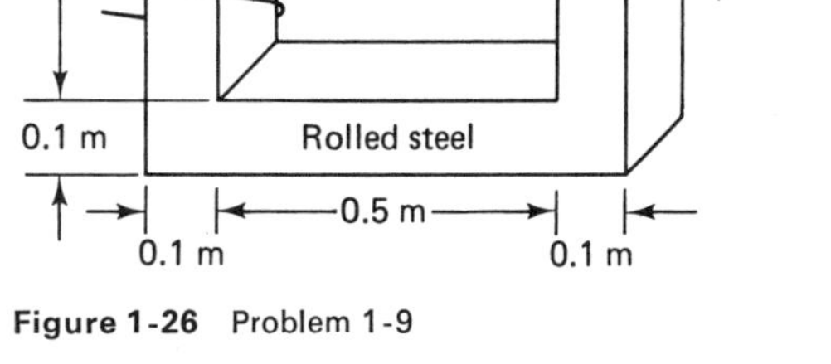

Figure 1-26 Problem 1-9

1-10. The dimensions of Fig. 1-27 are $BE = 0.2$ m, $BCDE = BAFE = 0.4$ m, $ag_2 = 2.5$ mm, and a uniform cross section of 2×10^{-3} m². $\Phi_{ag_2} = 3 \times 10^{-3}$ Wb. Find Φ_1, l_{ag_1}, and $\mathfrak{F}$.

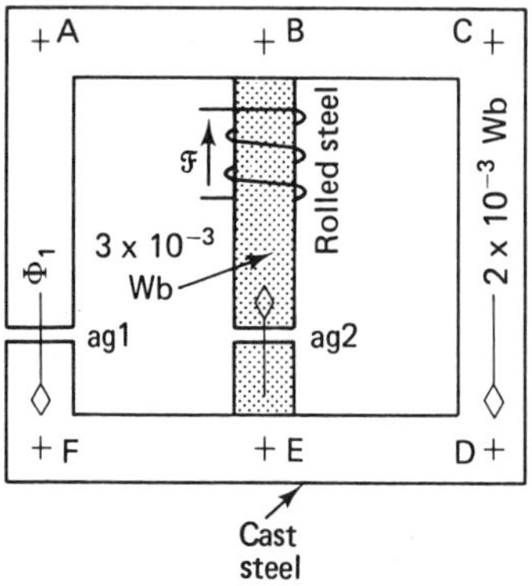

Figure 1-27 Problem 1-10

1-11. The flux in the air gap of Fig. 1-28 is 64.5 kilolines and $\mathfrak{F}_1 = \mathfrak{F}_2$. Find $\mathfrak{F}_1$ and the ratio N_1/N_2.

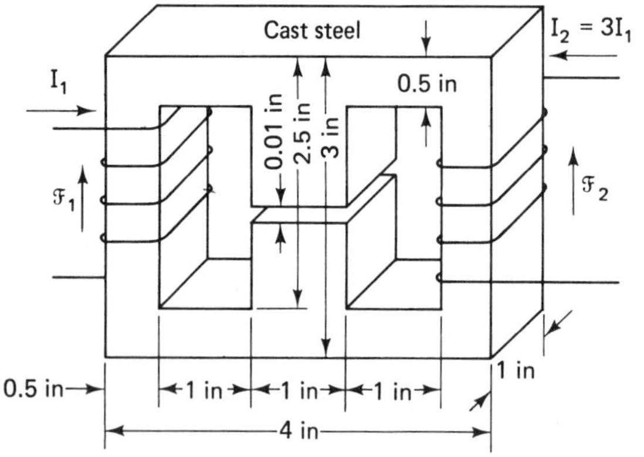

Figure 1-28 Problem 1-11

1-12. The following dimensions apply to Fig. 1-29.

	AB and EF (ea)	BC and DE (ea)	CD	AF
Length	0.118 in	7 in	20 in	9 in
Area	56 in²	40 in²	14 in²	18 in²

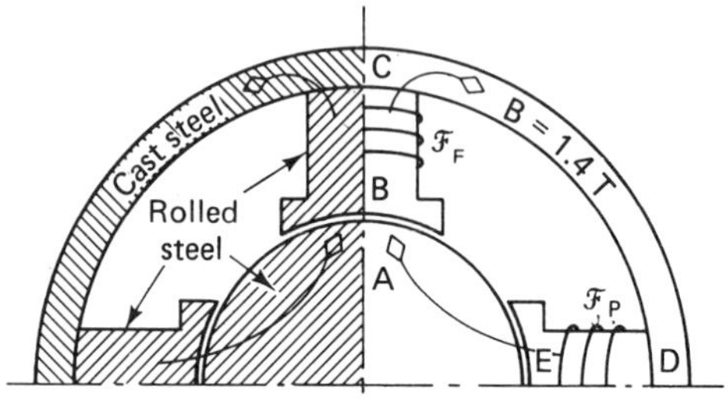

Figure 1-29 Problem 1-12

Find $\mathfrak{F}_p$, the mmf per pole.

1-13. In Fig. 1-30 flux is confined to the area delineated by short dashed lines. Spacing of the crosses indicates relative flux density. In each instance, the solid conducting loop is moved to the position where it is shown as a dashed loop. For each figure, determine whether a voltage is induced during the move-

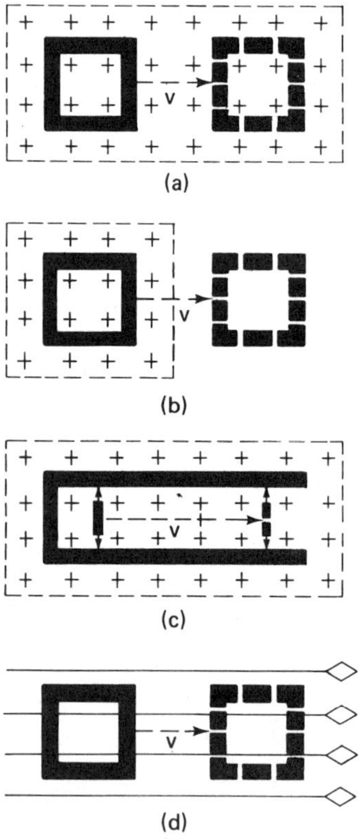

Figure 1-30 Problem 1-13

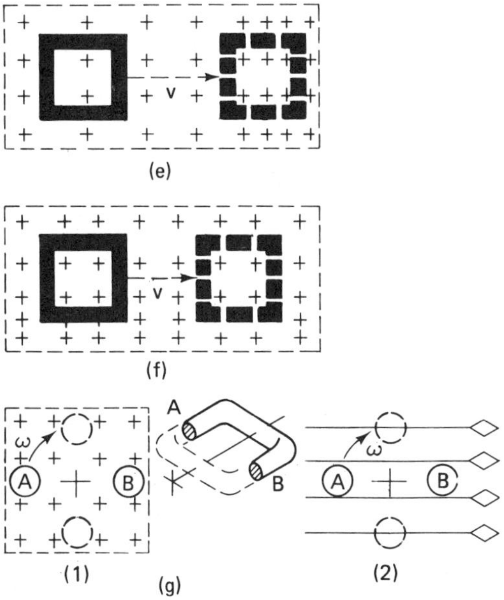

Figure 1-30 Continued

ment and, if so, the direction of the resulting current.

1-14. A magneto of the type used to supply the spark in small single-cylinder gasoline engines is illustrated in Fig. 1-31.

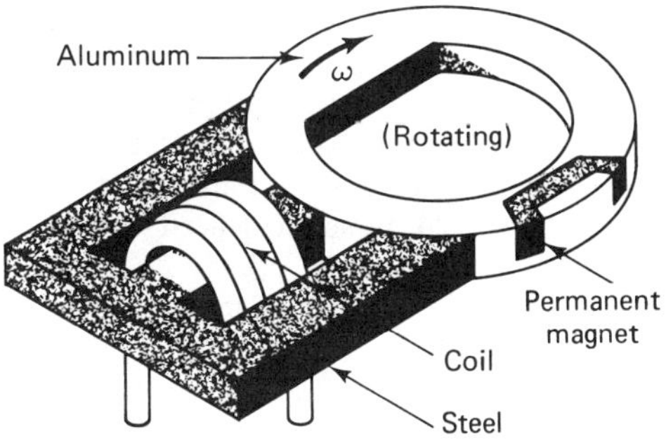

Figure 1-31 Problem 1-14

(a) Explain the generation of the voltage.
(b) Is the voltage ac or dc? Why?
(c) What parts, if any, should be laminated? Why?
(d) Will a point on the flywheel, diametrically opposite the magnet, attract iron filings? Why?

1-15. The flux through a loop is time-variant, as shown in Fig. 1-32. Sketch the induced voltages (time function) and show the magnitudes.

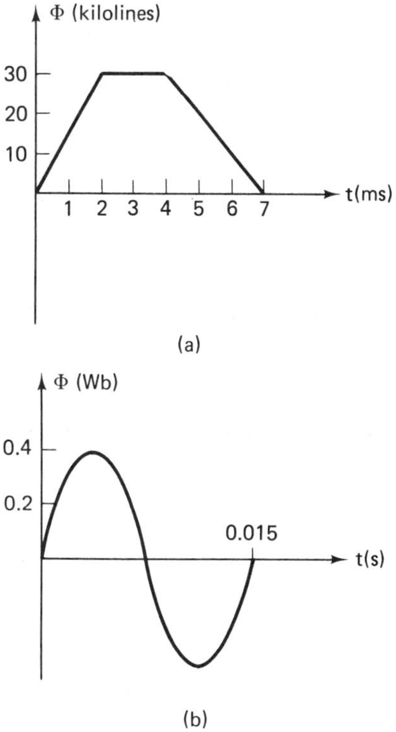

Figure 1-32 Problem 1-15

1-16. Two long, parallel conductors are freely suspended. Sketch the flux fields and determine the direction of any magnetic forces acting on the conductors if (a) both are carrying currents in the same direction, and (b) the currents are in opposite directions.

1-17. The relay of Fig. 1-33 has a core and armature constructed from cast iron. A flux density of 0.2 T through the coil produces a tractive force that closes the air gap and, once closed, maintains closure. The two coils are connected in series. (a) What current is required to close the gaps? (b) After closure, to what value may the current be reduced

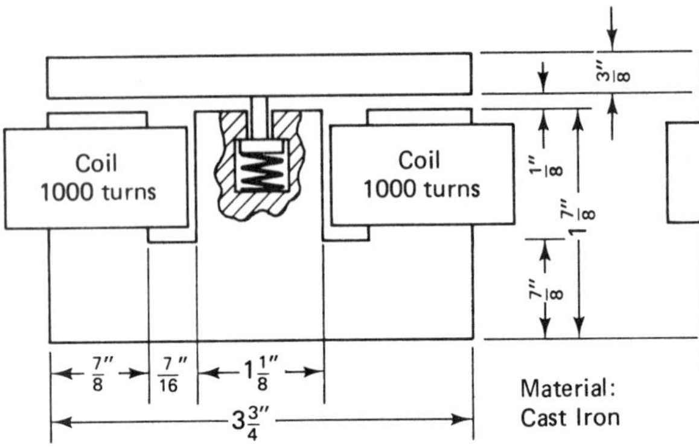

Figure 1-33 Problem 1-17

without the armature returning to the open position? (c) What is the tractive force in each air gap that causes closure?

1-18. The linear motor shown in Fig. 1-34 consists of a movable conductor that maintains electrical contact with the stationary U-shaped conductor of infinite length, whose sides are separated by 0.5 m and enclose a uniform magnetic field of 0.8 T. A current generator maintains a constant current in the closed loop. After an accelerating period, a force of 10 N maintains a 10-m/s steady-state velocity of the moving conductor. (a) What is the magnitude of the current? (b) What voltage is induced in the closed loop?

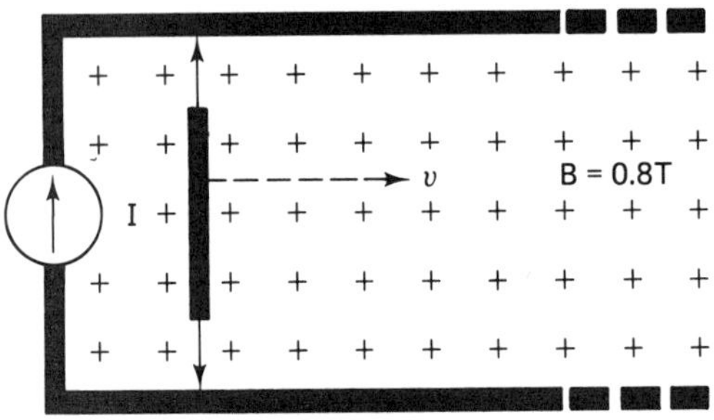

Figure 1-34 Problem 1-18

1-19. From the relationship $B = \mu H$, express μ in terms of the units weber, meter, and ampere.

1-20. A restraining spring holds the needle of the meter movement of Fig. 1-35 at the left scale index. One hundred turns of wire are coiled around a movable core to which the needle is attached; the core has a diameter of 2 cm, an axial length of 2 cm, and is centered between the faces of two poles with an air-gap flux density of 0.15 T. Angular deflection of the coil is proportional to the current in the windings; 120° full-scale deflection is caused by a current of 1 mA.

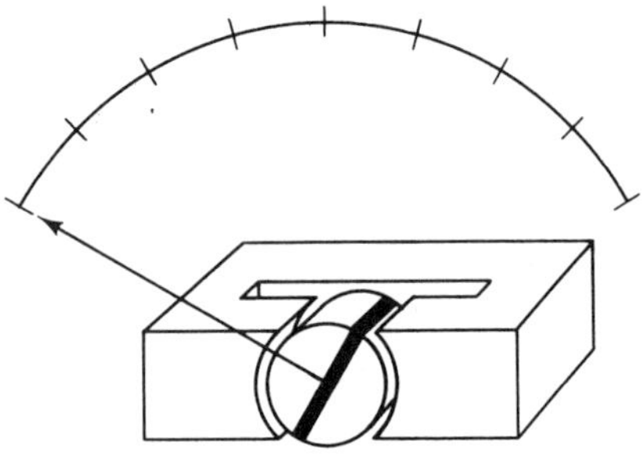

Figure 1-35 Problem 1-20

(a) What is the angular deflection for a current of 300 μA?

(b) At full-scale deflection, (1) what is the torque developed in one turn, and (2) what is the net torque produced by the coil?

1-21. (a) An engine develops a torque of 230 lb-ft at 1600 r/min. What is the hp? (b) The hp at 4200 r/min is 145. What is the torque?

1-22. A constant tangental force is applied to the cylinder of Fig. 1-36.

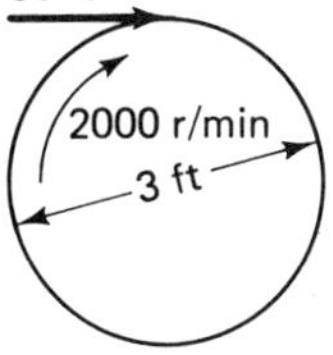

Figure 1-36 Problem 1-22

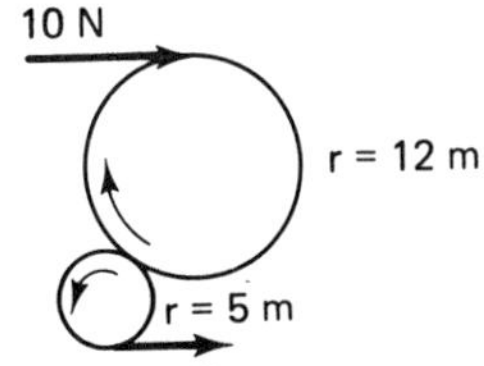

Figure 1-37 Problem 1-23

(a) What is the torque and the developed horsepower?

(b) What is the torque in N·m, angular velocity in rad/s, and power in W?

1-23. (a) If there are no losses in the system of Fig. 1-37, what is the magnitude of F? (b) If the magnitude of F is measured and found to be 9 N, what is the torque loss in the system? (c) If the large cylinder rotates at 100 rad/s, the small cylinder rotates at 240 rad/s. For the measured F of 9 N, what is the power input to the large cylinder, power output of the small cylinder, and power loss in the system?

1-24. In SI units, $F_m = Bli$ (Paragraph 1-10). Show that $F_m = Bli/11{,}300{,}000$ with F in lb-ft, B in lines/in^2, and l in inches.

2

DC Machinery

SECTION 2-1. DC GENERATOR

Most power generated and used in the United States is in ac form, but dc is essential to many applications such as control systems, elevators, and electrochemical processing. In part, this requirement is met by converting ac to dc, formerly with mercury-arc rectifiers and, with the evolution of the thyristor (silicon-controlled rectifier), increasingly with the latter. But the need often is met with dc generators; many are operational in industry and, despite improved thyristors, will continue to be preferred for some applications in the predictable future. This section develops the concepts and characteristics of the dc generator.

2-1. GENERATOR PRINCIPLES Mechanical generators exploit the phenomena of induction. If flux through a conductive loop varies sinusoidally, a sinusoidal voltage is induced in the loop (Paragraph 1-7). The magnetic field need not be time-variant; relative motion between the loop and a constant field can achieve the necessary flux variation. In the dc generator, the loop rotates within a constant, static field.

A simplified structure of a basic dc generator is represented in Fig. 2-1. A continuous conductor is wound around the pole cores, evenly divided between, and providing N turns on each of the poles. A current in this field winding estab-

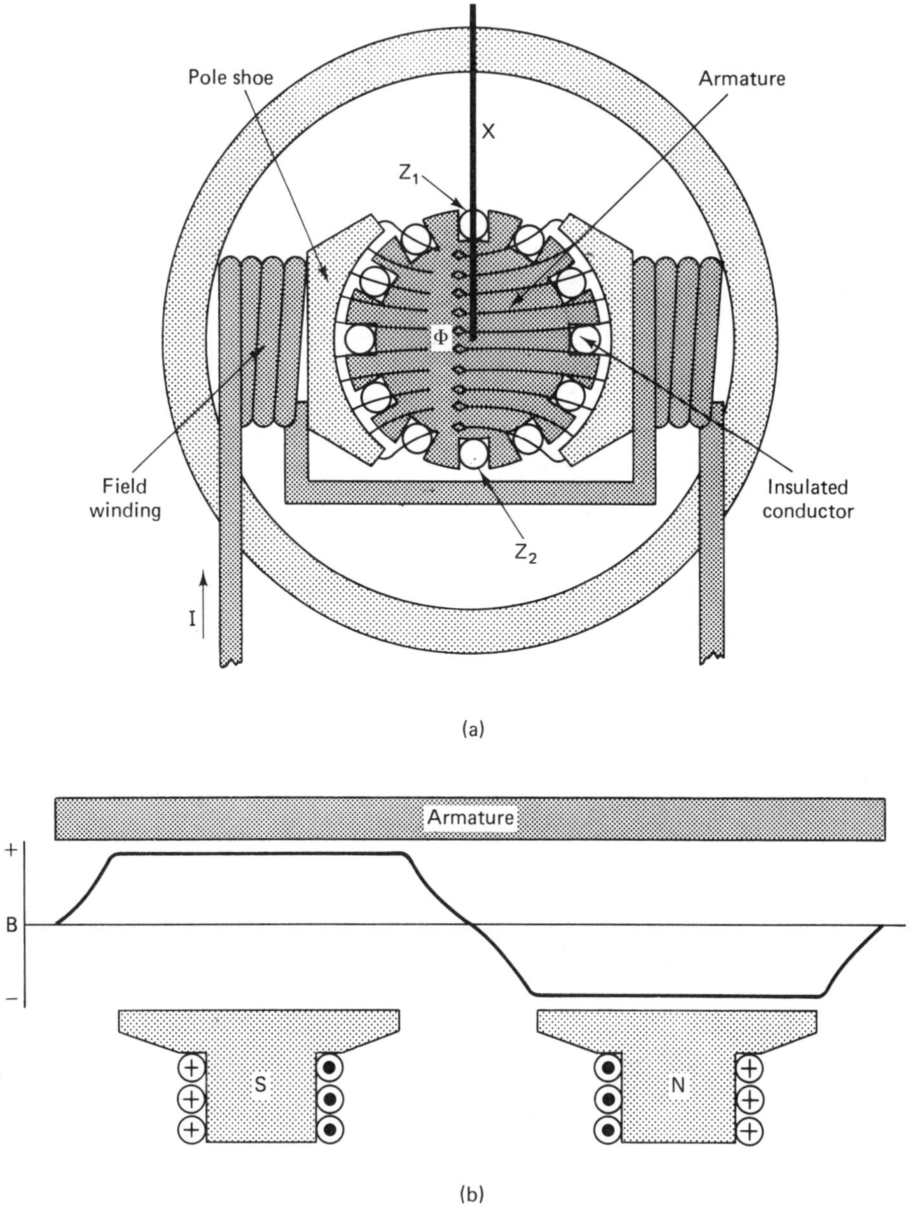

Figure 2-1 Basic Structure of a DC Generator

lishes an mmf of NI amperes on each pole, creating a flux Φ per pole. This flux crosses an air gap between each pole face and the surface of a cylindrical armature.

The armature, with insulated conductors located in slots around its periphery, rotates about its axis. The focal point for energy conversion is the circumference of the armature. It is here that the conductors, paired to form loops, are located. It is at this point that the strength and distribution of the flux fixes the magnitude and wave shape of the voltage induced in the loops. For this reason, the air-gap flux is of fundamental importance. The general pattern of this flux and its paths through the armature is shown in Fig. 2-1(a), but a more definitive picture is desirable. Assuming that the machine is cut along line X and folded outward, the structure can be viewed in the linear form of Fig. 2-1(b). In the area between the armature and the poles, the flux density, B, is plotted as a spatial function. Density is positive at points where it exits the armature and negative where it enters. Note that the flux density is constant across the pole faces.

In Fig. 2-2, a rectangular coil of length l, width w, and area A is placed in a field of uniform density. The loop sides Z_1 and Z_2 appeared as diametrically opposed conductors in Fig. 2-1, but the sides m and n were not represented.

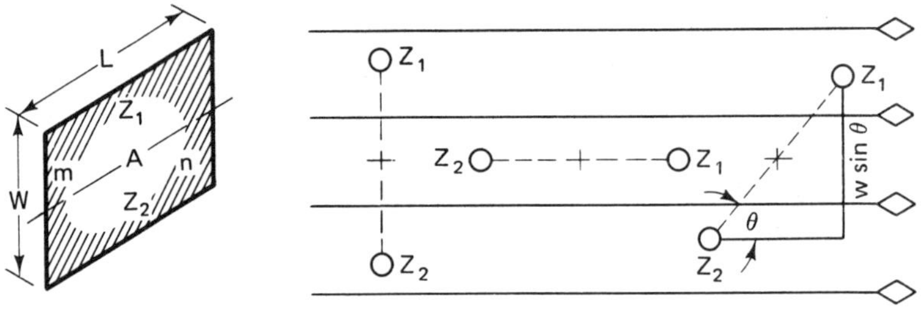

Figure 2-2 Loop Flux as a Function of Angular Orientation

Oriented vertically, the flux through the loop is BA, but when the loop is horizontal, no flux passes through it. At any angle θ, defined in the figure, the flux is $Bl(w \sin \theta) = BA \sin \theta$.

If the loop now rotates about its axis at a uniform angular velocity, θ equals ωt and the loop flux is $BA \sin \omega t$. This induces a sinusoidal voltage e having the average value $4f\Phi$ (equation 1-1). Substituting $\omega/2\pi$ for f, $e = 2\omega\Phi/\pi$.

In one complete revolution of the loop of Fig. 2-2, one full sinusoidal voltage cycle is generated. But most dc generators have not two but four or more poles (always an even number), arranged around the circumference with alternating polarity. In the four-pole machine of Fig. 2-3, two sinusoidal cycles are generated per armature revolution. (Repositioning of Z_1 and Z_2 is discussed in Paragraph 2-9.) A cycle is generated during the movement past each pair of poles—thus, for P poles, there are $P/2$ cycles per revolution. The resulting frequency is $(\omega/2\pi)(P/2) = \omega P/4\pi$. At this frequency, $e = P\omega\Phi/\pi$.

The loop now is interrupted and leads are brought out to two slip rings (Fig. 2-4). The loop and the rings rotate as a unit, with static carbon brushes riding on the rings. If a load is connected to the brushes, the loop voltage will appear across the load. Note that this is an ac voltage.

Next, replace the slip rings with a single split ring (Fig. 2-5). During the positive portions of the voltage cycle, the + brush is in contact with coil side Z_1,

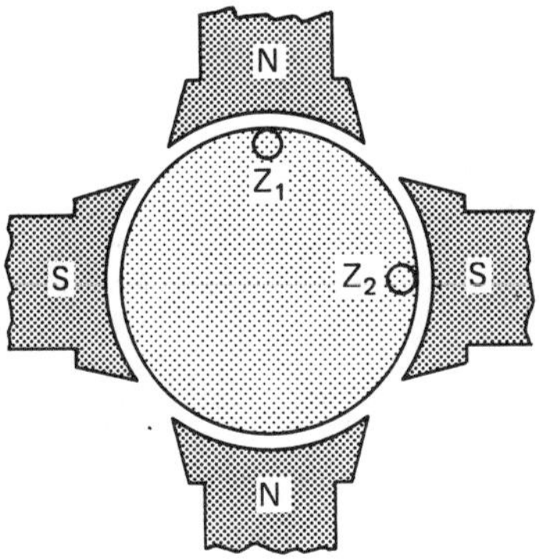

Figure 2-3 Four-Pole Generator

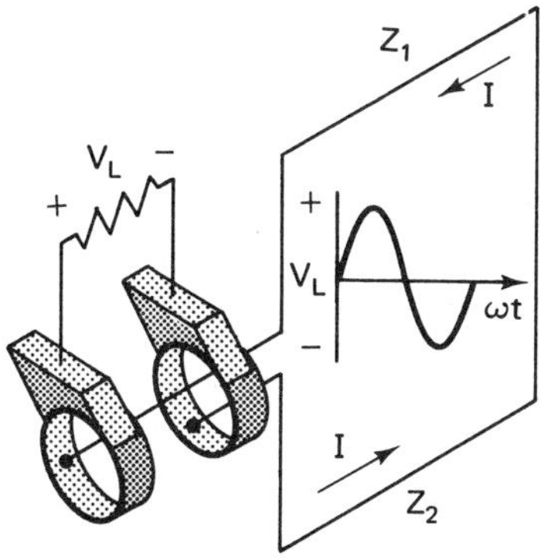

Figure 2-4 Slip Rings

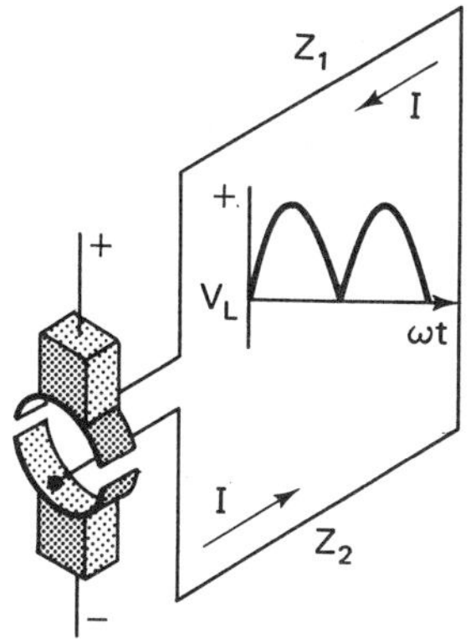

Figure 2-5 Split Rings

but as the rotation produces a voltage polarity change in the conductors, the + brush will contact side Z_2. This load voltage is a rectified sine wave.

A second loop and its split ring now are added, electrically displaced from the original by 90° (Fig. 2-6). The voltages generated in the two loops are identical but separated in phase by 90°. If the brushes are interconnected as shown, the voltage between the + and − brushes will be the sum of the two rectified waves. Successively doubling the number of loops, each time locating the additions so as to maintain a uniform angular distribution around the armature circumference and with a series connection of all loops at the split rings, results in load voltage waveforms as shown in Fig. 2-7 for 4, 8, and 16 loops. The voltage is not constant, but as the number of loops increases, the difference between maximum and minimum values diminishes. Fluctuations are above and below the horizontal lines, each of which has the value $n_c e$, where n_c is the number of series loops for that plot. From the illustrated patterns it may be concluded that if n_c is very large, the fluctuation (ripple) is very small—essentially, it is a steady dc voltage.

While the approach to dc generator construction described above is theo-

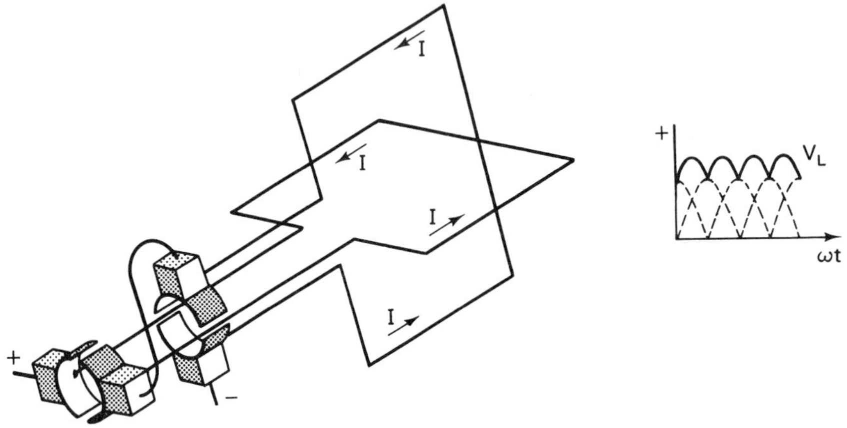

Figure 2-6 Two-Loop Generator

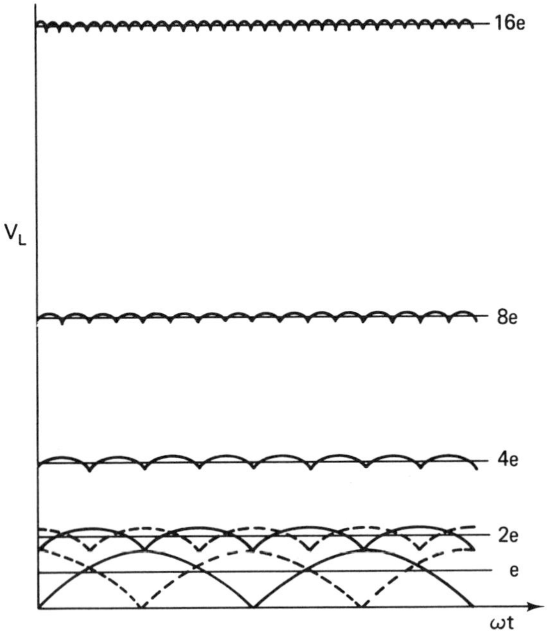

Figure 2-7 Load Voltage Waveforms for Multiple Loops

retically possible, it is impractical. An inordinate number of split rings and brushes would be required to accommodate a realistic number of loops. Connections that form only one series path limit the terminal voltage (i.e., load voltage) of a unique machine to but one possible value and restrict the load current to the relatively low quantity that a single conductor can safely carry; greater voltage and current options are required.

Practical methods for interconnecting loops and extracting power from the armature are developed in Section 2-2, but certain aspects are pertinent at this point. These features may be summarized as follows:

1. The number of conductors inserted in the armature is represented by Z. A pair of compatibly positioned conductors are required for each loop, resulting in $Z/2$ loops within the armature.

2. Instead of interrupting each loop to bring out leads, several series loops with a single pair of leads may be formed by continuously winding a single conductor into n_t turns (Fig. 2-8). This is termed a *coil.* Within the armature there are n_c identical coils.

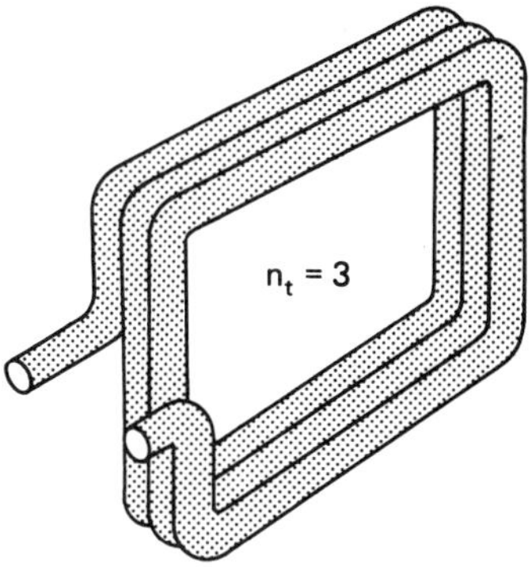

Figure 2-8 Coil

3. The coils are evenly divided among a parallel paths, a always being an even number. Each path has n_c/a coils connected in series.

4. The electrical quantities for each coil are e_c, i_c, and p_c. The voltage, e_c, equals $n_t e$, where e is the average voltage per turn. The current, i_c, also is the current in each of the parallel paths. The power, p_c, equals $e_c i_c$.

From these relationships and definitions, the electrical characteristics of the armature are developed.

Generated Voltage (E_g). The generated voltage is the voltage across the series–parallel network of coils.

$$E_g = \frac{n_c}{a}(n_t e)$$

$$= \frac{(n_c n_t)e}{a}$$

But $n_c n_t = Z/2$ and $e = P\omega\Phi/\pi$; therefore,

$$\boldsymbol{E_g = ZP\omega\Phi/2\pi a} \quad [\text{SI}] = \boldsymbol{ZPS\Phi/60a} \quad [\text{US}] \qquad (2\text{-}1)$$

In these equations, the elements Z, P, and a are design characteristics, built into the machine by the manufacturer, that cannot be changed without extensive rebuilding, but the user *can* control the speed at which the generator is driven and the flux per pole. Therefore, the equations commonly appear as

$$\begin{aligned} E_g &= k\omega\Phi \quad \text{[SI]} \qquad (k = ZP/2\pi a) \\ &= kS\Phi \quad \text{[US]} \qquad (k = ZP/60a) \end{aligned} \tag{2-2}$$

Armature Current (I_a). The armature current is the total current drawn from the coil network.

$$I_a = ai_c$$

Developed Power (P_d). This is the total electrical power developed within the armature.

$$\begin{aligned} P_d &= E_g I_a \\ &= \frac{n_c e_c}{a}(ai_c) \\ &= n_c(e_c i_c) \\ &= n_c p_c \end{aligned}$$

In the expressions above, the generated voltage and armature current are functions of the number of parallel paths. But note that the developed power is independent of the paths. Structurally, a desired number of paths may be created with relative ease, and the manufacturer can apply different arrangements to otherwise identical machines. Consequently, two machines with the same power ratings may have widely disparate voltage and current ratings; for example, a general-purpose generator may be rated 18 kW 250 V 72 A while its nearly identical twin, possibly intended for arc welding, may be rated 18 kW 30 V 600 A.

EXAMPLE 2-1: A dc generator has six poles, 780 conductors, and six parallel armature paths. It rotates at an angular velocity of 40π rad/s. The flux per pole is 0.0392 Wb. (a) What is the generated voltage? (b) If the speed is altered to 900 rpm but the flux is unchanged, what is the voltage? (c) If, at 900 rpm, the flux is changed to 4.35×10^6 maxwells, what is the new generated voltage? (d) With a permissible current per coil of 50 A, what is the rated P_d for the voltage of part (c)?

SOLUTION:

(a) $E_g = P\Phi Z\omega/2\pi a = (6)(0.0392)(780)(40\pi)/(2\pi)(6) = 612$ V

(b) $E_{g(b)}/E_{g(a)} = k\Phi\omega_{(b)}/k\Phi\omega_{(a)}$; therefore,

$$E_{g(b)} = \frac{\omega_{(b)}}{\omega_{(a)}}E_{g(a)} = \frac{2\pi \times 900/60}{40\pi}(612) = 459 \text{ V}$$

(c) $E_{g(c)}/E_{g(b)} = \Phi_{(c)}/\Phi_{(b)}$; therefore,

$$E_{g(c)} = \frac{\Phi_{(c)}}{\Phi_{(b)}} E_{g(b)} = \frac{4.35 \times 10^6 \times 10^{-8}}{0.0392}(459) = 509 \text{ V}$$

(d) $P_d = E_g I_a = E_g(ai_c) = (509)(6)(50) = 152.7 \text{ kW}$

Both speed and flux are variables in the expression for generated voltage. For general-purpose machines, constant speed commonly is presumed. Practically, speed-control devices on the prime mover often justify the presumption, with variations that are small enough to be negligible. On the other hand, relatively small speed variations may be a vital consideration—for example, in paralleling compound generators (Paragraph 2-8).

The topic of variations in flux is covered in Paragraph 2-2. Speed variance is considered in Paragraph 2-3.

2-2. FORMS OF THE DC GENERATOR

There are several basic types of dc generators, each with its unique characteristics. All have an armature and one or both of two types of field windings, shunt and series.

A *shunt* field consists of a large number of turns carrying a relatively small current. The current may be supplied by an external dc source (separate excitation) or by the generator itself (self-excitation).

A *series* field has relatively few turns but a large current, supplied by the armature of the generator.

When both fields are present, the series field usually is wound on top of the shunt field to improve heat dissipation.

The graphic symbols for the armature and fields are shown in Fig. 2-9 [ANSI Y32.2–1975]. Note the distinction in number of loops between the shunt- and series-field symbols.

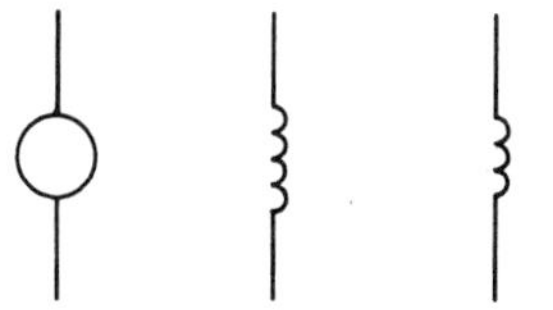

Figure 2-9 Graphic Symbols (ANSI)

Separately Excited Shunt Generator. This generator has a shunt field drawing current from an external source. The composite symbol and electrical parameters of the generator are shown in Fig. 2-10.

Focusing on the field, V_f is an external voltage applied to the winding and R_f the inherent resistance of the winding. The *field current*, I_f, is V_f/R_f.

The armature circuit has resistance R_{ac}, the equivalent resistance of the network consisting of the brushes and the armature conductors. Although no resis-

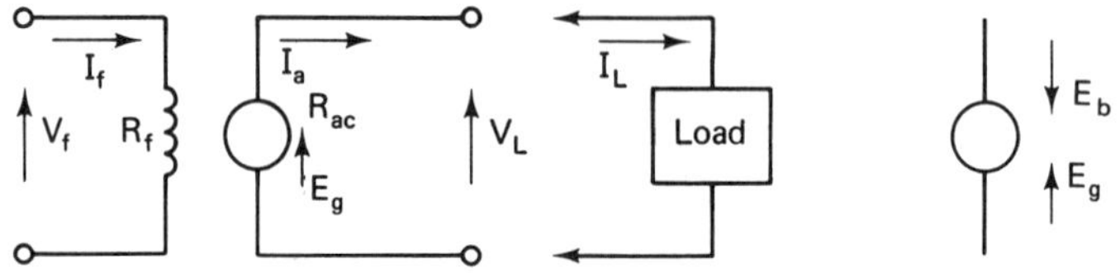

Figure 2-10 Separately Excited Shunt Generator

tor symbol appears on the schematic, it is understood that R_{ac} is present and must be included in electrical calculations.

Resistance of carbon brushes is nonlinear, varying with pressure, humidity, current density, and temperature; for this reason, instead of employing a constant R_{ac}, the network may be treated alternatively as two components: R_a, the equivalent resistance of the armature conductors, and E_b, a constant voltage drop across the brushes.

The voltage, E_g, is generated within the armature. Unloaded, no armature current flows and E_g appears between the line terminals, but under load a *line current* (**I_L**), equal in this case to I_a, flows through the armature and causes a voltage drop, I_aR_{ac}, across the armature circuit resistance; consequently, except at no load, V_L is less than E_g ($V_L = E_g - I_aR_{ac}$).

Power relationships in the generator are emphasized in Fig. 2-11. The power input consists of the electrical power supplied to the field and the mechanical power, P_m, delivered by the prime mover.

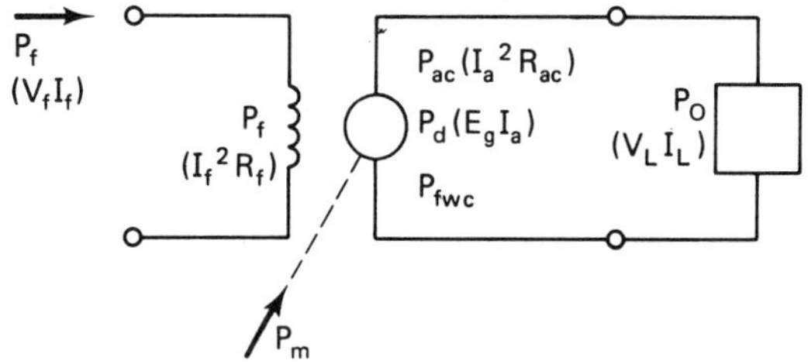

Figure 2-11 Power Relationships

Copper losses (P_{cu}) occur in each resistance within the machine. Note that in the case of the field, P_f is both a power input and a copper loss; that is, $P_f = V_fI_f = I_f^2R_f$. The copper loss in the armature circuit is $P_{ac} = I_a^2R_{ac}$.

All machines have rotational losses that are partly electrical and partly mechanical in nature. These are frictional, windage, and core losses, P_{fwc} (Paragraph 1-13), all of which are proportional to the speed of the armature.

The power output (P_o) is the electrical power delivered by the generator to the load.

$$P_o = V_LI_L$$

Power quantities within a machine in a steady state are related by the equation $P_{in} = P_o + \sum P_{losses}$. For the separately excited shunt generator,

$$P_f + P_m = P_o + P_f + P_{ac} + P_{fwc}$$

Deleting P_f from both sides of the equation and substituting electrical expressions for power wherever possible,

$$\begin{aligned} P_m &= V_L I_a + I_a^2 R_{ac} + P_{fwc} \\ &= (V_L + I_a R_{ac}) I_a + P_{fwc} \\ &= E_g I_a + P_{fwc} \end{aligned}$$

Earlier, $E_g I_a$ was defined as the developed power, P_d. This is power that has been converted from mechanical to electrical form. Note that, except for P_f, P_d consists of the power output plus the electrical losses in the machine.

Especially in larger machines, the amount of power that must be expended to produce the desired power output is an economic concern; output power is a necessary expense to which the cost of power losses is added. *Efficiency*, a measure of merit, is applicable to any device. Efficiency (η) is the ratio of the power output to the power input, usually expressed as a percentage: $\%\eta = 100P_o/P_{in}$. For a given power output, the operational cost of a machine with a high efficiency is less than that of a low-efficiency machine.

For the separately excited generator,

$$\%\eta = \frac{100P_o}{P_d + P_f + P_{fwc}}$$

Self-Excited Shunt Generator. The self-excited shunt generator (Fig. 2-12) differs from the separately excited generator only in the source of the field current. The shunt field is connected not to an external source but to the generator armature, as illustrated. Note that the voltage across the field is the line voltage, V_L.

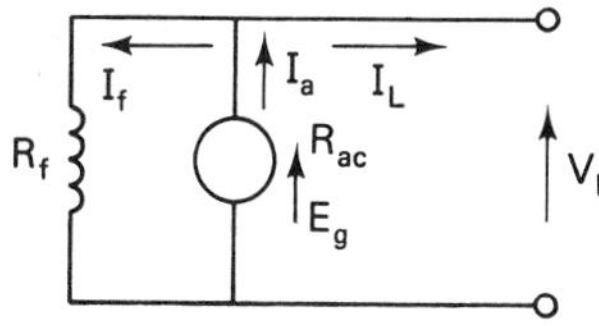

Figure 2-12 Self-Excited Shunt Generator

From Ohm's and Kirchhoff's laws, the following relationships exist:

$$I_a = I_f + I_L$$

$$I_f = \frac{V_L}{R_f}$$

$$E_g = V_L + I_a R_{ac}$$

All the input power is mechanical. The power balance is

$$\begin{aligned} P_{\text{in}} &= P_o + \sum P_{\text{loss}} \\ &= V_L I_L + V_L I_f + I_a^2 R_{ac} + P_{fwc} \\ &= V_L(I_L + I_f) + (I_a R_{ac}) I_a + P_{fwc} \\ &= (V_L + I_a R_{ac}) I_a + P_{fwc} \\ &= E_g I_a + P_{fwc} \\ &= P_d + P_{fwc} \end{aligned}$$

In order for any form of self-excited generator to function, residual magnetism must be present when the machine is at rest.

The saturation curve of a generator is shown in Fig. 2-13. The generated voltage is plotted as a function of the field current. Since $E_g = k\Phi S$, for a constant speed $E_g \propto \Phi$; further, with a constant effective cross section of the magnetic circuit, $E_g \propto B$. Similarly, $I = Hl/N$, where l/N is a machine constant; therefore, $I_f \propto H$. The E_g–I_f and B–H curves are identical except for the indexing of the abscissa and ordinate.

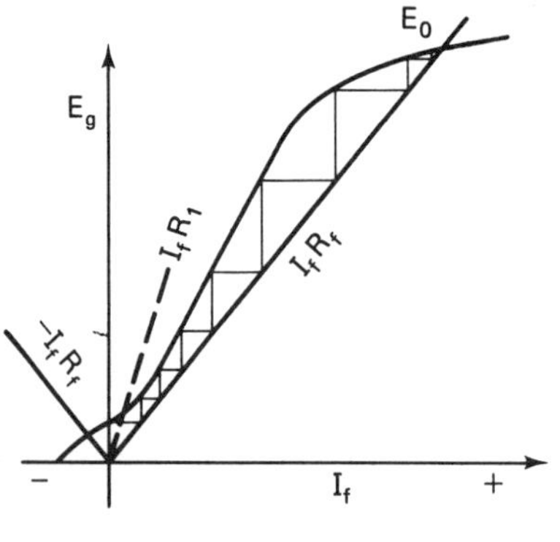

Figure 2-13 Voltage Buildup

When armature rotation begins, a small voltage will be generated due to residual flux; sequentially, a small field current flows, increasing the flux, and the generated voltage increases. The higher voltage leads to further increases in field current, flux, and generated voltage. The process continues as illustrated in Fig. 2-13 until equilibrium is achieved. On the graph, the plot of the field-voltage line, $I_f R_f$, intersects the magnetization curve at the point E_o. With no load connected at the time the generator is started, the line voltage builds to this point.

Should the field resistance be such that the plot is the dashed line, $I_f R_1$, the desired buildup of the terminal voltage will not occur. Also, there can be no buildup if the shunt field connections are reversed; in Fig. 2-12, if the top and bottom field connections to the armature were interchanged, both the coil current and the flux it creates would be reversed, also. This corresponds to the $-I_f$ portion of the curve in Fig. 2-13, resulting in a terminal voltage less than that which is due to the residual alone.

There is a third possibility that can preclude voltage buildup. In Fig. 2-14, during rotation of the coil Z_1–Z_2 from vertical to the angle ωt, the flux through the loop is decreasing; by Lenz's law, the direction of the induced current must produce a flux opposing the decrease—that is, the current attempts to maintain the original strength and direction of the flux through the loop. The right-hand rule identifies the recorded direction of currents in the conductors. While traversing the next quadrant [Fig. 2-14(b)], the flux through the coil is increasing, flux created by induced currents must oppose the increase, and the currents are as shown.

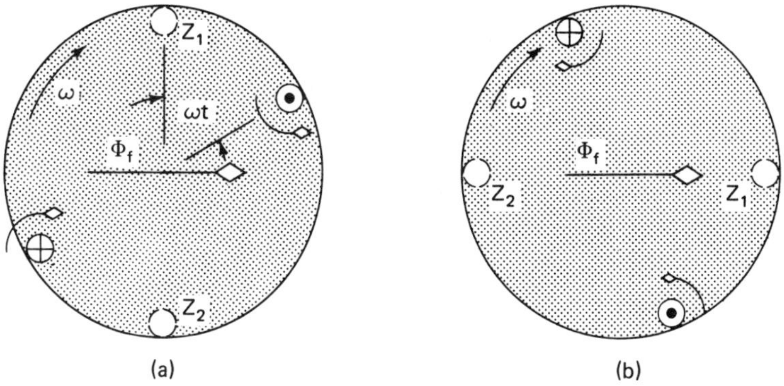

Figure 2-14 Direction of Induced Currents

During the remaining half of the revolution, the circumstances are identical to Fig. 2-14(a) and (b) but with conductors Z_1 and Z_2 interchanged.

In Fig. 2-15 the directions of both the rotation and the flux are identical to those of Fig. 2-14, resulting in currents as shown. The axis of the field flux is the *direct axis*; perpendicular to the direct axis is the *quadrature axis*. All conductor currents to the right of the quadrature axis have one polarity, while those to the left are opposite. No voltage is induced in conductors on the quadrature axis (corresponding to 0° and 180° on the sine curve); for this reason, it also is referred to as the neutral axis.

In Fig. 2-16, the direction of Φ_f is the same as that of Fig. 2-15 but the rotational direction is reversed. Applying the preceding rationalization results in a determination that current polarities are the reverse of that of Fig. 2-15. The reversal of currents between Figs. 2-16 and 2-15 also means reversal in polarity of the generated voltage, E_g, and the field current, I_f. Thus, if buildup occurs in the first case, it will not occur in the second.

A newly constructed machine initially is magnetically neutral; further, one that has been idle for an extended period may lose its residual. In order for these machines to build up a voltage, a residual must be established. Procedurally, an

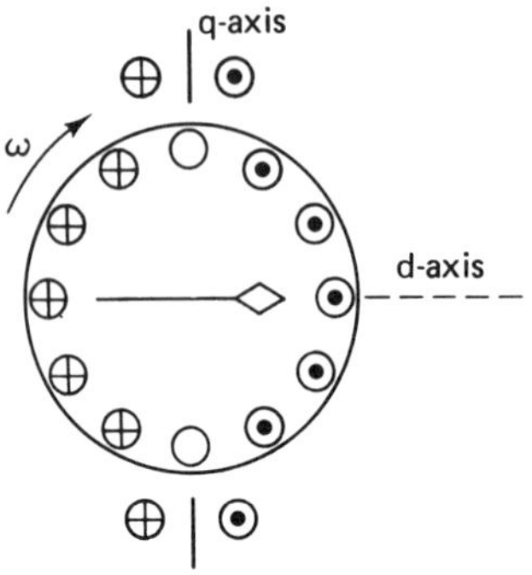

Figure 2-15 Direct and Quadrature Axes

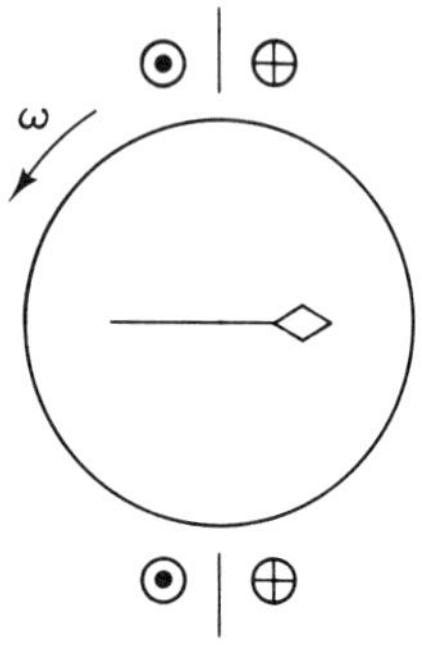

Figure 2-16 Reverse Rotation

external voltage is momentarily applied across the field, then removed. This process is called "flashing the field."

Compound Generator. In the shunt generator, the line and the generated voltages are related by the equation $V_L = E_g - I_a R_{ac}$. Even if E_g remains constant throughout the loading range, the line voltage will decrease as the armature current increases (Fig. 2-17). A decrease may be acceptable in some applications but undesirable in others. The compound generator provides compensation for the $I_a R_{ac}$ drop. Since the drop varies with the armature current, the correction also must be governed by this current. The remedy is addition of a series field to each pole. The term "series" implies that the field is connected in series with the armature and carries the armature current; in practice, the field may be in series with the line, carrying I_L, but I_f is small in contrast to I_a under usual loading conditions and the difference between I_a $(= I_L + I_f)$ and I_L is slight.

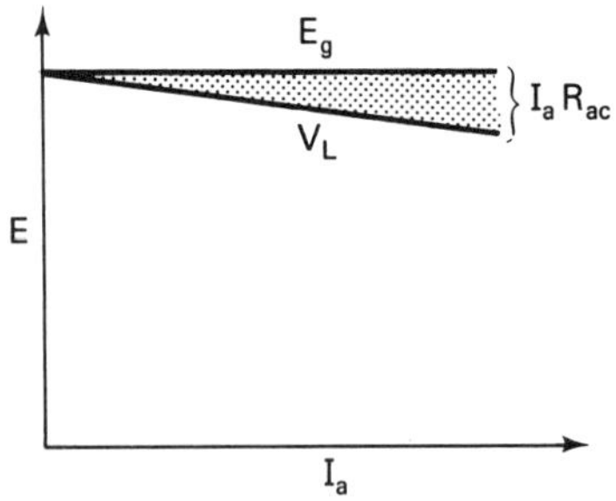

Figure 2-17 Armature Voltage Drop

In Fig. 2-18, the shunt and the series windings on one pole are illustrated. Each contributes a component to the total flux in the pole, $\Phi_f = \Phi_{sh} + \Phi_{se}$; Φ_{sh} is relatively constant, but Φ_{se} varies with the armature current. On the magnetization curve, the voltage corresponding to $\mathfrak{F}_{sh}$ is due to Φ_{sh}, alone; adding $\mathfrak{F}_{se}$ causes an increase (ΔE_g) in the generated voltage. ΔE_g may be equal to, greater than, or less

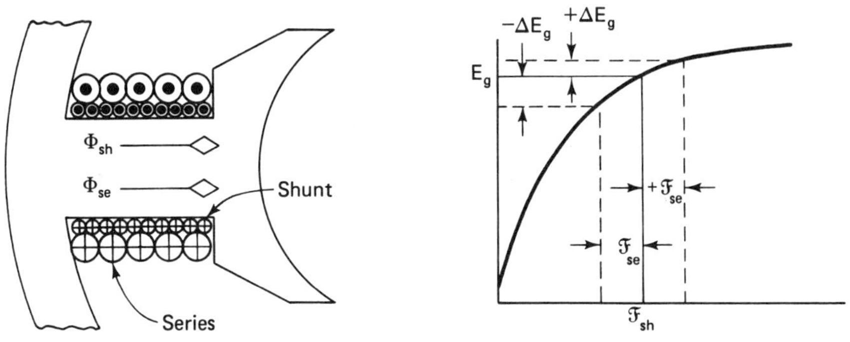

Figure 2-18 Shunt- and Series-Field Windings

than $I_a R_{ac}$, depending upon the iron characteristics, current, and number of series turns.

It is also possible to reverse the polarity of the current in the series winding by interchanging the field connections, with the result that the total flux in the pole is $\Phi_f = \Phi_{sh} - \Phi_{se}$. On the magnetization curve this condition is represented by $-\mathfrak{F}_{se}$ and $-\Delta E_g$.

When the series field flux adds to that of the shunt, the generator is cumulatively compounded; conversely, if it subtracts, the machine is differentially compounded. The norm is cumulative compounding—differential is rarely employed.

The schematics and parameters of the compound generator are shown in Fig. 2-19. The terminology "long" or "short" refers to the connection points of the shunt field and identifies the current through the series field as either I_a or I_L. Unless qualified, the term "compound" hereafter will reflect the long shunt connection. Schematically, there is no distinction between differential and cumulative compounding.

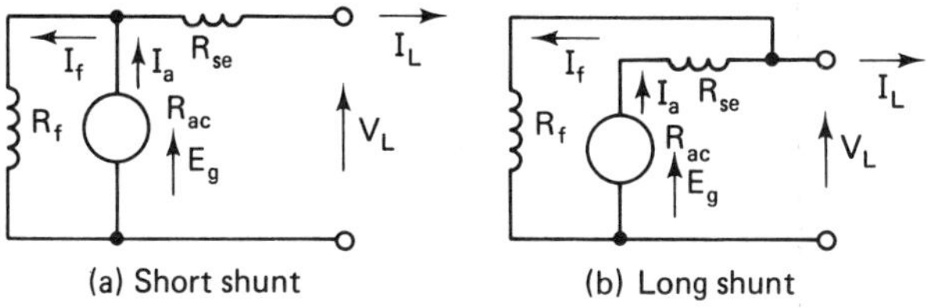

Figure 2-19 Compound Generator

In all forms of the generator, flux is the key to the magnitude of E_g. Logically, provisions to control E_g are linked to adjustment of the flux. Control is imposed upon the shunt field; by adjusting the current through the field, the flux is varied. Note, however, that only in the atypical circumstance of an operating range confined to the linear portion of the magnetization curve is flux directly proportional to current; normally, operation is beyond the knee of the curve.

The common method for control of E_g is insertion of a variable resistor (rheostat) in series with the shunt field.

EXAMPLE 2-2: The generated voltage of the compound dc generator of Fig. 2-20 is 265 V and the line current is 100 A. What is the line voltage?

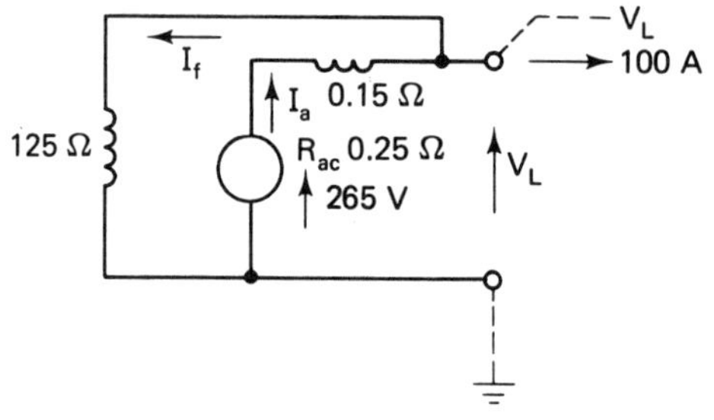

Figure 2-20 Example 2-2

SOLUTION: Assuming a ground at the lower end of the armature, V_L appears as the potential at the indicated node. Writing the nodal equation

$$\frac{V_L}{R_f} + \frac{V_L - E_g}{R_{se} + R_{ac}} + I_L = 0$$

then

$$V_L = \frac{R_f[E_g - I_L(R_{se} + R_{ac})]}{R_{se} + R_{ac} + R_f}$$

$$= \frac{125[265 - (100)(0.15 + 0.25)]}{0.15 + 0.25 + 125}$$

$$= 224.3 \text{ V}$$

The power equation for the compound generator differs from that of the self-excited shunt only in inclusion of the copper loss in the series field.

Commonly, compound generators are designed to be overcompounded. After installation, the degree of compounding is optimized by inserting a diverter resistor (R_d in Fig. 2-21) in parallel with the series winding, thereby reducing the

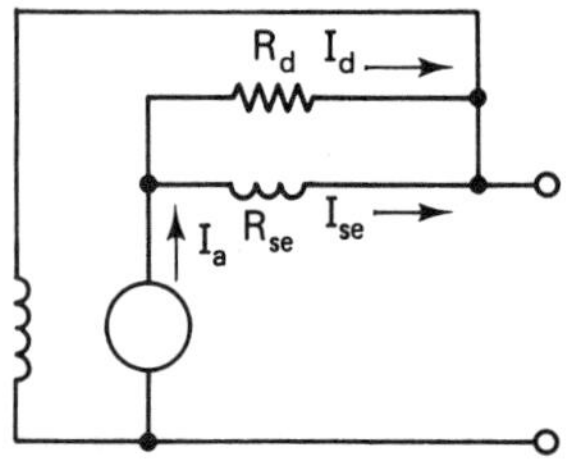

Figure 2-21 Diverter

series-field current to a value that produces the terminal characteristics required by the user.

Series Generator. The schematic of a series generator is shown in Fig. 2-22. There is no shunt field; the flux is created by the load current, and E_g will rise dramatically as this current increases until saturation occurs. This is not a general-purpose form of the generator but is employed in limited special applications; the series generator is most likely to appear as a line-voltage booster in an electrical traction system (e.g., railways).

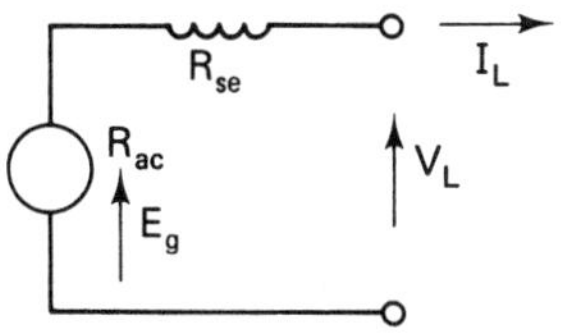

Figure 2-22 Series Generator

2-3. EFFECT OF SPEED ON E_g The discussion of the voltage characteristics in Paragraph 2-2 tacitly assumed that the speed of the generator drive remained constant throughout the loading range. If the speed changes but the flux remains constant, the generated voltage will vary directly with speed. The saturation curve of a self-excited shunt generator, known for one speed, may be determined for any other speed by the speed ratio.

EXAMPLE 2-3: The saturation curve of a separately excited shunt generator at 1200 r/min appears in Fig. 2-23. Find the curve for 900 r/min.

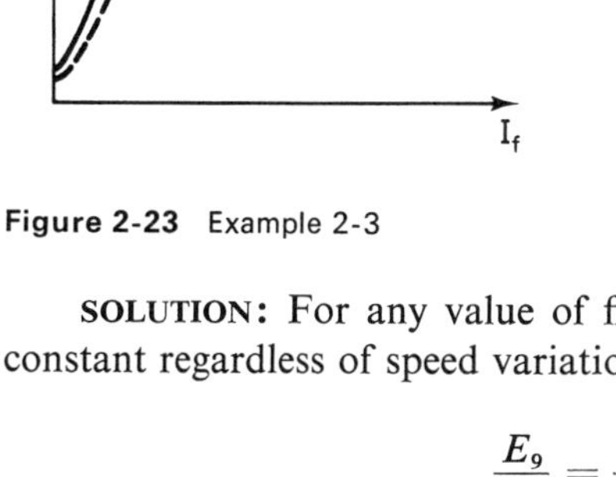

Figure 2-23 Example 2-3

SOLUTION: For any value of field current, the strength of the flux remains constant regardless of speed variations.

$$\frac{E_9}{E_{12}} = \frac{k\Phi(900)}{k\Phi(1200)} = 0.75$$

Therefore,

$$E_9 = 0.75E_{12}$$

The dashed curve is the plot for 900 r/min.

Because of a countertorque developed in the generator, the speed of unregulated movers normally will decrease with loading. This will affect the $V_L - I_L$ characteristic curves. For all forms of the generator, decreased speed will result in a lower line voltage for a given line current than would occur at constant speed.

2-4. ARMATURE REACTION With a load attached to a generator and current flow in the armature conductors, an armature flux (Φ_a) is superimposed upon the field flux (Φ_f). Neglecting the small armature current drawn by the field of a self-excited generator, at no load I_a is zero, but graduated loading results in increasing armature currents and increasing Φ_a. The field flux lies on the direct axis (Fig. 2-24), but the armature reaction flux is on the quadrature axis; this circumstance is termed *cross-magnetization*. Some special forms of the dc generator exploit cross-magnetization (Paragraph 2-17), but in a conventional machine it degrades performance unless countermeasures are provided.

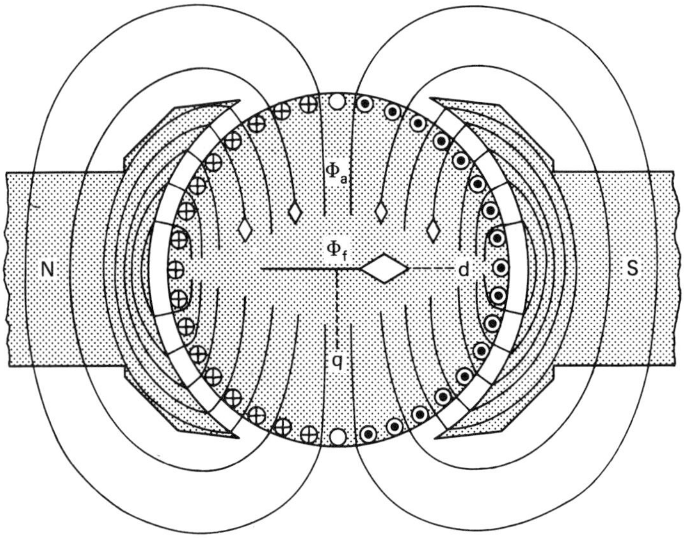

Figure 2-24 Armature Reaction

The pattern of the armature flux is illustrated in Fig. 2-24. In the south pole, Φ_a opposes Φ_f in the upper half but augments it in the lower half. This circumstance is inverted and repeated in the north pole. Translated into flux density distribution, B_f, B_a, and their sum ($\sum B$) are plotted in Fig. 2-25 with the assumption that the

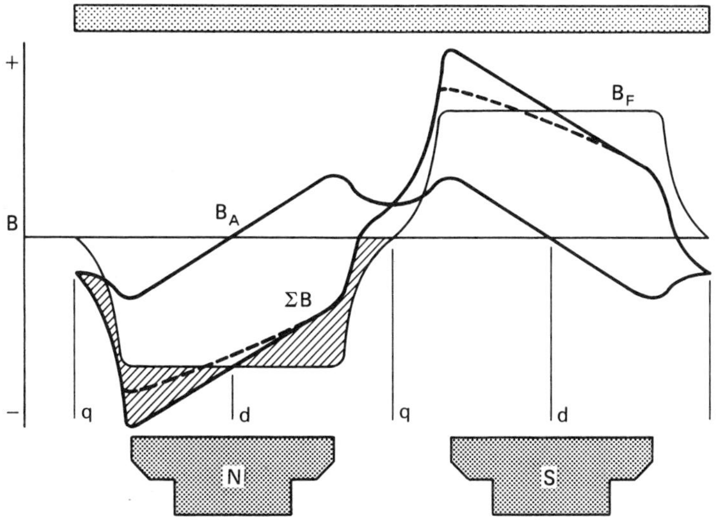

Figure 2-25 Air-Gap Flux Density

magnetic circuit is linear. The crosshatched areas are equal and it appears that the net flux would remain Φ_f. But the actual condition of nonlinearity of the saturation curve must be considered. The effect can be deduced by first reverting to an electrical circuit analogy.

A series circuit with current $I = E/R$ is shown in Fig. 2-26(a). If the resistor were split along the centerline to form two parallel resistors, each resistor would have a cross-sectional area one-half that of the original resistor and, therefore, an ohmic value $2R$. Inserting equal emfs V into each of the parallel branches but with opposite polarities [Fig. 2-26(b)], $I = I_1 + I_2 = (E - V)/2R + (E + V)/2R = 2E/2R = I$, the same current that resulted in Fig. 2-26(a).

In the magnetic circuit equivalent, E becomes $\mathfrak{F}_f$ and V is $\mathfrak{F}_a$. Referring to Fig. 2-26(c), the net mmf in one of the paralleled branches is $\mathfrak{F}_f + \mathfrak{F}_a$, and in the other it is $\mathfrak{F}_f - \mathfrak{F}_a$. Adding $\mathfrak{F}_a$ results in an increase, $+\Delta B$, but subtraction of $\mathfrak{F}_a$ results in the decrease, $-\Delta B$. The magnitude of the increase is less than that of the decrease, and the actual flux density distribution is approximately that shown by the dashed amendment to the summation curve of Fig. 2-25. Therefore, the net flux is decreased by armature reaction.

Probably more important than the drop in E_g due to flux reduction is the distortion of the voltage waveform attributable to the distorted pole flux. At the point where B peaks, the instantaneous voltage induced in a coil is greater than the amplitude of the sinusoidal e_c resulting from normal distribution. The ends of the coil are connected to copper commutator segments separated only by a very small air gap; voltages generally larger than 30 to 40 V across this gap will cause arcing that not only is destructive to the brushes and commutator segments but also may lead to armature current flow from the positive to the negative brush, jumping

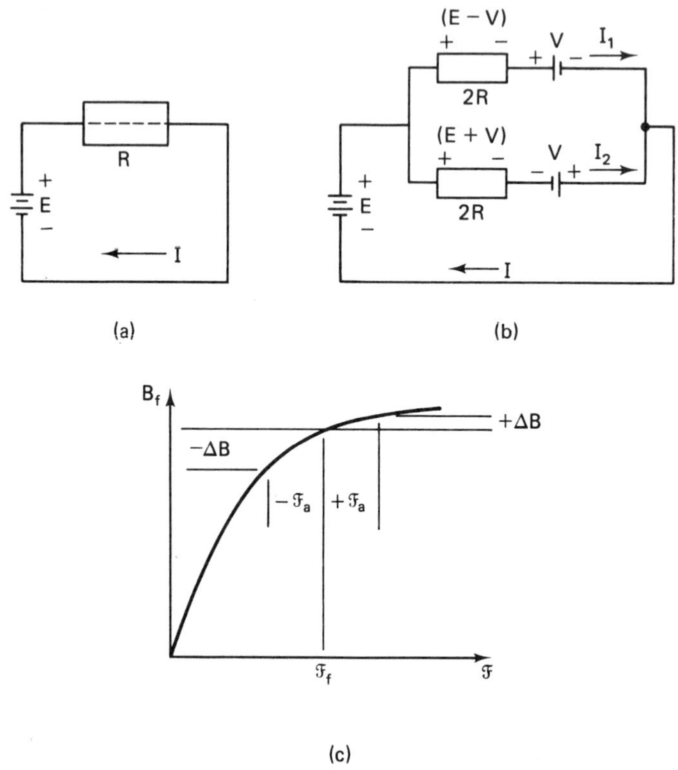

Figure 2-26 Electrical Circuit Analogy

from segment to segment through the air gaps, and short-circuiting the load. This condition is termed *flashover*.

Another detrimental aspect of armature reaction is the establishment of flux in the neutral (or commutating) zone, the areas in the vicinity of the intersection of the quadrature axis and the armature circumference. In Fig. 2-24, note that no current appears in conductors at these points. As the armature rotates, a conductor crosses the quadrature axis and the direction of the current in that conductor must reverse. The problems and remedies associated with this reversal (commutation) are discussed in greater detail in Paragraph 2-10; for the moment, note that armature reaction flux in the neutral zone will induce conductor voltages that aggravate the commutation problem.

Uncorrected armature reaction also causes a shift in the position of the quadrature axis. The vectorial combination of Φ_a and Φ_f in Fig. 2-24 results in the net flux Φ_r shown in Fig. 2-27. The axis of the flux has shifted from the *d*-axis to d'; concurrently, the *q*-axis shifts to q'. The direction of the shift of these axes always is in the direction of rotation of the armature, but since the magnitude of Φ_a in Fig. 2-24 varies with I_a, the amount of displacement is not constant. This

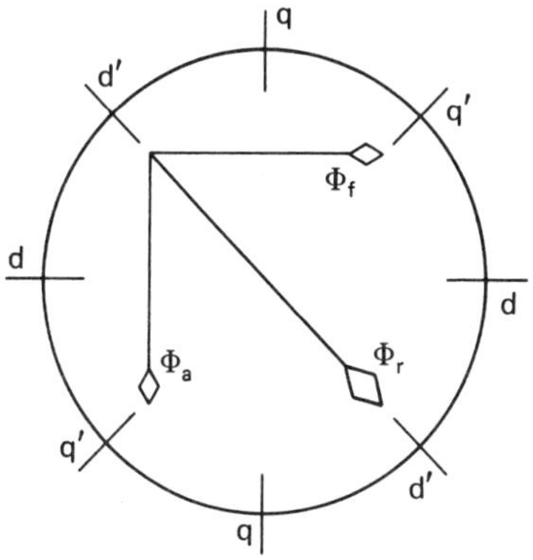

Figure 2-27 Resultant Flux

displacement also has a detrimental impact on commutation; optimally, the brushes are electrically situated on the q-axis, but if they remain fixed while the q-axis shifts with each load change, this requirement seldom is met. Remedial measures to combat armature reaction require (1) elimination of the flux in the neutral zone and (2) restoration of the net flux to the direct axis.

If small, slender poles are centered between the main field poles of the generator, they can be excited to produce a flux in opposition to the armature reaction flux (Fig. 2-28). Located on the quadrature axis, they can free the neutral zone of all flux. These poles are termed *commutating poles* or, alternatively, interpoles. The magnitude of the required correction is proportional to the armature current; therefore, the commutating windings are placed in series with the armature. The number of turns in the winding is designed to provide the proper strength flux to

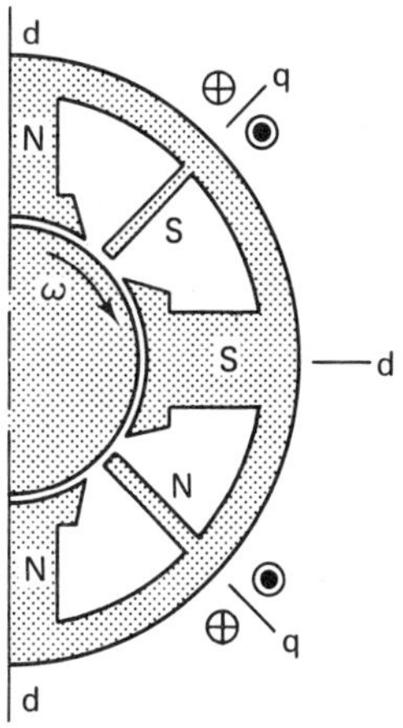

Figure 2-28 Commutating Poles

neutralize Φ_a under all load conditions. (Actually, the discussion in Paragraph 2-10 develops reasons for providing an interpole flux that is greater than the zonal armature reaction flux.)

While interpoles eliminate flux in the commutation zone, they do not correct the distorted field pattern. This condition may be corrected by the addition of *compensating* windings. Unlike the shunt, series, or interpole windings, compensating windings are not wound around a salient pole core but rather are inserted directly through the shoe of the main poles (Fig. 2-29). These windings carry armature current and are designed to provide an mmf equal to, but opposing, that of the armature conductors directly opposite the pole face. They can be regarded as producing a flux equal to, but in the opposite direction of, the armature flux; by superposition, the resultant flux is zero.

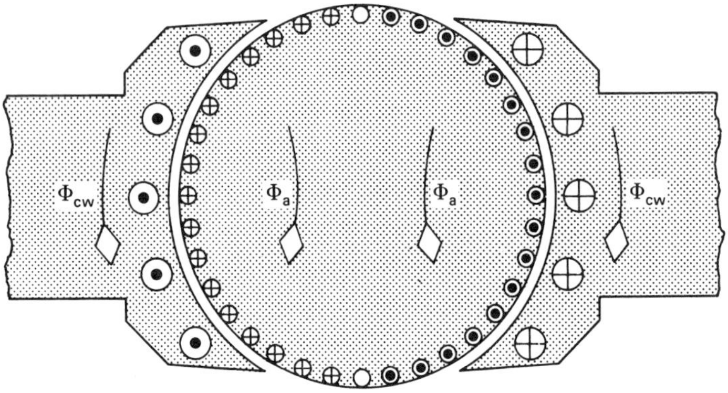

Figure 2-29 Compensating Windings

Generally, 65 to 70% of the surface of the armature cylinder is directly opposed by surfaces of the pole faces. Letting p represent this percentage, the number of armature conductors directly opposite a single pole face is $pZ/100P$. Since the armature conductors each carry the current I_a/a, the armature reaction mmf is $pZI_a/100Pa$. The compensating windings carry I_a; letting Z_{cw} represent the number of required conductors in each pole face and equating the mmfs,

$$Z_{cw}I_a = \frac{pZI_a}{100Pa}$$

Therefore,

$$Z_{cw} = \frac{pZ}{100Pa}$$

The schematic symbol for either a commutating (interpole) or compensating winding is shown in Fig. 2-30(a). In this text the subscripts *ip* and *cw* identify the type where necessary. When both windings are present, the symbol is not dupli-

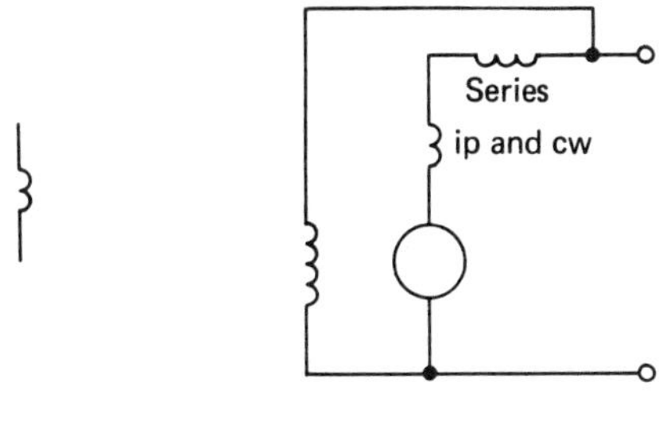

(a) Commutating and/or compensating field (b) Compound generator

Figure 2-30 Graphic Symbol

cated but appears only once; Fig. 2-30(b) is a compound generator with interpole and compensating windings.

While compensating windings are the most effective means for correcting the decrease in flux due to armature reaction, they are costly. Less effective but cheaper methods do not neutralize the armature mmf but rather (1) aim at reducing the armature reaction flux by increasing the reluctance in its path, (2) encourage greater field flux densities in the center of the pole face where distortion due to armature reaction is minimal, and (3) ensure that the ratio of field flux to armature reaction flux is high. Figure 2-31 illustrates two types of pole construction designed to accomplish (1) and (2).

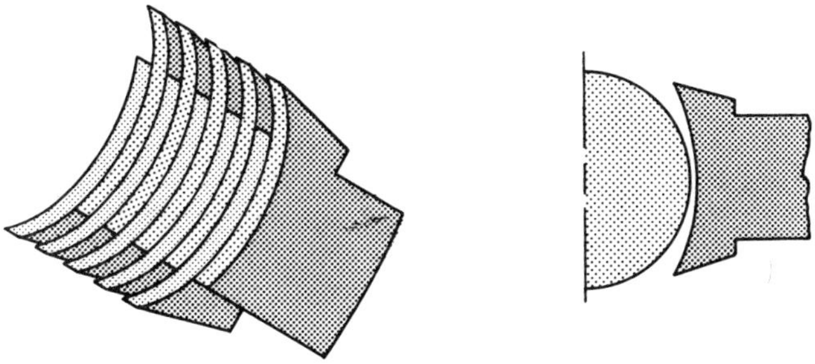

Figure 2-31 High-Reluctance Pole Tips

2-5. LINE-VOLTAGE CHARACTERISTICS

The comparative V_L vs. I_L characteristics of the several types of generators are considered in selecting a particular type for a specific application. These characteristics can be deduced from the electrical and magnetic parameters, including the effect of armature reaction.

In subsequent discussions, V_L is the voltage appearing between the terminals of the generator. The no-load voltage, V_0, is the terminal voltage when the line current, I_L, is zero. The nameplate of the generator has recorded the rated power

(P_R) and rated voltage (V_R); rated current I_R ($= P_R/V_R$) appears on the nameplate of modern machines but may be missing on older ones.

Figure 2-17 illustrated the drop-off in V_L of a shunt generator with the qualifying assumption that E_g remains constant. *But the assumption is invalid.* A realistic assessment must consider two factors, both of which act to reduce E_g as load current increases. First, since $I_f = V_L/R_f$ in the self-excited shunt, the decrease of V_L due to loading is accompanied by a decrease in I_f and the field flux. Second, increasing armature reaction adds to the reduction. The corrected plot is represented in Fig. 2-32.

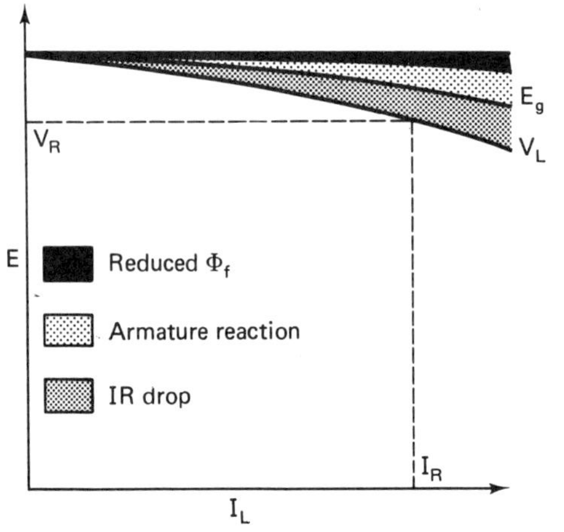

Figure 2-32 V_L I_L Characteristics—Shunt Generator

The cumulative-compound generator is subjected to additional influences. The generated voltage is $k\omega(\Phi_{sh} + \Phi_{se} - \Delta\Phi_a)$. Φ_{se}, not present in the shunt generator, is responsible for the distinct characteristic of the compound generator. It not only adds directly to the net flux but also affects the drop in Φ_{sh}, since V_L droops less than it does in the shunt. The magnitude of Φ_{se} is a function of the mmf, $N_{se}I_a$, with N_{se} a design option and I_a controllable by a diverter. Consequently, the compound generator may exhibit any of three V_L vs. I_L characteristics (Fig. 2-33).

If the series-field mmf is such that the rated full-load line voltage is equal to the no-load voltage, the generator is *flat-compounded.* A stronger series field will cause *overcompounding*, in which V_R is greater than V_0, while a weaker field results in *undercompounding*, in which V_R is less than V_0. In the latter case, the magnitude of the decrease is less than that of the same machine without the series field, operating as a shunt generator. Intuitively, it might appear that flat-compounding is preferable, but this is not a mandate. For example, economic and design considerations may optimize a winding that is undercompounded; the speed of the prime mover may decrease with loading; or separation of the generator and load sites

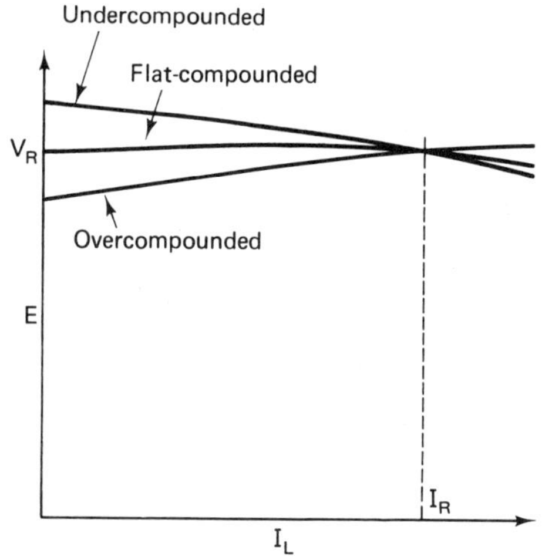

Figure 2-33 V_L/I_L Characteristics—Compound Generator

may argue for overcompounding in order to compensate for voltage drops in the transmission line. Reiterating an earlier comment, not uncommonly the generator delivered by the manufacturer is overcompounded and is adjusted to the desired degree of compounding by means of a diverter, after installation.

In the series generator, the no-load voltage is entirely dependent upon residual magnetism. Under load, the series-field flux, tempered by armature reaction, determines the voltage. The comparative characteristics of the several generator types are shown in Fig. 2-34.

An underlying assumption in the development of the V_L–I_L curves was that the speed of the generator is constant. Exceptionally, broad speed variations may be the norm (e.g., an automobile generator). The design of a generator is based upon a particular speed, recorded on the nameplate, that cannot be ignored with impugnity. Mechanically, the construction must provide adequate heat dissipation, which requires a minimum speed, and withstand the centrifugal forces created at high speeds. Electrically, the design establishes the permissible generated voltage. User option is confined to a limited speed range. Within the limits, the principal effect of a change in speed is in the generated voltage.

2-6. VOLTAGE REGULATION

Voltage regulation (VR) is a measure indicative of the drop of V_L between no load and full load. Usually expressed as a percentage, it is the ratio of the voltage drop to rated V_L: that is, $\%\text{VR} = 100(V_0 - V_R)/V_R$. For an overcompounded generator, the %VR will be negative.

The %VR usually is based on the condition of constant speed throughout the load range. However, it may be expressed for a motor–generator set not on the basis of constant speed but rather on V_0 and V_R resulting from a drop in the motor speed with loading in addition to the electrical and magnetic effects.

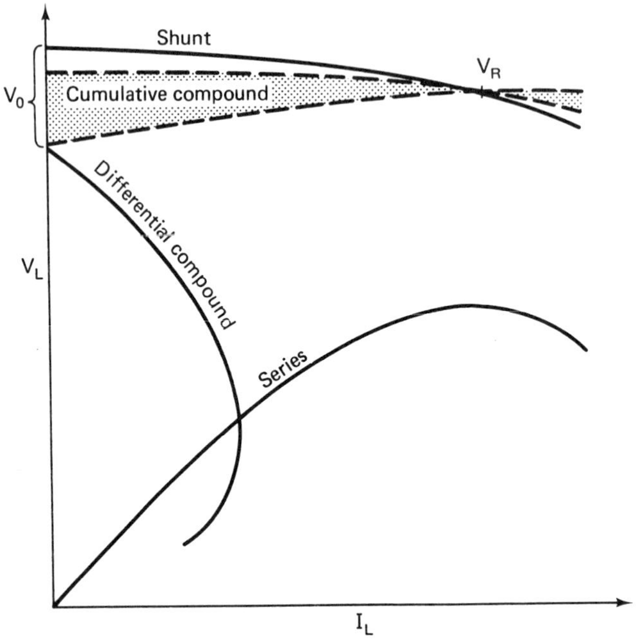

Figure 2-34 Comparative Characteristics

EXAMPLE 2-4: A cumulative compound generator is rated 240 V 24kW 1800 r/min. Parameters of the generator are:

$$N_f = 2000 \qquad N_{se} = 6$$

$$R_f = 205\ \Omega \qquad R_{ac} = 0.035\ \Omega \qquad R_{se} = 0.065\ \Omega$$

The open-circuit saturation curve at 1800 r/min with only the shunt winding excited is shown in Fig. 2-35. The effect of armature reaction is negligible.

(a) What is the %VR when the prime mover maintains a constant speed of 1800 r/min?

(b) What is the %VR when the no-load speed is 1800 r/min but decreases to 1710 r/min when the generator delivers rated full-load current?

SOLUTION:

(a) V_0**:** The $I_f \times R_f$ voltage line is plotted on the graph. The intersection of this line with the saturation curve is the no-load voltage, 204 V. [The field-current component of the voltage drop across the armature and series field, $I_f(R_{ac} + R_{se})$, is about 0.1 V and is neglected in this and subsequent calculations.]

V_R **(1800 r/min):**

$$I_R = \frac{P_R}{V_R} = \frac{24\times 10^3}{240} = 100\ \text{A}$$

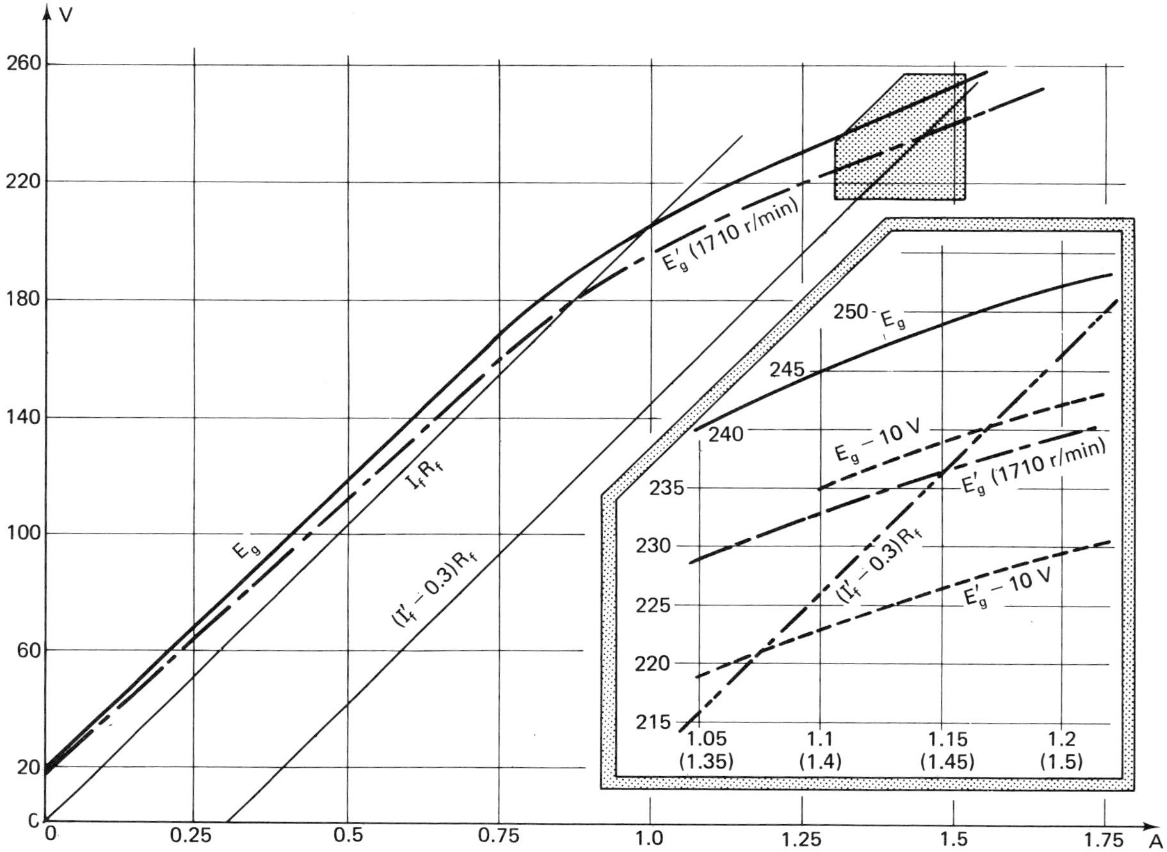

Figure 2-35 Example 2-4

The voltage drop across the armature and series field at full load is

$$I_R(R_{ac} + R_{se}) = (100)(0.035 + 0.065) = 10 \text{ V}$$

On the graph, a line segment is plotted parallel to the E_g curve at $E_g - 10$ V. V_R must fall on this line.

The effect of the series field is introduced by finding the shunt-field current increment, ΔI_f, that provides the same mmf as the series field under full-load conditions.

$$\frac{\Delta \mathfrak{F}_{sh}}{\mathfrak{F}_{se}} = \frac{N_f(\Delta I_f)}{N_{se} I_{se}} = 1$$

$$\Delta I_f = \frac{N_{se} I_{se}}{N_f} = \frac{(6)(100)}{2000} = 0.3 \text{ A}$$

The full-load line voltage results from an equivalent shunt-field current of $I_f + 0.3$ A. But V_L still must satisfy the Ohm's law relationship $V_L = I_f R_f$. Letting I'_f represent the equivalent field current at full load, $I_f = I'_f - 0.3$ A. A plot of $(I'_f - 0.3)R_f$ is the locus of the possible V_R.

The intersection of $(E_g - 10 \text{ V})$ and $(I'_f - 0.3)R_f$ fulfills both of the required conditions and $V_R = 240$ V.

$$\%\text{VR} = \frac{V_0 - V_R}{V_R} \times 100 = \frac{204 - 240}{240} \times 100 = -15\%$$

(b) Since the no-load speed is unchanged, V_0 remains 204 V.

For the full-load speed of 1710, the original saturation curve must be adjusted. For any value of I_f,

$$\frac{E'_g}{E_g} = \frac{1710}{1800}$$

Therefore,

$$E'_g = 0.95 E_g$$

The dashed-line curve is the plot of E'_g. V_R is found as in part (a)—that is, plot $(E'_g - 10 \text{ V})$ and find its intersection with $(I'_f - 0.3)R_f$. V_R is 221 V.

$$\%\text{VR} = \frac{204 - 221}{221} \times 100 = -7.7\%$$

Note that the plots of $V_L I_L$ for the two circumstances show less overcompounding when the prime-mover speed decreases with loading.

2-7. TORQUES IN THE GENERATOR

In Paragraph 1-10, electromagnetic forces were discussed and an expression for the torque developed by a current loop lying in a magnetic field was rationalized ($T = Blid$).

In the generator, armature currents flow because of the voltage induced by rotation of the conductors in the pole flux. But as current-carrying elements in the same flux field, these conductors will experience an electromagnetic force. Referring to Fig. 2-36, the generator of Fig. 2-15 is expanded. In order for the generator to rotate in the indicated direction, a clockwise torque must be applied by the prime mover to the generator shaft. However, when the directions of the armature currents and the flux field are analyzed from the viewpoint of magnetic forces, the conclusion is that a counterclockwise torque is developed—that is, a developed torque (T_d) opposes the applied input torque (T_i). Additionally, friction, windage, and core effects contribute a small countertorque (T_{fwc}). If rotation is at a uniform angular velocity, the net torque on the rotor must be zero. Thus, $T_i = T_d + T_{fwc}$.

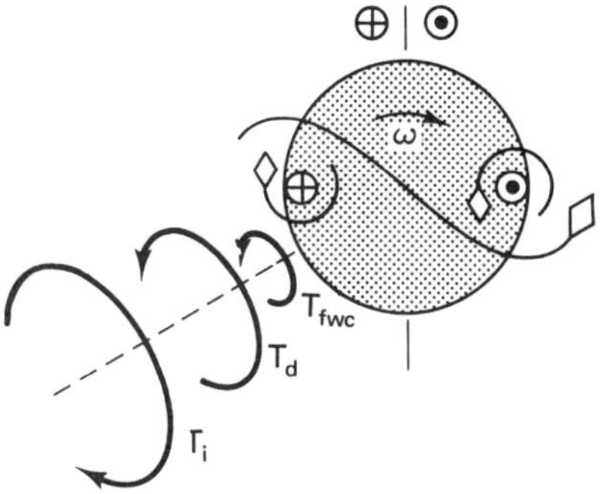

Figure 2-36 Generator Torques

Developed torque is more fully explored in Section 2-3. However, at this point, it is interesting to note how the generator demands greater power input from the prime mover in order to meet increased load requirements. For simplicity, assume that ω, B, and P_{fwc} are constant. The developed electrical power, $E_g I_a$, is directly proportional to I_a, the armature current. The loop torque, *Blid*, has one variable, $i\ (= I_a/a)$. Therefore, T_d (= number of loops $\times$ torque per loop) is proportional to I_a. An increase in loading demands an increase of input torque from the prime mover and, thereby, an increased mechanical power input (ωT). For all self-excited forms of the generator, the power balance is

$$\omega(T_d + T_{fwc}) = E_g I_a + P_{fwc}$$
$$= P_o + \sum P_{cu} + P_{fwc}$$

2-8. PARALLELING GENERATORS

A requirement to parallel generators may stem from any of a number of causes: plant expansion may require additional generating capacity; periodic peaks in the power demand perhaps are most economically accommodated by temporary paralleling of two generators rather than continuous operation of one large-capacity generator; an on-line generator may need replacement for repair or maintenance without interrupting the flow of power to the line loads. Whatever the reason, generators may be paralleled providing they are compatible and proper procedures are followed.

In the following discussion, the term *normal excitation* means that the current in the shunt field is adjusted by means of the rheostat to the level that produces rated terminal voltage when the generator is delivering rater power. At lesser field currents the generator is *underexcited*—greater currents, *overexcited.*

Two self-excited shunt generators, G1 and G2 in Fig. 2-37, are connected to the line by switches 1 and 2. Their rated voltages, but not rated currents, are identical. With respect to the assumed ground, the terminal voltage of each generator, V_1 and V_2, appears as the potential on the generator side of its switch; the potential on the line side of both switches is V_L, the line voltage. When either switch is closed, the terminal voltage of the corresponding generator equals the line voltage—for example, $V_1 = V_L$ when Sw 1 is closed.

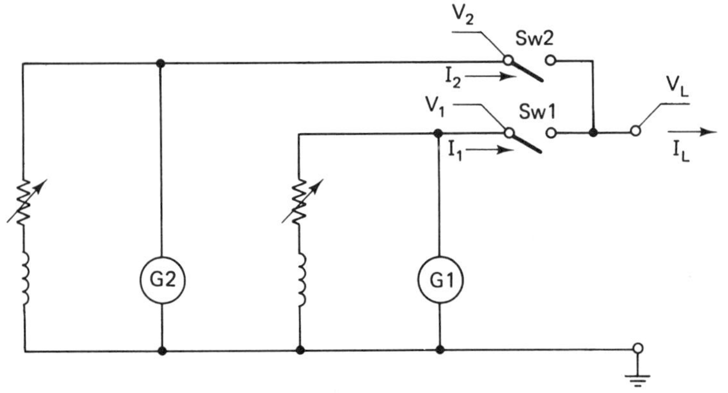

Figure 2-37 Paralleled Shunt Generators

I_L is the line current. It is equal to $I_1 + I_2$, the sum of the load currents of the generators. With a switch open, the corresponding generator load current is zero; with both switches closed, I_1 and I_2 normally are positive, but in unique circumstances, subsequently discussed, one or the other (but not both simultaneously) may be zero or negative.

In Fig. 2-38, the characteristic load curve of G1 with normal excitation is plotted to the right of the vertical axis (V_1–I_1). The characteristic of G2 is plotted to the left for several different excitation levels; note that, unlike usual coordinate scales, indexing to the left of the vertical axis is positive.

If Sw 1 is closed and Sw 2 is open, the line current will be supplied by G1 only. An equivalent load resistance, R_A, is connected to the line; the load line, $I_L R_A$, for this resistance is plotted, intersecting the V_1–I_1 curve and establishing the circuit operating point at V_{1R} and I_{1R}; that is, G1 is delivering its rated voltage and current.

G2 is now underexcited to produce the characteristic curve V_{2F}–I_{2F}. Note that its open-curcuit terminal voltage is equal to the line voltage (V_R); consequently,

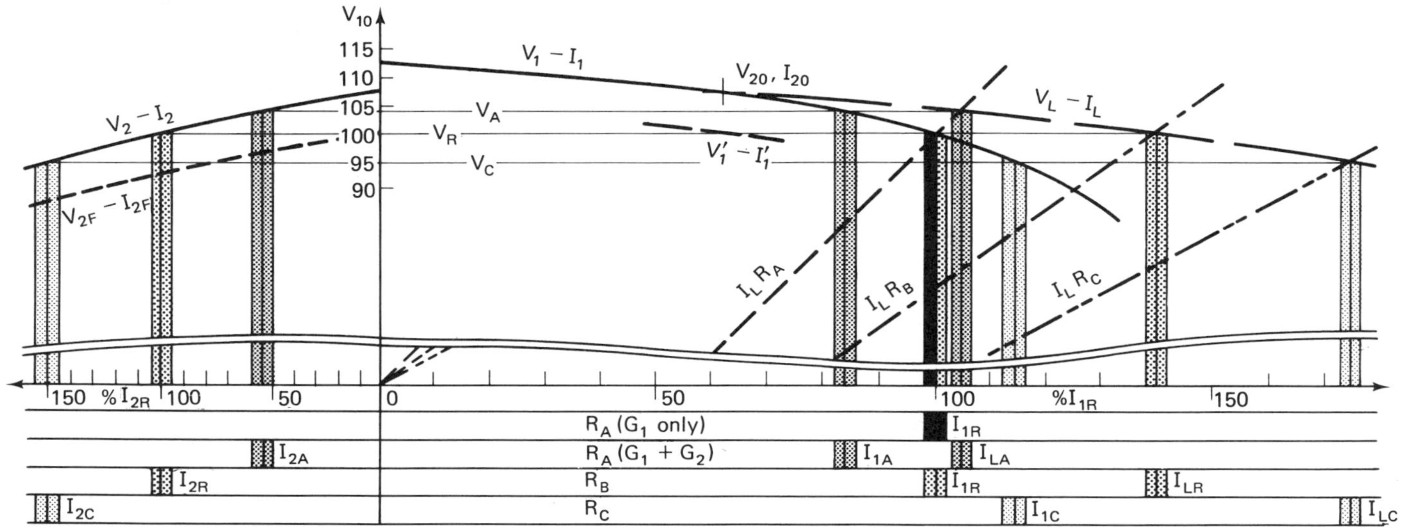

Figure 2-38 Load Division in Paralleled Generators

if Sw 2 is closed, I_2 will be zero and the condition of G2 is termed *floating*. G1 continues to supply the line at the previous current level.

Let the rheostat of G2 now be adjusted for normal excitation, resulting in the curve V_2–I_2. The load resistance remains unchanged. To determine the new operating point, the curve V_L–I_L is plotted; for each voltage level, I_1 and I_2 are added to obtain the corresponding I_L. The intersection of the V_L–I_L and $I_L R_A$ curves identifies the line voltage V_A and line current I_{LA}. G2 is providing the current I_{2A} while G1 has dropped to I_{1A}. Both the line voltage and the line current have increased; these increases can be negated if, concurrent with the increase to normal excitation in G2, the excitation of G1 is reduced to produce the characteristic V'_1–I'_1. Properly coordinated, it is possible to gradually and simultaneously make these changes so as to maintain constant line voltage and current throughout the transition; this procedure is followed in bringing an additional generator on line. However, note that for the assumed load and generator characteristics of the example, G2 will be loaded to rated capacity; no additional line power is available without overloading G2.

With normal excitation in both G1 and G2, the line power may be increased to that for the equivalent load resistance R_B without overloading either generator. This is the full-load condition of the system; a requirement for increased capacity has been met by bringing a second generator on line, and the system performs satisfactorily as long as the line voltage does not rise above the point V_{20}. If this occurs, G2 no longer will function as a generator but will change to a motor (I_2 negative).

One basic consideration in evaluating a system is stability. Although designing is meticulous, the possibility of disruptive disturbances cannot be eliminated—events such as breakdown of a prime mover or accidental short-circuiting of the line. Also, during normal operation, transient adjustments occur: for example, when a change is made in the load. A *stable* system is one that is not driven to a disastrous state by a disturbance and will return to steady-state operation. Inherently, paralleled shunt generators are stable. To illustrate, assume normal excitation and operation at rated power for both generators, with system levels at V_R and I_{LR}. Further, assume that the load includes a motor that momentarily stalls, resulting in an equivalent load resistance R_C and operating levels V_C and I_{LC}. When the motor regains its operating speed, the generating system, on its own, will return to the normal full-load condition. It has experienced a severe short-duration overload, but the system sustains and recovers from the abnormality.

Unlike paralleled shunt generators, compound generators, if overcompounded, are unstable. In Fig. 2-39 the characteristic curves of two generators are shown; each curve is based upon independent operation. In the absence of any disturbance they would exhibit the V_L–I_L characteristic as plotted. For an equivalent load resistance R_A, the operating point is V_A–I_{LA}. Now assume that a momentary stutter in the prime mover of G2 results in a drop in speed for which the characteristic curve, adjusted by speed ratios, is V_{2D}–I_{2D}. What will happen to the system?

Instinctively, there is a temptation to plot a new line characteristic curve,

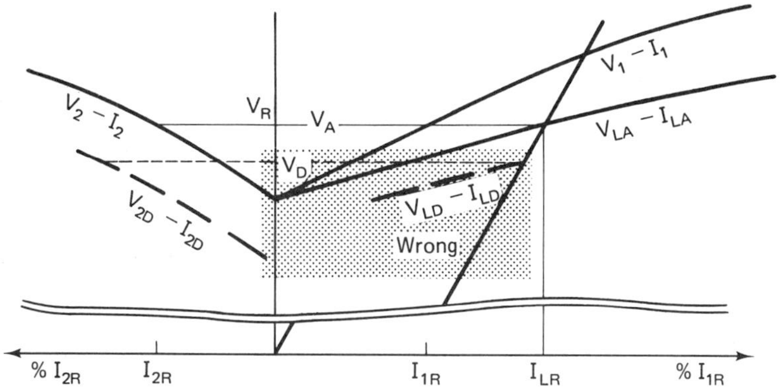

Figure 2-39 Paralleled Compound Generators

V_{LD}–I_{LD}, and claim its intersection with the $I_L R_A$ load line as the new operating point. *But this is wrong!* It ignores several basic facts. The actual consequences of the disturbances are as follows:

1. Immediately after the drop in V_2, V_1 and V_L remain at the level V_A; since the load is unchanged, the line current, I_{LA}, continues.

2. The load current of each generator is $(E_g - V_L)/(R_{ac} + R_{se})$ (for simplicity, the small shunt-field current is disregarded in this discussion). The drop in G2 that, in independent operation, would produce the open-circuit voltage V_{2D} inherently results from a drop in E_{g2}; this causes a decrease in I_2. Therefore, I_1 $(= I_L - I_2)$ immediately increases.

3. The drop in I_2 reduces the series-field mmf, causing a decrease in E_{g2}, while the increase in I_1 causes V_1, V_L, and I_L to rise. This leads to a further reduction in I_2, triggering successive repetitions of the events of paragraph 2 and this paragraph.

4. At some point, E_{g2} decreases to a level less than the corresponding V_L, with the result that I_2 becomes negative and G2 converts from generator to motor. As a motor it will attempt to drive its prime mover and, in the process, demand an increased current input. All currents, of course, must be supplied by G1. [$I_1 = I_L - (-I_2) = I_L + I_2$.]

5. When the disturbance is removed, the system cannot return to the original operating level. The imbalance remains; restoration will not drive I_1 back to the level I_{1A} and V_L does not return to V_A. The current, I_2, cannot return to the plotted V_2–I_2 curve, and the consequences of the preceding paragraphs continue.

To preclude this instability in paralleled generators, an *equalizer* is installed between the two machines. The equalizer is a very low resistance conductor that interconnects the juncture between the armature and the series field of the machines (Fig. 2-40).

With the equalizer present, V_E is common to both machines. I_1 and I_2, the currents in the respective series fields, may or may not equal the armature currents

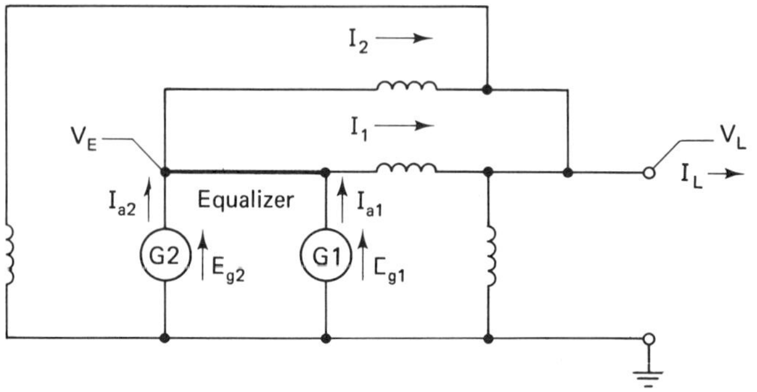

Figure 2-40 Equalizer

of the corresponding generators. The ratio, I_1/I_2, always equals $R_{\text{se }2}/R_{\text{se }1}$, and the armature current of either machine is $(E_g - V_E)/R_{ac}$, with E_g and R_{ac} unique to each machine. Repeating the disturbance in Fig. 2-38, but with the equalizer in place, the consequences are as follows:

1. *The drop in E_{g2} is less pronounced.* $E_g = k\omega(\Phi_{\text{sh}} + \Phi_{\text{se}})$. Without the equalizer, Φ_{se} (a function of I_2) would drop, but I_2 and Φ_2 remain essentially constant with the equalizer in place. I_{A2} will drop less than without the equalizer, and G1 will assume a lesser increased share of I_L.

2. *V_E and V_L will not rise.* Both V_E and V_L are dependent upon E_{g1}. Since I_1 and $\Phi_{\text{se }1}$ do not increase, E_{g1} will not increase. To the contrary, an increase in I_{A1} results in a reduction of V_E because of the increased I_aR_{ac} voltage drop. In turn, this leads to some reduction in V_L and I_L.

When the disturbance is removed, E_{g2} and I_{a2} increase. Each generator regains its share of the line current and the initial operating point is restored.

The problem in paralleling may be eliminated by the use of *stabilized shunt-wound* generators. These are slightly undercompounded generators designed to minimize the voltage regulation while assuring a definite drooping characteristic that negates the need for equalizers.

SECTION 2-2. ARMATURE WINDINGS AND COMMUTATION

The building block for tailoring electrical terminal characteristics of a generator is the coil of the armature winding with its rated power, current, and voltage. The number of coils fixes the power rating, and the series–parallel arrangement determines the terminal voltage and current. Although design is a topic well beyond the scope of this text, it is appropriate to consider construction from the viewpoint of electrical influences.

The nonrotating member of the machine is termed the *stator*; the revolving portion is the *rotor*. The stator of the dc machine basically consists of a large electromagnet and circuit components required to transfer electrical power from the armature to the terminals; the rotor is the armature in which the power is developed. Transfer of power from the rotor to the stator is accomplished by means of the *commutator*, a rotor component, and the *brushes*, a stator element.

This section is concerned with methods whereby power is extracted from the armature. A basic understanding of two topics, armature windings and commutation, serves the purpose.

2-9. ARMATURE WINDING

The construction of the rotor is illustrated in Fig. 2-41. The armature core is formed from stamped steel laminations that include slots into which sides of the coils are inserted. The illustration has 12 slots.

Each slot in the illustration houses two coil sides (numbers 10 and 4′ in slot number 10). For clarity, only three complete coils are shown; there would be 12 in this machine, inserted so that all prime-numbered sides are in the inner portion of the slot.

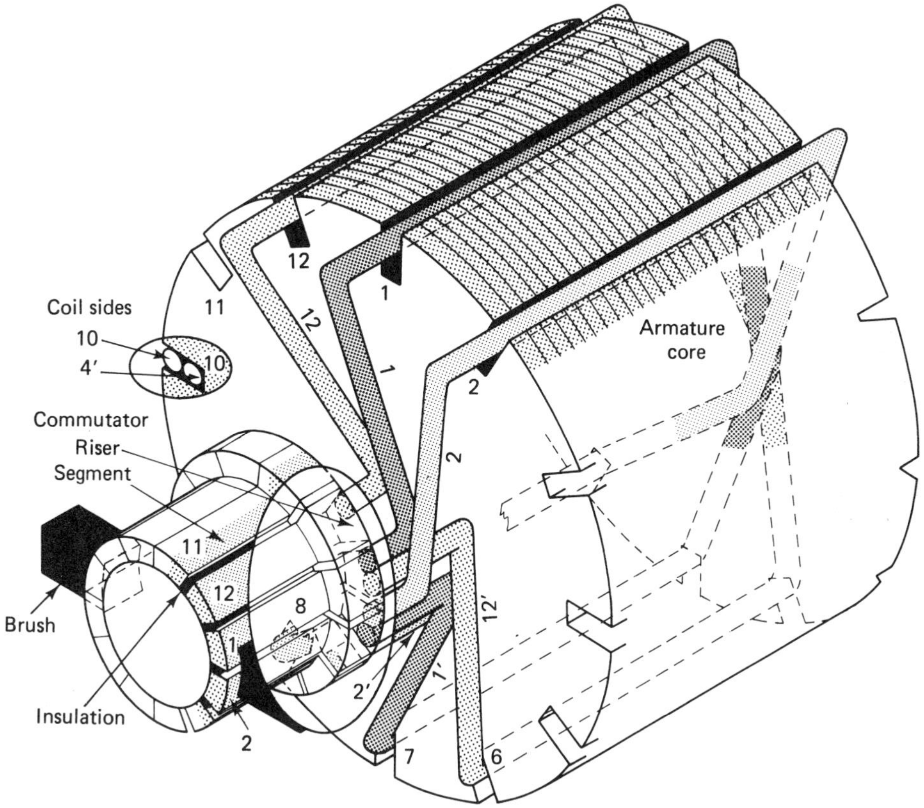

Figure 2-41 Rotor Construction

Focusing on coil number 12, it begins with an electrical connection to the riser of commuter segment 12, leads to coil side 12 in slot 12, continues across the back of the armature core to coil side 12′ lying in slot 6, and terminates in a connection to commutator segment 1.

The commutator has 12 copper segments, each completely insulated from other segments and from the armature. The riser of each segment is the connecting point for one side of each of two coils. The commutator revolves during operation; there is no relative movement between coils and commutator segments.

Brushes, commonly constructed of carbon and graphite, do not rotate; they are under spring pressure to ensure maintainance of an electrical contact with the rotating commutator. For the machine illustrated, two brushes are required.

While Fig. 2-41 graphically illustrates the physical construction, it is cumbersome for representing the complete rotor. The essential features are represented in Fig. 2-42(a). All coil sides are shown, together with connecting leads to the commutator segments; the rear portions of the coils, not pictured, are understood to provide continuity between primed and unprimed sides of a numbered coil. For legibility the brushes are placed in contact not with the outside but with the inside of the commutator. Inclusion of the quadrature axis not only permits ready identification of polarities in each coil side but also identifies this as the armature of a two-pole machine.

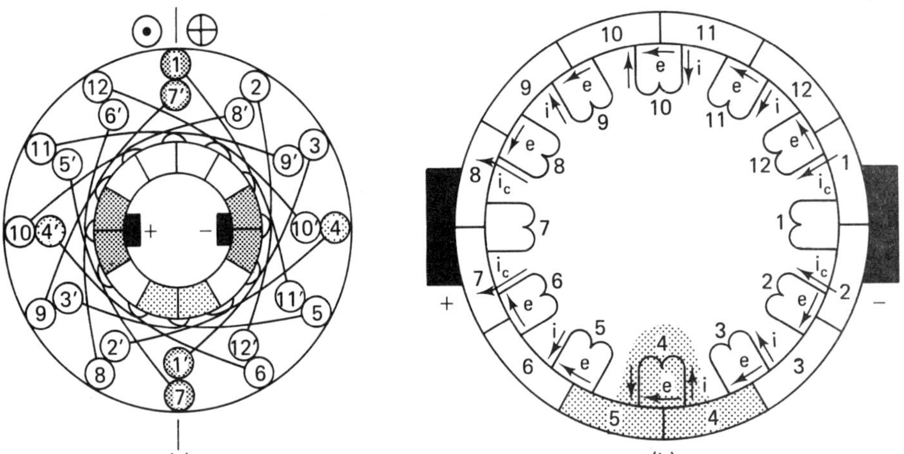

(a) (b)

Figure 2-42 Simplex Lap Winding

Figure 2-42(a) can be redesigned to more clearly depict the electrical circuit pattern of the machine by simple eversion to Fig. 2-42(b). In all the illustrations, the unprimed coil side is connected to the correspondingly numbered commutator segment. Considering coil 4, the current direction, identifiable by location with respect to the quadrature axis in Fig. 2-42(a), is from 4 to 4′; this is carried over to Fig. 2-42(b). Since this is a generator, current flow is from lower to higher potential;

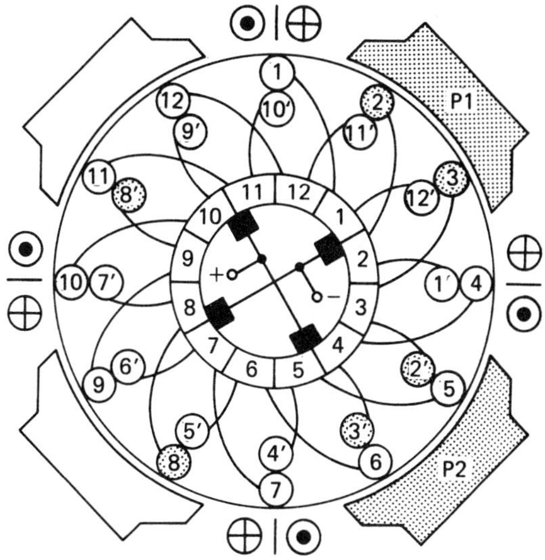

(a) Winding

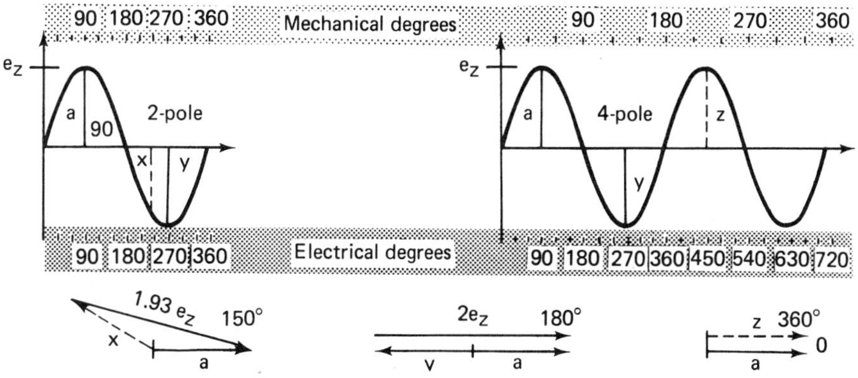

(b) Placement of coil sides

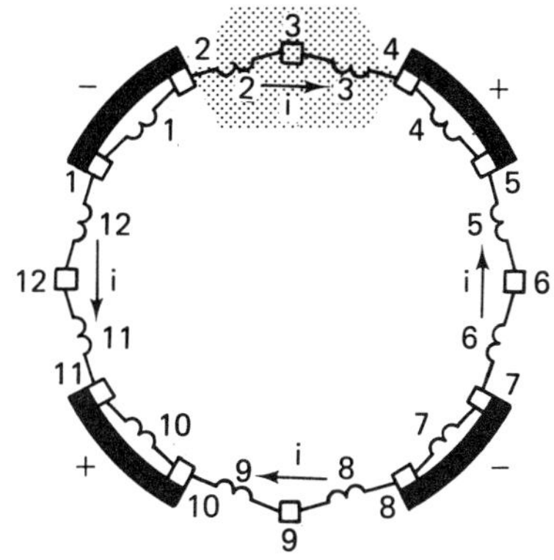

(c) Parallel paths

Figure 2-44 Simplex Lap Winding (Four-Pole)

only 90° in the four-pole machine. The reason for this change is explicable with the aid of Fig. 2-44(b).

In Paragraph 2-1 it was noted that $P/2$ full sine waves are generated in one complete revolution of the armature. The waves for the voltage e_Z induced in one conductor during one revolution are shown for a two- and a four-pole machine. The coil voltage is the vector difference between the instanteous voltages in its sides. If the coil sides were separated by 150 mechanical degrees in a two-pole machine, the conductor voltages also would be separated by 150 electrical degrees, leading to a peak coil voltage of $e_Z\underline{/0^\circ} - e_Z\underline{/150^\circ}$, or $1.93e_Z$; with the separation increased to 180°, the peak voltage is at a maximum of $2e_Z$.

In the four-pole machine, a sine wave is generated during revolution through 180 mechanical degrees; thus, if the coil sides were separated by 180 mechanical degrees, the conductor voltages would be in phase and the vector difference would be zero. To obtain 180-electrical-degree separation, the conductors must be mechanically separated by 90°. Practically, there are mechanical considerations such as shorter end connections and more compact coil placement that favor separation of the sides by less than 180 electrical degrees; the lower limit is about 145°, with an e_c of approximately $1.9e_Z$.

With the winding of Fig. 2-44(a) redrawn to accent the circuitry [Fig. 2-44(c)], it is seen that, again, the coils and commutator segments form one continuous closed loop. Similarly, a duplex would have two, and a triplex three, loops. However, in tracking around a loop in the two-pole machine, there are two points at which current reversal is encountered, whereas there are four reversal points in the four-pole machine. The number of such reversals is equal to the number of poles; to tap the power, a brush must be located at each point, resulting in P brush locations. If the brushes of corresponding polarity are interconnected, each closed loop has P parallel paths; hence, a machine with P poles and an m-plex lap winding will have mP parallel armature paths.

The principal advantage of lap windings is the relatively large number of parallel paths that may be created. There are practical limits on the permissible magnitude of the current in a single coil; consequently, a generator with a large current rating must have a relatively large number of paths.

But the lap winding has one serious flaw. In Fig. 2-44(a) and (c), coils 2 and 3 are the series elements in one parallel path; both of these coils are in the quadrants of poles P1 and P2. Similarly, each of the remaining three paths consists of series coils all of which are uniquely influenced by only two of the four poles. In an ideal machine, with identical flux densities opposite each pole face, no problems would arise. However, in the practical machine, uniformity seldom, if ever, exists; tolerances in features such as resistance of the pole windings, reluctance of ferromagnetic components, and lengths of the air gaps result in imbalances in flux density. If one pole had a somewhat greater density than that of the other three, coils under its influence would generate higher voltages than other coils and the voltages across the paths would be imbalanced, with potentially destructive circulating current flowing through the coils, commutator segments, and brushes. To minimize these currents, equalizers may be added; an equalizer is a direct low-resistance connection between points in the circuit that should be at the same potential. For example, coils 2 and 8 of Fig. 2-44(a) are identically situated from an electromagnetic viewpoint, and an equalizer can be installed between corresponding points of the two coils, either at the commutator or at the rear of the armature. Six equalizers would be required in this machine: 1-7, 2-8, 3-9, 4-10, 5-11, and 6-12.

The *wave* winding avoids this problem by ensuring that the coils in any series path are distributed across *all* the poles; an imbalance will affect all paths equally. To achieve this distribution, the ends of a coil are connected to commutator segments separated by approximately 360 electrical degrees. Beginning at commutator segment 11 in Fig. 2-45 and tracing through coil 11 to segment 6, then through coil 6 to segment 1, a trace of approximately one lap around the commutator terminates on a segment whose number is the starting segment number plus two—a *duplex-wave winding*. An alternative duplex wave, equally valid, would begin on segment 11 and track through coil 11 to segment 4 to coil 4 to segment 9. Similarly, if the trace around the commutator ends on immediately adjacent segments, the winding is simplex. A simplex wave is not possible for the illustrated machine; winding from segment 11 through coil 11 to segment 5 and through coil 5 would terminate on segment 11, the beginning segment, and there would have to be six separate loops, each containing only two coils.

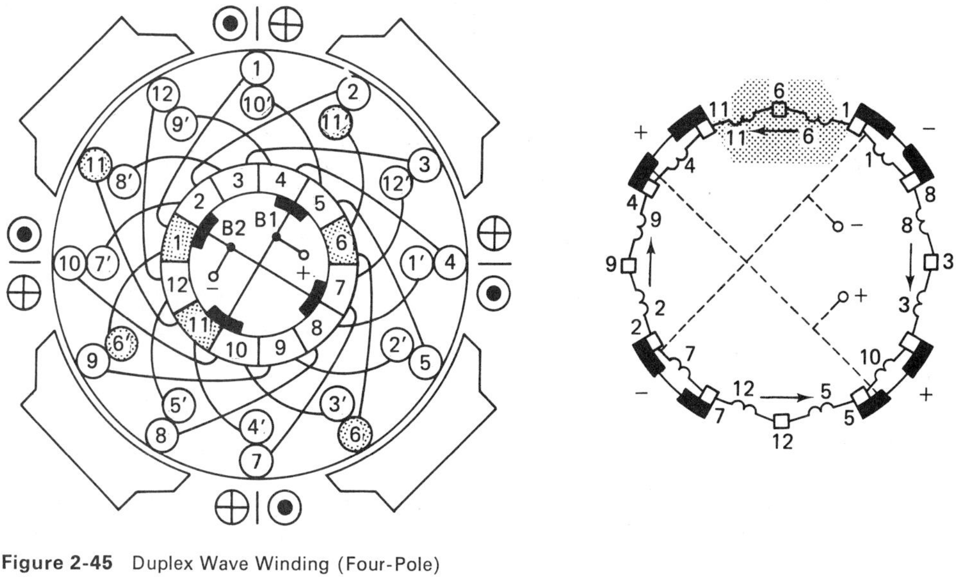

Figure 2-45 Duplex Wave Winding (Four-Pole)

The number of parallel paths in a wave winding always is $2m$; it would appear that a large plex would permit any desired number of paths, but in actual machines considerable difficulty has been experienced with higher plexes and the winding is generally limited to triplex or less.

If only brushes B1 and B2 were retained in Fig. 2-45, satisfactory operating conditions still would be achieved; continuity between the two current paths feeding into a common node would be provided by a coil undergoing commutation. This is a minor advantage of wave windings—only two brushes are required, regardless of the structural aspects. But there also are limitations on the permissible magnitude of the current flow through a single carbon brush; even on relatively small machines, several brushes are used at each brush location. Consequently, the two-brush option frequently is disregarded.

Figures 2-41 through 2-45 illustrate a machine having two single-turn sides per slot. A practical machine will have multiple turns per coil and the slot may house more than two coil sides. A six-turn, lap-wound coil is represented in Fig. 2-46(a). Note that the voltage between commutator segments 1 and 2 will be six times the instantaneous coil voltage; this voltage may be large enough to cause arcing between segments. To retain the same net voltage but reduce the voltage between adjacent segments, the six-turn coils may be split into two three-turn coils, connected to adjacent segments as shown in Fig. 2-46(b). For either a wave or a lap winding, a commutator segment provides the connecting point for two coil sides—therefore, the number of segments equals the number of coils.

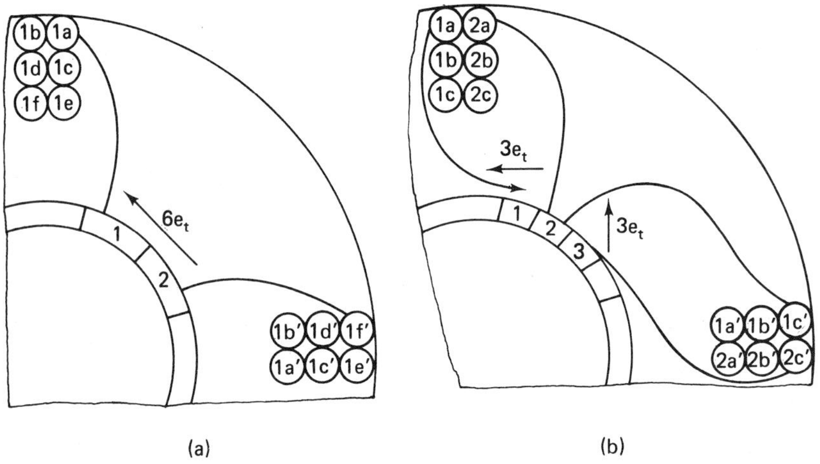

Figure 2-46 Voltage Between Segments

The illustrated simplex-lap and duplex-wave windings may be combined in a single machine (Fig. 2-47). Each slot houses four coil sides, two lap and two wave. But note that four coil sides are connected to each commutator segment. In this winding, the wave coils serve as equalizers between corresponding electrical points of the lap winding. The two windings are in parallel (lap coils 2–3, wave coils 2–9); this requires that the number of series coils, and therefore the number of parallel paths, must be the same in both windings. Thus, $2m_w = m_L P$ and the plex of the wave winding must be $m_L P/2$ (in practice, m_L, the lap plex, normally is

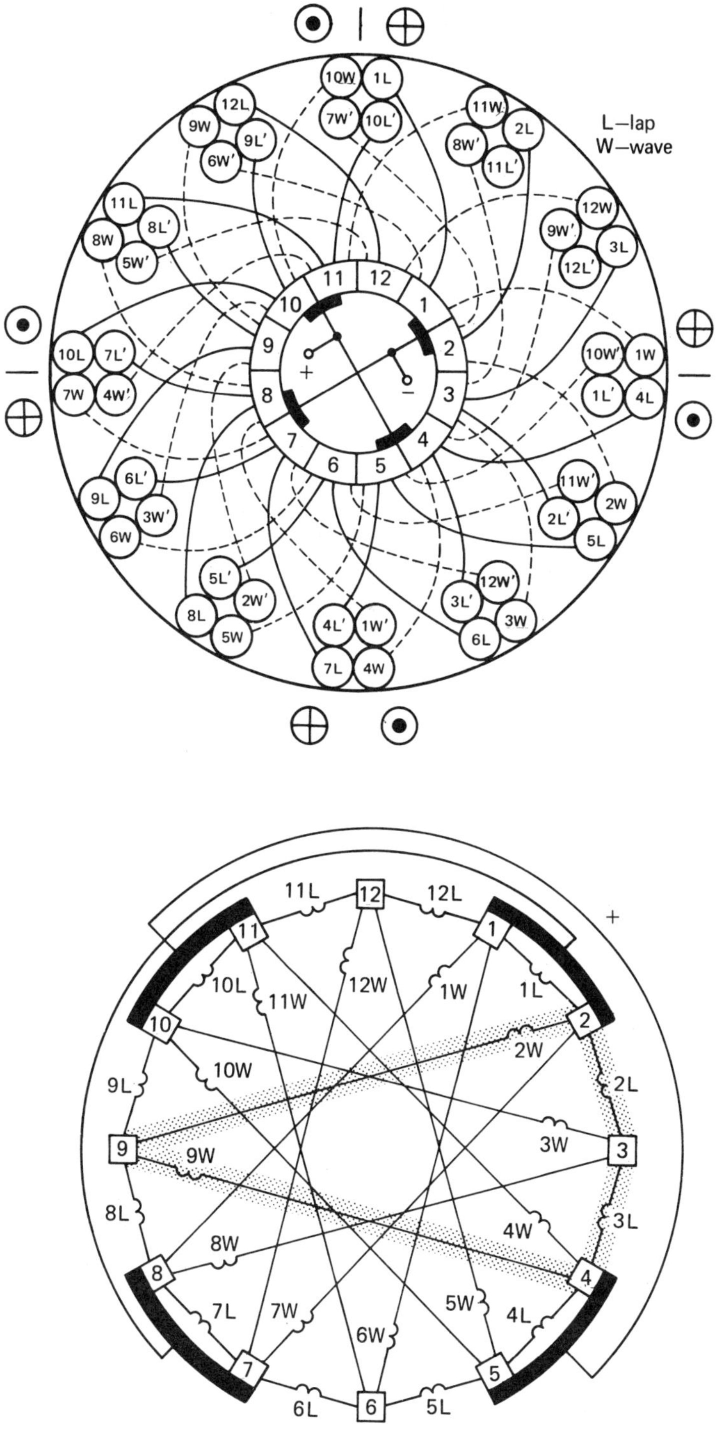

Figure 2-47 "Frog-Leg" Winding

one). The aggregate number of parallel paths is the sum of the paths of the two windings—in the illustration, there are eight. This type of winding is termed *frog-leg.*

In the preceding discussion of windings, the 12-slot, 12-segment illustrations are oversimplifications; practical machines have greater numbers of components.

1. *Number of slots*: Electrical and mechanical considerations are compromised in determining the number of slots in the armature. The total slot space must accommodate the aggregate mass of the conductors, whose size and quantity are determined by the desired electrical characteristics. This space must be distributed among the slots while retaining teeth of adequate mechanical strength and providing sufficient iron for the magnetic circuit.

2. *Number of coils*: Any number of coil sides can be inserted in each slot. Electrically, the number of turns in the armature is related to the desired terminal voltage and line current. These turns must be allocated among coils to provide the required number of parallel paths while avoiding excessive voltage potentials between adjacent commutator segments. For lap and wave windings, each commutator segment is the connecting point for two coil sides; since each coil has two sides, it follows that the number of segments must equal the number of coils. For a frog-leg winding, the required number of segments equals one-half the number of coils.

2-10. COMMUTATION

While a coil is crossing the quadrature axis, moving from one to another armature path, the current through the coil must undergo a full reversal of polarity. The reversal is not instantaneous, but the current gradually decays from a positive maximum to zero followed by buildup to a negative maximum.

If only the contact area between brushes and commutator segments influenced the current pattern, commutation would be ideal (Fig. 2-48). In the illustration, each space spans 4° of the circumference and each segment spans 26°. Commutation begins when the coil rotates through 4° from the 15° orientation; at that point segment 2 begins to make contact with the brush. Throughout the next 22° of rotation, both segments 1 and 2 are in brush contact. As the contact area of segment 1 decreases, the area of segment 2 increases a like amount. The angular velocity and the rate of change in areas is uniform.

Resistance is inversely proportional to area; current division is inversely proportional to resistances and, therefore, directly proportional to contact areas. Consequently, the coil current entering segment 2 would divide between two paths to the brush, one direct and the other through the commutating coil, with linear current increase in the former and decrease in the latter until, at 0° coil orientation, the coil current would be zero. A similar change in the division of current entering segment 1 would occur during the succeeding 11° rotation, and the current in the coil would conform to the dashed IDEAL line shown on the graph.

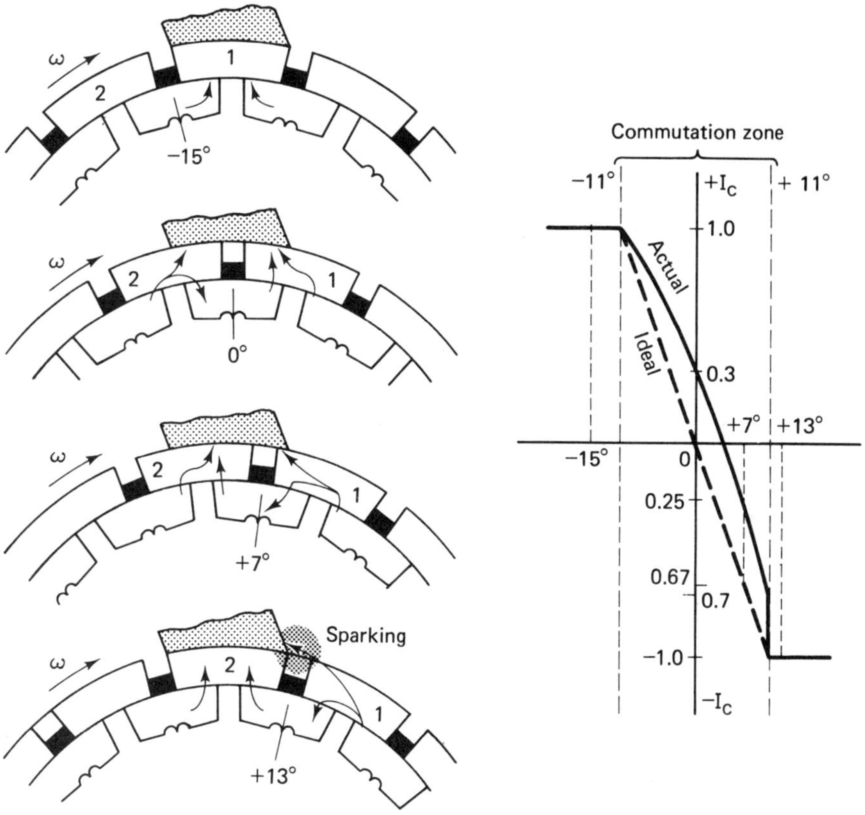

Figure 2-48 Commutation

However, the ideal ignores the resistance of the coil and commutator segments. Of greater importance, when the current in a coil is time-variant, self-inductance causing a reactive voltage is present. Consequently, the change in current in the commutating coil will lag the ideal. The ACTUAL curve of the illustration shows a current of $+0.3I_c$ in the coil at 0°, a point where the current ideally would be zero. At 11°, the point where current reversal should be complete, instead of $-I_c$, the actual coil current is only $-0.7I_c$. Consequently, immediately after the break of contact between the brush and segment 1, a current of $0.3I_c$ must jump the air gap from segment 1 to the brush. The accompanying spark is highly erosive to both the brush and the commutator segment.

In a simple inductive circuit, current lag is fixed by a time constant, $\tau = L/R$. Although there is no constant τ applicable to the commutating coil, increased circuit resistance will reduce the amount by which the actual lags the ideal. The principal resistance in the commutating circuit is the brush-contact resistance; increasing this resistance improves the commutation. However, a more effective remedy is to neutralize the *reactance* voltage of the coil by inducing a *rotational* voltage of equal but opposite polarity. This requires a flux of proper polarity and magnitude on the quadrature axis.

In Paragraph 2-4 (Fig. 2-28), interpoles were discussed from the viewpoint of eliminating armature reaction flux on the quadrature axis. If these poles are excited to a level beyond that required to neutralize armature reaction, the resulting flux will be of the proper polarity to develop a counterreactive voltage. This method is used in all but small machines.

SECTION 2-3. DC MOTOR

2-11. MOTOR FUNDAMENTALS Countertorque in a generator was discussed in Paragraph 2-7. In Fig. 2-49(a) the representation of flux and armature current as it appears in Fig. 2-36 is combined with the electrical schematic of a shunt generator. Two distinctive features of the generator are: (1) E_g is *greater* than V_L, and (2) T_i, the input torque, is *greater* than T_d, the developed torque.

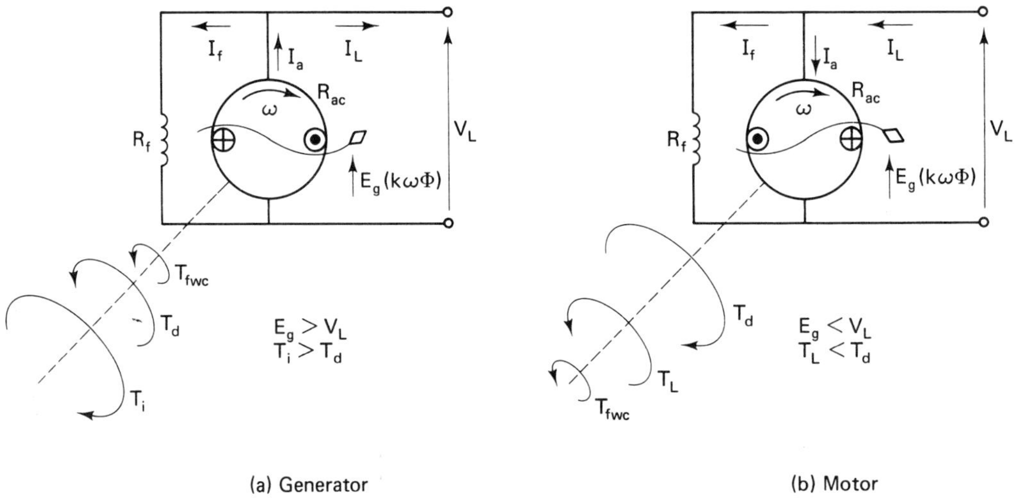

Figure 2-49 DC Machines

Assume that a power source with a terminal voltage greater than E_g is connected to the generator. The line and the armature currents reverse but the field current does not [Fig. 2-49(b)]. The direction of the field flux is unchanged, and the electromagnetic torque developed within the armature now will be in the opposite direction. The direction of rotation is unchanged, and T_{fwc}, which always opposes rotation, will not change. But the machine no longer is a generator—it is a motor. A passive load, driven by the motor, exerts a countertorque T_L that opposes rotation, and the torque equilibrium is $T_d = T_L + T_{fwc}$. Thus, in contrast to the generator, the distinctive features of the motor are: (1) E_g is *less* than V_L, and (2) T_L is *less* than T_d.

Mechanical power was discussed in Paragraph 1-12. Summarizing,

$$P_{\text{mech}} = \omega T \text{ watts} \quad [\text{SI}]$$
$$= \frac{2\pi ST}{33{,}000} \text{ horsepower} \quad [\text{US}] \qquad (1 \text{ hp} = 746 \text{ W})$$

At the point of energy conversion between electrical and mechanical forms, the system model is ideal, with equality between electrical and mechanical power. The converted power earlier was identified as developed power, P_d; the relationships are

$$\omega T_d = E_g I_a \quad [\text{SI}]$$
$$\frac{2\pi S T_d}{33{,}000} = \frac{E_g I_a}{746} \quad [\text{US}]$$

From these equalities, the mechanical quantities can be developed:

$$\boldsymbol{T_d} = \frac{E_g I_a}{\omega} = \frac{(k\Phi\omega)I_a}{\omega} = \boldsymbol{k\Phi I_a} \quad (k = ZP/2\pi a) \quad [\text{SI}]$$
$$\boldsymbol{T_d} = \frac{33{,}000}{2\pi(746)}\frac{E_g I_a}{S} = \frac{7.04(k\Phi S)I_a}{S} = \boldsymbol{7.04k\Phi I_a} \quad (k = ZP/60a) \quad [\text{US}] \tag{2-3}$$

$$\boldsymbol{\omega} = \frac{E_g I_a}{T_d} = \frac{E_g I_a}{k\Phi I_a} = \frac{\boldsymbol{E_g}}{\boldsymbol{k\Phi}} \quad [\text{SI}]$$
$$\text{S} = \frac{7.04 E_g I_a}{T_d} = \frac{7.04 E_g I_a}{7.04 k\Phi I_a} = \frac{\boldsymbol{E_g}}{\boldsymbol{k\Phi}} \quad [\text{US}] \tag{2-4}$$

From the viewpoint of performance, the characteristics of principal interest in the generator are efficiency and voltage regulation; in the motor, they are efficiency, speed regulation (SR), and load torque variations. Note that the load torque, T_L, is not T_d but rather $(T_d - T_{fwc})$.

Example 2-5: A 200-hp 240-V dc shunt motor has a rated speed of 850 r/min and rated line current of 670 A. The friction, windage, and core losses are 5.5 kW (assumed to be constant throughout the operating range). $R_{ac} = 0.02\ \Omega$ and $R_f = 30\ \Omega$. Find (a) full-load output horsepower, (b) full-load developed and load torque, (c) full-load efficiency, (d) no-load speed (S_o), and (e) per cent speed regulation (%SR).

SOLUTION:

(a) The nominal output horsepower is 200 but the actual output, computed from known data, may differ.

$$\text{hp}_R = \frac{P_d - P_{fwc}}{746}$$
$$= \frac{E_g I_a - P_{fwc}}{746}$$

$$I_a = I_L - I_f$$
$$= 670 - \frac{240}{30} = 662 \text{ A}$$
$$E_g = V_L - I_a R_{ac}$$
$$= 240 - (662)(0.02) = 226.8 \text{ V}$$
$$\text{hp}_R = \frac{(226.8)(662) - 5.5 \times 10^3}{746} = 193.9 \text{ hp}$$

(b) $T_d = 7.04 k \Phi I_a = 7.04 \frac{E_g}{S} I_A$

$$= \frac{(7.04)(226.8)(662)}{850} = 1244 \text{ lb-ft}$$
$$T_R = \frac{33{,}000 \text{ hp}_R}{2\pi S}$$
$$= \frac{(33 \times 10^3)(193.9)}{2\pi(850)} = 1198 \text{ lb-ft}$$

(*Note*: Although two apparently different methods are illustrated in computing T_d and T_R, both originate from the same source and the methods are interchangeable. For the unwary, a minor trap is failing to distinguish between developed and output torque or power.)

(c) $\%\eta_R = \frac{100 P_o}{P_{\text{in}}}$

$$= \frac{(100)(746)(\text{hp}_R)}{V_L I_R}$$
$$= \frac{(100)(746)(193.9)}{(240)(670)} = 90.0\%$$

(d) The no-load speed of a motor is fixed by the equality $P_d = P_m + P_{fwc}$. Since P_m is zero at no-load, $P_d = P_{fwc}$. Thus,

$$P_{fwc} = E_g I_A$$
$$= E_g \frac{V_L - E_g}{R_{ac}}$$
$$E_g^2 - V_L E_g + P_{fwc} R_{ac} = 0$$

Therefore,

$$E_g = \tfrac{1}{2}(V_L \pm \sqrt{V_L^2 - 4 P_{fwc} R_{ac}})$$
$$= \tfrac{1}{2}(240 \pm \sqrt{240^2 - (4)(5.5 \times 10^3)(0.02)} = 239.6 \text{ V}$$

Assuming that armature reaction is negligible, for the shunt motor $\Phi_o = \Phi_R$.

$$\frac{S_o}{S_R} = \frac{E_{go}}{E_{gR}}$$

$$S_o = \frac{S_R E_{go}}{E_{gR}}$$

$$= \frac{(850)(239.6)}{226.8} = 898 \text{ rpm}$$

(e) Speed regulation is comparable to voltage regulation in the sense that it is the ratio of the drop in speed (no-load to full-load) to the rated speed.

$$\%\text{SR} = \frac{100(S_o - S_R)}{S_R}$$

$$= \frac{(100)(898 - 850)}{850} = 5.65\%$$

2-12. SPEED–TORQUE CHARACTERISTICS

The alternative forms of the dc motor are counterparts of the dc generator—shunt, compound, and series—in which the distinction in types is solely attributable to which type of field windings, shunt and series, are present in the machine. In construction and in schematic representation, dc generators and motors are identical. The general pattern applied to schematic representations in this text maintains input power on the left and output power on the right; for this reason, line terminals of the motor usually (but not inflexibly) will appear to the left of the field and armature circuits.

Shunt Motor. The schematic for the shunt motor was illustrated in Fig. 2-49(b). The speed and torque characteristics over the loading range may be deduced from equations 2-3 and 2-4 using either load power or armature current as the controlled variable. The curves illustrating individual characteristics are based upon armature currents.

A presumption throughout this discussion is that line voltage remains constant. Therefore, in a shunt motor the field current will remain constant throughout the loading range; however, this does not necessarily mean that the field flux is constant since armature reaction, if uncorrected, causes reductions with loading. A *stabilized* shunt motor is one in which a light series field is added to compensate for the effect of armature reaction.

Neglecting armature reaction, the denominator in the speed equation, $\omega = E_g/k\Phi = (V_L - I_a R_{ac})/k\Phi$, is constant but a drop in speed, varying with the armature current, will occur as the load increases. From the torque equation, $T_d = k\Phi I_a$, a constant flux results in direct proportionality between developed torque and armature current. The plot of speed and torque as functions of the armature current appears in Fig. 2-50.

When the effects of armature reaction are introduced, the constant field flux is reduced by armature reaction flux and the term $k\Phi$ now becomes $k(\Phi_f - \Delta\Phi_a)$. In the speed equation, $\omega = (V_L - I_a R_{ac})/k(\Phi_f - \Delta\Phi_a)$, a decrease in the denomi-

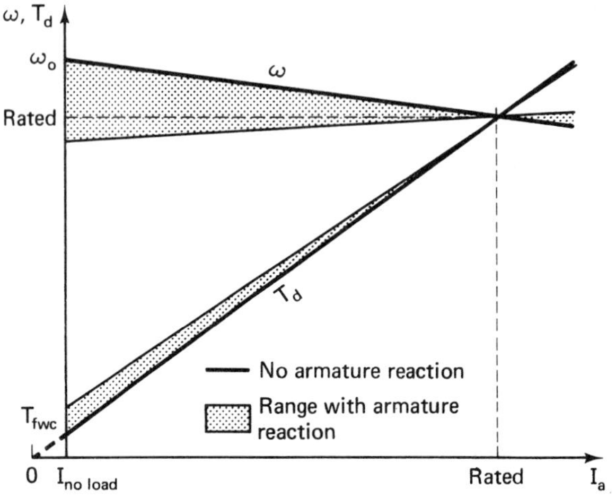

Figure 2-50 Speed and Torque as Functions of I_a in a Shunt Motor

nator influences ω to rise, but this effect is concurrent with increased armature current, which, in the numerator, acts to decrease ω. The actual change in ω is dependent upon the magnitudes of the armature resistance and armature reaction flux and upon the point on the saturation curve at which the motor is operating. For most motors, ω will decrease with load, but to a lesser degree than without armature reaction; however, in some machines the full-load speed may equal or even exceed the no-load speed, a potentially unstable condition that may be precluded by the addition of stabilizing windings. In the torque equation, $T = k(\Phi_f - \Delta\Phi_a)I_a$, the effect of armature reaction is to reduce torque.

The shaded areas of Fig. 2-50 are representative of the region within which the speed and torque curves might lie with uncompensated armature reaction. This figure should be viewed merely as illustrative of the nature of possible variations; specifically (1) most shunt motors have drooping characteristics—that is, $\omega_R < \omega_o$, and (2) the %SR lies generally between 2 and 8%. The dc shunt motor is categorized as a constant-speed motor, a category defined as motors with constant or near-constant speed over the loading range.

Series Motor. The schematic of the series motor appears in Fig. 2-51. The unique speed and torque characteristics result from the highly variable flux Φ_{se} produced by the armature currents. Currents have a double effect in both the speed and the torque formulas. Assuming operation on the linear portion of the saturation curve, $\Phi = k_\Phi I_a$. The speed equation is

$$\omega = \frac{V_L - I_a(R_{ac} + R_{se})}{kk_\Phi I_a} = \frac{V_L}{kk_\Phi I_a} - \frac{R_{ac} + R_{se}}{kk_\Phi}$$

As I_a approaches zero, ω approaches infinity. Thus, the no-load speed is high, limited only by friction, windage, and core losses. Mechanically, the no-load speed must be restricted in order to avoid destruction of the machine by centrifugal

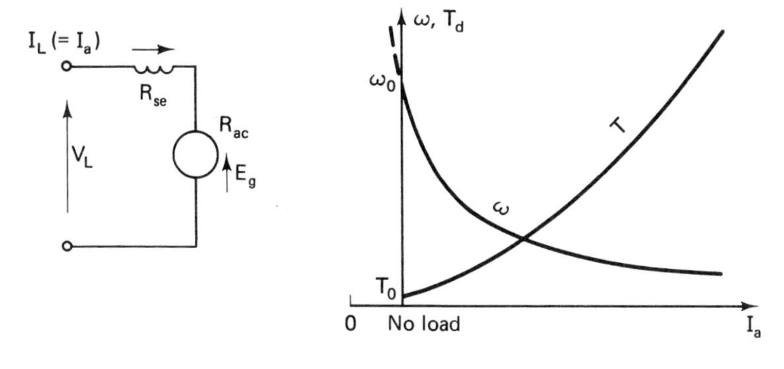

(a) Schematic (b) Speed-torque characteristics

Figure 2-51 Series Motor

forces. Frequently, particularly in large series motors, use is limited to applications in which the machine never is fully unloaded.

As I_a increases, the speed drops—rapidly for small loads but more moderately as the load increases—and the curve is that shown in Fig. 2-51. The torque equation is $T = kk_\Phi I_a^2$ and the resulting curve is shown in the figure.

When the nonlinearity of the saturation curve is introduced, at low currents Φ remains directly proportional to I_a, but beginning at the break of the upper knee, $\Phi < k_\Phi I_a$. In the speed equation, the denominator increases less rapidly and the speed, therefore, does not decrease as much as it would in a linear relationship. Similarly, beyond the knee of the saturation curve, the torque increases at a lesser rate. However, the shapes of the ω and T curves are not substantially changed from those of Fig. 2-51.

Compound Motor (Fig. 2-52). In the compound motor, shunt-field current is constant and normally produces a flux beyond the linear portion of the saturation curve. Consequently, the flux increment due to the series field is *not* directly

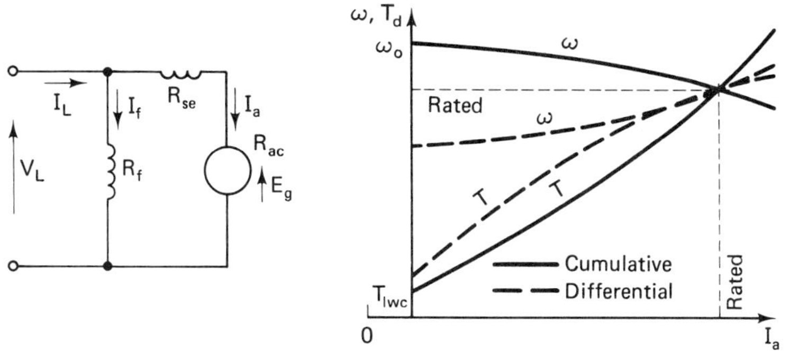

(a) Schematic (b) Speed-torque characteristics

Figure 2-52 Compound Motor

proportional to I_a. The series increment, $\Delta\Phi_{se}$, may add to the shunt flux (cumulative compounding) or subtract from it (differential compounding).

The characteristic equations are $\omega = [V_L - I_a(R_{ac} + R_{se})]/k(\Phi_f \pm \Delta\Phi_{se})$ and $T_d = kI_a(\Phi_f \pm \Phi_{se})$.

Comparing the cumulative compound and a comparable shunt motor, with increasing armature currents (1) the numerator of the speed equation decreases more rapidly in the compound, and (2) the denominator increases for the compound while the shunt remains constant (armature reaction neglected). Both of these differences contribute to a greater drop in speed in the compound motor. But torque in the compound is greater than that of the shunt by an increment, $k\Phi_{se}I_a$.

With the differential compound, circumstances are reversed. The decrease in net flux causes ω to rise as I_a increases, and torque is less than that of a comparable shunt motor. Because of the rise in speed as loading increases, this form of the motor is unstable, particularly when driving a load such as a generator in which the load power increases with speed; increased loading of the generator demands an increased I_a in the motor, leading to a higher motor speed, a sequential increase in E_g and P_d of the generator, and a new demand for a larger I_a in the motor—a "runaway" chain of circumstances. A differentially compounded motor rarely is employed—its one tempting feature is that careful adjustment of the degree of compounding allows identical no-load and full-load speeds, but there are alternative means of providing this feature if it is required. Differential compounding hereafter is discounted and the term compound is understood to mean cumulative compounding.

The comparative characteristics of the three motors appear in Fig. 2-53. Several changes have been made in the indexing; speed in r/min ($= 60\omega/2\pi$) is used in lieu of ω, horsepower output is the abscissa of the speed graph, and load

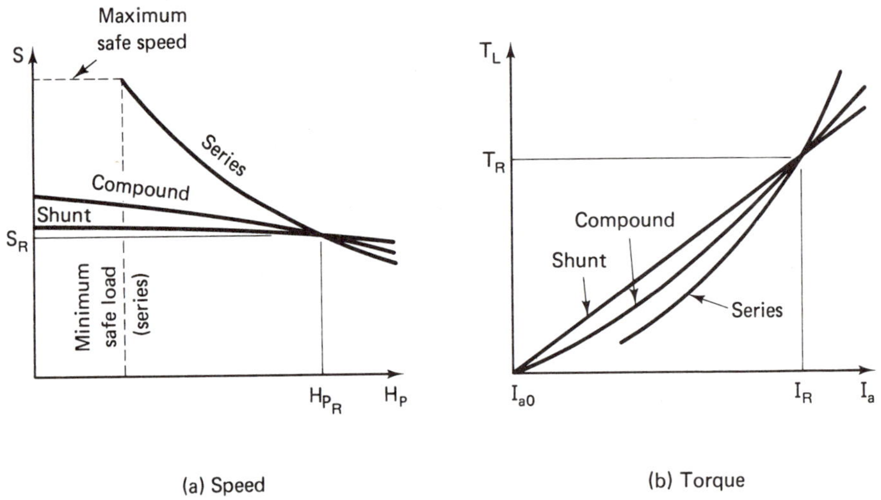

Figure 2-53 Comparative DC Motor Characteristics

torque has replaced developed torque. These graphs provide clues to the preferred type motor for specific applications. The speed characteristic of the series motor is noncompetitive with compound and shunt motors, but commonly either of the latter two may be satisfactory. In overload torque, the compound overshadows the shunt, but speed regulation favors the latter; final selection may hinge on which of these features is to be emphasized.

2-13. STARTING DC MOTORS

A problem common to all motors is the magnitude of starting currents. The armature current of a shunt motor is $(V_L - E_g)/R_{ac}$; when starting, $E_g = 0$ and $I_a = V_L/R_{ac}$. At full load a representative motor may have an E_g of 95% V_L, resulting in a rated armature current of $0.05(V_L/R_{ac})$; when started, the initial current would be 20 times this rated current. Excepting small motors, currents of this order usually are unacceptable not only in the motor itself, but also in the power source and the interconnecting system of transmission lines, switches, protective devices, and so on. Measures are required to reduce the current during the starting period. From the defining equation for the current, the alternatives are reduction of the line voltage or addition of resistance in series with the armature. Either of these approaches results in the application of a reduced voltage, V_A, across the terminals of the armature; the difference between the two is in the magnitude of the field current.

During the starting period, the motor is in a transient state. Electrically and mechanically, the physical parameters are undergoing change. The armature and field circuits are not purely resistive; every electromagnetic device inherently is an *RL* circuit. Under steady-state conditions, a dc *RL* circuit acts as though only resistance were present, but when a voltage initially is impressed the current is $I(1 - \epsilon^{-t/\tau})$. The circuit of Fig. 2-54(a) represents the field circuit of the shunt motor; the steady-state current is $I_f = V_L/R_f$ and the transient current is $i_f = I_f(1 - \epsilon^{-t/\tau_f})$, where τ_f, the time constant, is L_f/R_f. Practically, the current is steady after five time constants. The transient period in the field circuit is small in comparison with the starting period, and the following discussion assumes that the field flux is at its steady-state value when the armature circuit initially is energized.

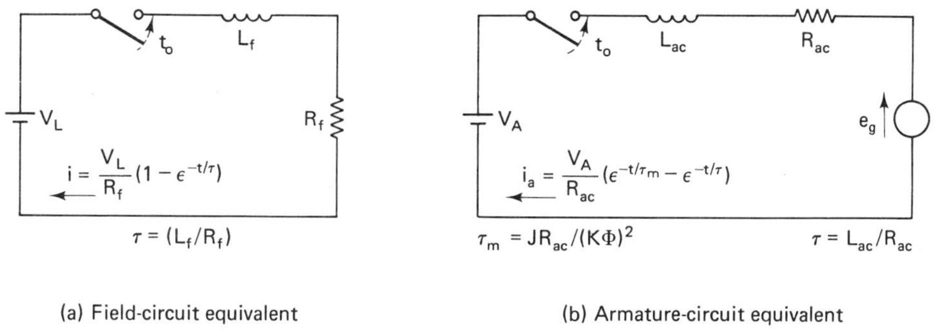

(a) Field-circuit equivalent (b) Armature-circuit equivalent

Figure 2-54 Transient-State Circuits

The electrical schematic for the armature during the starting period is shown in Fig. 2-54(b). V_a is not constant but normally is increased by steps between a relatively low initial and the final (usually V_L) value. For the period that V_a is constant, e_g is not—to the contrary, at any instant $e_g = k\omega\Phi$ and ω will be increasing as the motor accelerates. The decisive elements in determining ω_t, the angular velocity at time t, are the developed torque and the rotor moment of inertia, J. In linear motion, $F = ma$, where F is the net force operable upon a body, m the mass, and a the linear acceleration. The corollary in rotational motion is $T = J\alpha$ (α is angular acceleration). ω is a function of α ($\omega = \alpha t$). J is constant, but $T(= k\Phi i_a)$ changes with the armature current.

When the contributing factors are analyzed assuming that (1) $k\Phi$ and V_a are constant, (2) T_{fwc} and armature reaction are negligible, and (3) the motor is unloaded during the starting period, the armature current is

$$i_a = \frac{V_a}{R_{ac}}(\epsilon^{-t/\tau_m} - \epsilon^{-t/\tau})$$

$$\tau_m = \frac{JR_{ac}}{(k\Phi)^2}$$

$$\tau = \frac{L_{ac}}{R_{ac}}$$

In Fig. 2-55, i_a is plotted for a typical motor. Note that the electromechanical time constant is predominant—the electrical transient ($\tau = L/R$) is of short duration. Although the curve indicates that i_a does not actually peak at V_a/R_{ac}, it approaches this value as τ decreases. The speed of the motor during the starting period is typified by the plotted curve. If the motor is started under load, the general shape of current and velocity curves are unchanged, but the elapsed time for reaching steady state is increased.

The common method of starting a dc motor is insertion of resistance in series

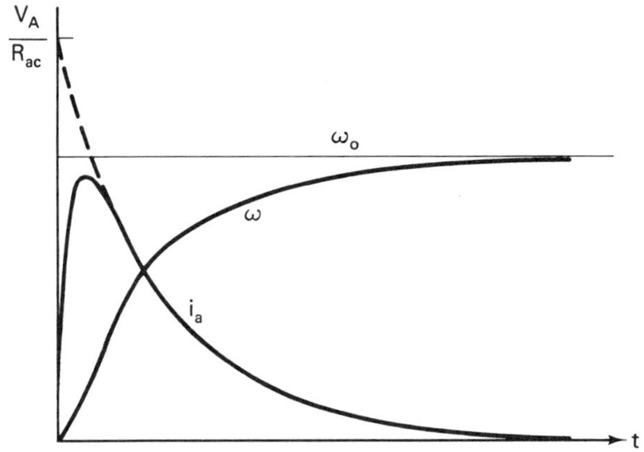

Figure 2-55 Transient Characteristics During Starting

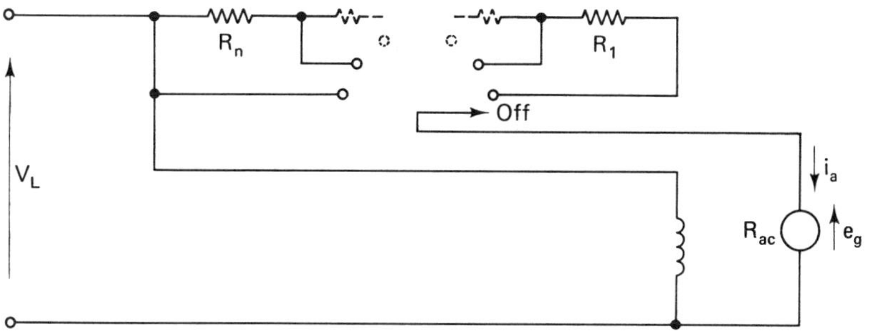

Figure 2-56 Starting Circuit

with the armature (Fig. 2-56). Initially, the magnitude of the resistance is such that the resulting current, $V_L/\sum R$, is limited to the allowable value. ($\sum R$ is the sum of R_{ac} and all resistances in series with it.) As the speed builds and the current decays to an appropriate level, one segment of the series resistance (R_1) is eliminated; after elimination, the remaining resistance must restrict $(V_L - e_g)/(\sum R - R_1)$ to the permissible current. The process continues until all starter resistance has been eliminated.

Example 2-6: A 200-hp 240-V 1800-r/min shunt motor has a full-load armature current of 660 A (I_{aR}). R_{ac} is 0.0147 Ω. At full load, $T_d = 595$ lb-ft. The motor is to be started by successive elimination of resistance; during starting, it is desired that $I_{aR} \lessapprox i_a \lessapprox 1.75 I_{aR}$.

(a) How many starting resistors are required and what is the ohmic value of each?

(b) What is V_a, the voltage across the armature immediately prior to elimination of each resistance?

(c) What is the speed of the motor at each instant when an increment of starting resistance is eliminated?

(d) What is the maximum and the minimum torque developed during the starting period?

Solution: A tabular form is convenient to aid computations and record results.

Step (n)	$\sum R$ [1]	R_n [2]	$e_g\|_{i=I_{aR}}$ [3]	$V_A\|_{i=I_{aR}}$ [4]	$S\|_{i=I_{aR}}$ [5]
0	0.208 Ω		102.7 V	112.4 V	803 r/min
1	0.1189	0.0891	161.5	171.2	1263
2	0.0679	0.0509	195.2	204.9	1526
3	0.0388	0.0291	214.4	224.1	1676
4	0.0222	0.0166	225.4	235.1	1762
5	(0.0127)	(0.0095)	(231.6)	(241.3)	(1811)

$$[1]\ \sum R_n = \frac{V_L - e_{g(n-1)}}{1.75 I_A} = \frac{240 - e_{g(n-1)}}{1155}$$

$$[2]\ R_n\colon \sum R_{(n-1)} - \sum R_n$$

$$[3]\ e_g\colon e_{g(n)} = V_L - I_{aR} \sum R_n = 240 - 660 \sum R_n$$

$$[4]\ v_A\colon v_{A(n)} = e_{g(n)} + I_{aR} R_{ac} = e_g + 9.7$$

$$[5]\ S_n = \frac{e_{g(n)}}{k\Phi} = \frac{e_{g(n)} S_R}{E_g}$$

$$= \frac{e_{g(n)}(1800)}{240 - (660)(0.0147)} = 7.82 e_{g(n)}$$

$$T_d\colon T_d \propto I_a$$

Therefore,

$$595 \lesseqgtr T_d \lesseqgtr (1.75)(595) = 595 \lesseqgtr T_d \lesseqgtr 1042 \text{ lb-ft}$$

Computations for steps 0 and 1:

$$\text{Step 0:}\quad \sum R_0 = \frac{240 - 0}{1155} = 0.208\ \Omega$$

$$e_{g(0)} = 240 - (660)(0.208) = 102.7\ \text{V}$$

$$v_{a(0)} = 102.7 + 9.7 = 112.4\ \text{V}$$

$$S_o = (7.82)(112.4) = 803\ \text{r/min}$$

$$\text{Step 1:}\quad \sum R_1 = \frac{240 - 102.7}{1155} = 0.1189\ \Omega$$

$$R_1 = 0.208 - 0.1189 = 0.0891\ \Omega$$

$$e_g = 240 - (660)(0.1189) = 161.5\ \text{V}$$

$$v_a = 161.5 + 9.7 = 171.2\ \text{V}$$

$$S_1 = (7.82)(161.5) = 1263\ \text{r/min}$$

In step 5, $\sum R < R_{ac}$. The required R_5 is, therefore, less than the recorded 0.0095 Ω; $R_5 = R_4 - R_{ac} = 0.0222 - 0.0147 = 0.0075\ \Omega$.

At the termination of the starting period, the speed, current, and torque can range between no-load and full-load values depending upon the load, if any, coupled to the motor at starting.

Starting Torque. In numerous applications, load is applied after the motor is started and starting torque is a minor interest. But when the motor starts under load, the starting torque becomes significant; it must be large enough to overcome counterforces of inertia and static friction and provide adequate torque to accelerate the load to the desired speed within an acceptable time period. The series motor excels in this characteristic; the large flux and armature current at starting combine to produce large torques. The unusual speed characteristic of the series motor is easily controlled by the use of series resistance, discussed in Paragraph 2-14.

Because of its impressive starting torque, a common application of the series motor is to lifting devices such as cranes. To a lesser degree, the compound motor can be used to fulfill requirements for substantial starting torque.

2-14. SPEED CONTROL

In size, constructional simplicity, cost, and durability, the *polyphase ac motor* is preferable to the dc motor. However, the latter is unexcelled in one characteristic that often is vital to a particular application: by simple methods, the speed of the motor may be precisely adjusted within relatively broad limits.

Base speed of an adjustable-speed motor is defined as the lowest rated speed at rated load and rated voltage at the temperature rise specified in the rating. This discussion will assume that the base speed occurs when both the field and the armature circuit are connected to the line and only inherent machine resistances (R_f, R_{ac}, R_{se}, R_{ip}, R_{cw}) are present in the system.

Clues to methods for altering the speed are apparent in the speed equation, $\omega = E_g/k\Phi = (V_L - I_a \sum R)/k\Phi$. $\sum R$ is the total resistance in the armature circuit, to include added external as well as inherent resistance; if R_{ac} is redefined as the net machine resistance of the armature circuit, including fields in series with the armature, and R_x the *external* resistance, the equation may be changed to $(V_a - I_a R_{ac})/k\Phi$, where $V_a = V_L - I_a R_x$. Both V_a and $k\Phi$ may be varied.

Armature Control (variable V_a, constant Φ) (Fig. 2-57). This circumstance can apply to shunt and compound, but not series, motors. At base speed, $V_a = V_L$. With operation restricted to rated voltage, $V_a \leqq V_L$. Reduction of V_a causes ω to drop. Two basic methods are broadly applied in achieving this reduction: (1) rated line voltage with a variable resistance in series with the armature or, less common, an adjustable voltage divider across V_L and the armature across the divider output, and (2) supplying the armature power from a separate source whose terminal voltage is controlled. For the latter, the source often is a field-controlled dc generator; the term *Ward–Leonard system* broadly applies to this arrangement, with numerous variations in components and circuitry.

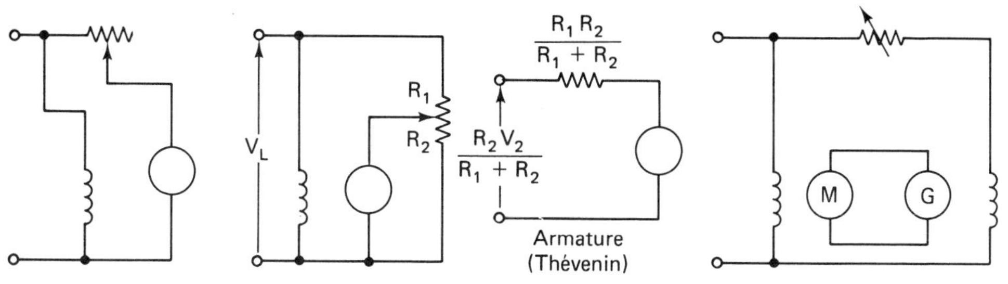

(a) Series resistance (b) Voltage divider (c) Ward–Leonard

Figure 2-57 Armature-Controlled Speed

Field Control (constant V_a, variable Φ) (Fig. 2-58). The highest possible I_f (and Φ) occurs with V_L applied directly to the field. Although I_f cannot be increased beyond this value, it may be *decreased* by means of a rheostat in series with the field. This reduces the flux and causes *above-base* speeds. An alternative to the rheostat is a controlled voltage applied directly to the field.

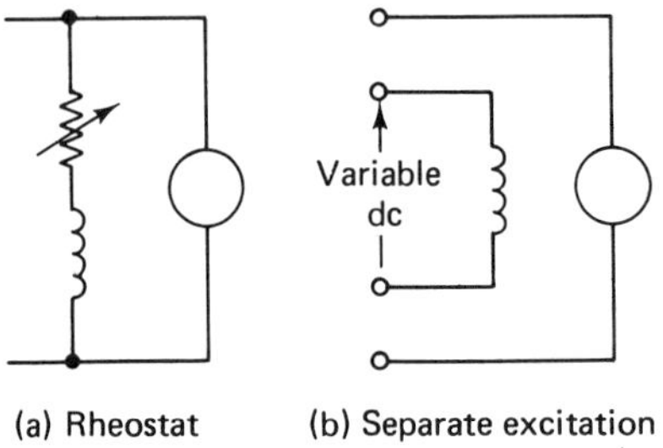

Figure 2-58 Field-Controlled Speed

Variable V_a, Variable Φ. While less common than armature or field control, this combination permits more precise adjustment of the speed. Depending upon the relative dominance of each element, the speed may be above, below, or equal to the base speed.

The speed of a motor is determined not alone by V_a and Φ. Interrelated elements include P_d, T_d, and I_a. The above discussion tacitly assumed that I_a either remained constant or was inconsequential in assessing the response to change in V_a or Φ. Is this true? Or is it possible that changes in V_a or Φ will trigger changes in I_a that invalidate the conclusions? For reconfirmation, consider first the premise that P_d may not exceed the rated developed power.

$$P_d = E_g I_a = (V_a - I_a R_{ac}) I_a$$

Rearranging and solving this equation,

$$I_a^2 - \frac{V_a I_a}{R_{ac}} + \frac{P_d}{R_{ac}} = 0$$

$$I_a = \frac{V_a/R_{ac} \pm \sqrt{(V_a/R_{ac})^2 - 4P_d/R_{ac}}}{2}$$

Discarding the plus sign (it would result in excessive currents) and simplifying the expression,

$$I_a = \frac{V_a}{2R_{ac}}\left(1 - \sqrt{1 - \frac{4R_{ac}P_d}{V_a^2}}\right)$$

With field control, V_a is constant. For any value of P_d, I_a is constant regardless of changes in $k\Phi$; *field control is characterized as constant power.* But $P_d = \omega T$, and if P_d is constant,

$$\omega_1 T_1 = \omega_2 T_2$$

Therefore,

$$\frac{\omega_1}{\omega_2} = \frac{T_2}{T_1} \quad \text{a reciprocal relationship}$$

The power and torque for above-base speed with field control are plotted in Fig. 2-59.

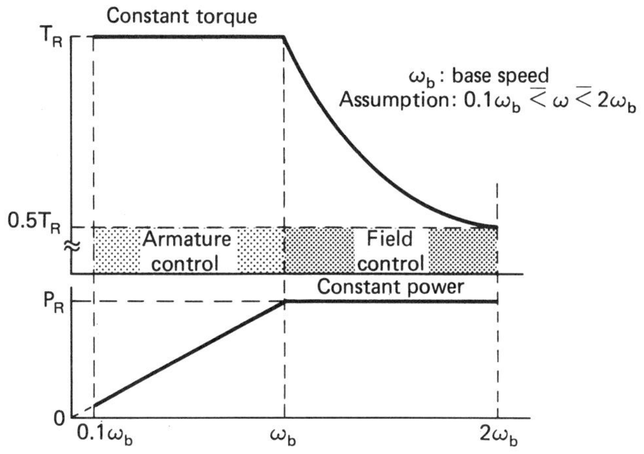

Figure 2-59 Torque and Power as Functions of Speed

A second premise is that I_a may not exceed the rated value. Since $T = k\Phi I_a$ and armature control in a shunt or compound motor has no impact on $k\Phi$, with a constant I_a, torque is constant and the developed power $(= \omega T)$ is directly proportional to the speed. (In the series motor, if Φ is directly proportional to I_a, both power and torque are proportional to I_a^2.) *Armature control is characterized as constant torque.*

When assessing the capabilities of control methods, several limitations are noteworthy:

1. *Field control*: The permissible speed is bounded by the ability of the machine to withstand the mechanical forces developed at high speeds; consequently, the size of a variable-speed motor is greater than that of a comparable constant-speed motor. Additionally, as the field is decreased, the armature reaction assumes greater significance; its effect is to further reduce the net flux and may lead to instability. A potential danger is accidental interruption of the field circuit with a weak residual in the magnetic circuit. The ratio of maximum permissible speed to base speed may be 8 : 1 for small motors, 4 : 1 in midrange, and 2 : 1 for larger motors.

2. *Armature control*: Although very low speeds are theoretically possible, reduction in speed also reduces the ability of the motor to dissipate heat. With series resistance as the control method, the system is inefficient because of the substantial power losses in this element. The lower limit on speed is about 0.1 base speed.

3. *Speed regulation*:

$$\text{SR} = \frac{\omega_0 - \omega_R}{\omega_R}$$

$$= \frac{E_{go}/k\Phi - (E_{go} - I_R R_{ac})/k\Phi}{(E_{go} - I_R R_{ac})/k\Phi}$$

$$= \frac{I_R R_{ac}}{E_{go} - I_R R_{ac}}$$

The rated I_R is identical for all speed settings and E_{go} is approximately equal to V_a. The SR with field control is essentially the same for all speed settings, but with armature control the SR increases as the speed setting is lowered.

Characteristics of the Load. To this point, the discussion of speed control has been exclusively in terms of motor characteristics; this can lead to serious misunderstanding. For example, suppose that a field-controlled motor is operated at base speed and rated power while driving a separately excited shunt generator, which, in turn, is supplying a fixed-resistance load. If the motor speed now is adjusted to 1.5 base speed, the generator E_g and I_a increase by a factor of 1.5, and the developed power increases by a factor of 2.25. Thus, Fig. 2-59, which shows constant power at above-base speeds with field control, is not truly representative; the fallacy is that the load does not conform to the restriction that I_a and P_d may not exceed their rated values.

For continuous loads—that is, loads that are driven over a significant time period in contrast to an elevator or crane that is frequently starting and stopping—there are three common responses to speed changes. First, if the steady-state load force that the motor must overcome is kinetic friction, the power demanded by the load is independent of velocity. A second type is characterized by viscuous friction with power directly proportional to the speed—P_{fwc} responds in this manner. Third, illustrated by the generator, load power may be proportional to the square of the speed.

2-15. ELECTRICAL BRAKING

If a motor is stopped by a simple disconnect means, inertia causes the rotor to coast to a halt over a time period whose duration well might be unacceptable in a particular situation. Also, in some circumstances the nature of the load is overhauling—that is, because of external influences on the load, it accelerates to the point where the load drives the motor; examples are an electric locomotive on a steep, downhill slope or a descending elevator. In both instances, a motor may be adapted to provide *electromagnetic braking*. The several methods of braking are based on the relationship between the direction of the developed torque and the directions of field flux and armature current. If the direction of the field flux of a motor is unchanged but the armature current is reversed, the developed torque will reverse.

Dynamic Braking [Fig. 2-60(a)]. As long as the armature is rotating and the field is energized, a speed voltage is generated. If the field circuit remains connected to the line but the armature is disconnected and reconnected to a resistor, the machine, in effect, changes from motor to generator, the armature current reverses, and the developed torque reverses and brakes the machine. The braking resistor must have an ohmic value that will limit the maximum armature current to the permissible magnitude. Maximum current occurs immediately upon switching the armature from the line to the resistor; at that moment, the value of the generated voltage is the E_g immediately prior to switching. The maximum braking torque occurs immediately after switching; as ω drops, E_g decreases and the braking torque decays. When the armature arrives at zero speed, no voltage is generated, no current flows, and no torque is developed.

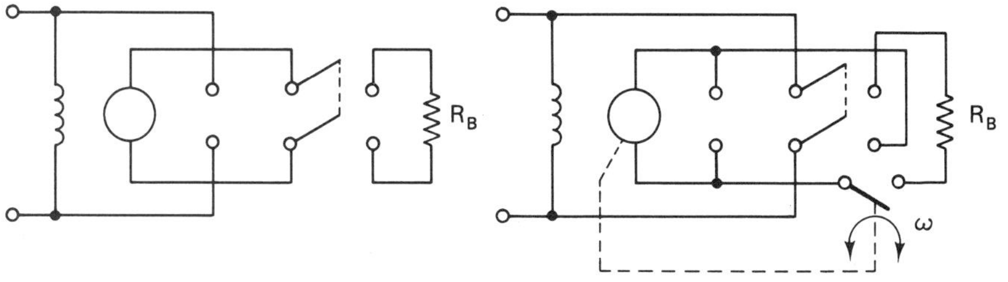

(a) Dynamic (b) Plugging

Figure 2-60 Electrical Braking

Plugging [Fig. 2-60(b)]. If the field remains energized but the armature is disconnected from the line and immediately reconnected with the polarity reversed, the current will reverse. Immediately after switching, E_g and V_L are additive in polarity; a braking resistor having an ohmic value of $(V_L + E_g)/I_{max}$, where I_{max} is the maximum allowable armature current, must be placed in series with the armature during the braking period. When the rotor comes to rest, the armature must be disconnected from the line; if not, the line would continue to supply armature current and the motor would accelerate in the reverse direction. The symbol in the illustration is a centrifugal switch; representing this switch in the open position signifies that it is open at zero speed but closed during rotation in either direction. The dashed line indicates mechanical coupling to the armature.

As E_g decays, the braking torque decreases, but, since the line is supplying armature current, a substantial torque continues throughout the period. Immediately prior to opening of the centrifugal switch the torque is approximately one-half of the maximum developed immediately after throwing the double-pole switch.

Regenerative Braking. With all forms of braking, energy is expended during the braking period. In plugging, this energy clearly is supplied by the power supply system. Dynamic braking taps *stored* energy that was supplied by the power system; during the starting period, kinetic energy is stored in the mass of the armature and this energy is expended during the braking period.

Regenerative braking is a form of dynamic braking in which energy is returned to the power supply system. It is possible only when the load is overhauling and requires that (1) E_g be greater than V_L, and (2) the source of the braking energy be other than the power system.

Normally, the load torque opposes rotation of the motor, but with overhauling loads it is in the direction of rotation. This added torque increases the rotor speed; with the increase in ω, E_g rises. If E_g is greater than V_L, the motor converts to a generator and power flow is from the generator to the power supply. Since the magnitude of I_a and T_d is proportional to $E_g - V_L$, the braking torque can be controlled by varying E_g, a condition readily imposed by field control.

Regenerative braking cannot be used to fully stop the motor; inherently, braking is possible only when the product $k\omega\Phi$ is greater than V_L. But for retardation to preclude buildup to excessive speeds, regenerative braking is available and, from the viewpoint of economics, simplicity of control, and durability, preferable to a mechanical braking system as the primary means for speeds at which it is effective.

2-16. STARTERS AND SPEED CONTROLLERS

The problems and elementary methods of starting and stopping the motor and adjusting the speed were discussed in Paragraphs 2-13 and 2-14. This section considers devices for accomplishing these objectives.

Starters may be manually or automatically controlled during the starting period. There are numerous varieties of each; those discussed are representative but not exclusive systems and methods.

A modern revolution in power systems was triggered by development of solid-state devices capable of accommodating large power. Solid-state control devices are excluded here but considered in Chapter Eight.

The system for starting a motor should:

1. Limit currents to permissible magnitudes.

2. Provide a high starting torque to reduce the acceleration time.

3. Protect against power interruptions that, upon restoration, would result in application of line voltage directly to the nonrotating machine.

4. Protect against low voltages that would cause overheating due to inability of the motor to adequately dissipate heat at the attendant lower speed.

5. Protect against excessive speed caused by accidental opening of the field circuit or excessive rheostat resistance.

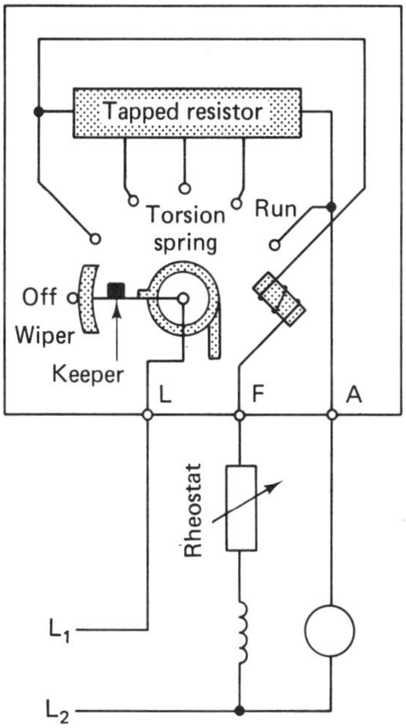

Figure 2-61 Three-Point Starter

A simple *three-point starter* is illustrated in Fig. 2-61. One line is common to the field and the armature. The other line enters the starter and is internally connected to a movable, conducting arm. A counterclockwise torsion spring forces the arm to the OFF position in the absence of other forces. On the arm is an iron keeper and, at the end, a copper wiper.

When the arm is moved from the OFF position to the first stud, line voltage is applied directly to the field circuit, but the full length of the tapped resistor is in series with the armature. As the motor accelerates, the arm is moved through intermediate studs to the final RUN position. At this point, the iron keeper contacts an electromagnetic coil; the starting arm will be held in the RUN position as long as the current in the coil provides a tractive force resulting in a torque greater than that of the torsion spring.

As the contact moves from stud to stud, increments of the tapped resistor are eliminated from the armature circuit and inserted in series with the field circuit. Additional field resistance required to raise the motor to rated full-load speed is provided by an external field rheostat.

The starter will protect against excessive current during the starting period only if the operator is judicious in timing the movement to successive studs. There is no inherent protection against abnormal currents if the motor is overloaded; protection is mandated by the National Electrical Code and is obviously essential to prevent damage to and prolong the life of the motor. It must be provided by fuses or circuit breakers at an appropriate point in the feeder circuit.

Protection against power interruption, low voltage, and low field currents is provided by the electromagnetic coil in series with the field; if the field current drops below the threshold holding current, the arm will release to the OFF position.

Starters of this type are commercially available; both motors and starters are standardized and the user need only match horsepower and voltage ratings for normal service.

An *above-and-below-base speed controller* is illustrated in Fig. 2-62. The movable arm has three wipers. In the first half of the movement arc, the outer wiper is in contact with a copper segment through which the line voltage is continuously applied to the field; for the last half of the movement, this wiper contacts studs connected to a tapped resistor and introduces resistance in the field circuit.

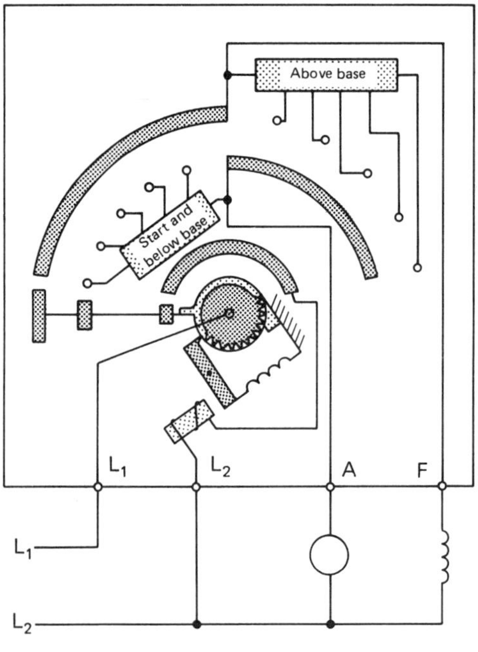

Figure 2-62 Above-and-Below-Base Speed Controller

The center wiper eliminates portions of a tapped resistor in series with the armature. The inner wiper energizes a magnetic coil whose force rotates a pawl to engage a ratchet in the base of the movable arm; the arm can be positioned on any desired stud and will remain there until the coil is deenergized or the arm is physically repositioned.

With the arm located at any point in the first half of the movement arc, the speed is below base. At the vertical, line voltage is applied directly to both the armature and the field, and the motor will run at base speed. Continuing clockwise movement introduces field control to produce above-base speeds.

Automatic starting is accomplished by means of a control circuit. There are numerous components offering a wide range of options in the operation of the circuit. The symbols for a few common elements are shown in Fig. 2-63. Usually, a system is activated by pushing a START button. Pushbuttons are momentary, spring-return switches, schematically represented in the normal condition, that is, when not physically depressed. The top switch closes a circuit when pushed but reopens it when released; consequently, when used to start a motor that is to run continuously, the control circuit must provide some means of maintaining circuit continuity.

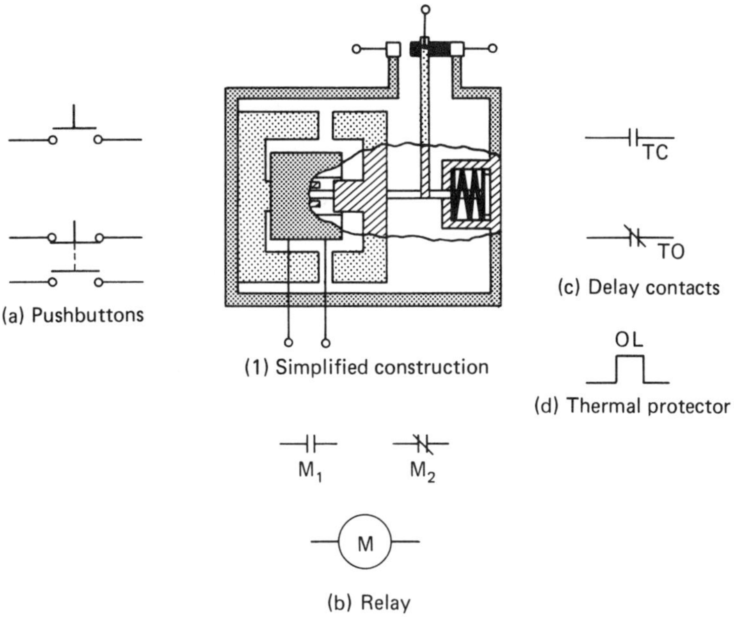

Figure 2-63 Control-Circuit Components

One pushbutton may control several switches in the circuit. A short-dashed line between any elements shows mechanical linkage; pushing the illustrated switch momentarily interrupts the top and closes the bottom line.

Pushing a START button is a command to the control system to commence operation. The system must be capable of both *receiving* and *obeying* the command. Further, it routinely is asked to monitor performance and automatically execute appropriate changes in the circuitry according to a preconceived plan. The basic element fulfilling these requirements is the *relay*.

A simplified sketch of one of many types of electromagnetic relays is illustrated. A coil is secured to an immovable, ferromagnetic, E-shaped core and yoke. A similar E-shaped element, the relay armature, is movable; when the coil is de-energized, a compression spring holds the armature to the right with an air gap between the opposing pole faces. But when a current of adequate strength flows in the coil, the tractive force of the flux closes the air gap by drawing the armature to the left. Coupled to the armature are relay armature contacts; two contacts with a common terminal appear on the illustration. Also shown are two fixed contacts, each with its terminal. In the illustrated mechanism, movement of the armature to the left interrupts circuit continuity on the right and establishes it on the left. This functioning is termed *break-make*, opening one circuit before closing a second. Alternatives to break-make are the converse, *make-break* (close one before opening the other) or simply *make* or *break*. A relay may have any number of fixed and movable contacts, allowing a number of branch circuits to be switched by energization of a single coil.

Also shown are the symbols for the relay. Since there are two functions in the relay, one sensory and the other switching, a circle is used for the first and two parallel lines for the second. Each of the symbols is placed on the control schematic at whatever point is most convenient for interpretation—although parts of a single assembly, proximity of the symbols on the schematic is neither required nor desirable. The control circuit will have more than one relay; identification is required to associate a contactor with its coil. The letter M commonly is used for the coil and contactors that establish the motor circuits.

Symbols on the control schematic are represented in the normal deenergized position, that is, with no electrical power applied at any point within the system. A change in state of the contacts from normal occurs at *pickup*; the pickup voltage or pickup current is the minimum voltage or current that provides sufficient tractive force to effect the change. Conversely, *dropout* occurs at dropout voltage or current, the maximum value at which the relay will release to its normal position. Since the magnetic circuit is different for the open and closed positions because of the air gaps, pickup and dropout values are not identical. In the illustration, at pickup M_1 will close and M_2 will open; at dropout, M_1 reopens and M_2 closes.

Frequently, there is a requirement for a *time delay* between energization of the coil and opening or closing of the associated contacts. Physically, an *RL* circuit, a timing motor, and a pneumatic retarder similar to that used on screen or storm doors are examples of delay devices. A single relay with but one coil can be constructed with instantaneous changes in some contacts, delayed opening in others, and delayed closing of the remainder. Both pickup and dropout may be delayed. As illustrated, TC denotes time-delayed closure and TO indicates delayed opening—for either, if desired, the actual delay time may also be shown.

Whether or not symbolically represented, an overload protector, usually a thermal device, will interrupt the circuit if, through failure to start or too-heavy loading, the resulting currents would cause excessive heating. The illustration is a thermal sensing component; the interrupting function would be shown at the appropriate point in the circuit by normally closed contacts with the identification OL.

The integration of these components in a functional system is illustrated in Fig. 2-64. The power circuit contains a shunt motor with a rheostat, preset for rated speed, and a tapped resistor that provides two starting resistances. Four contactors

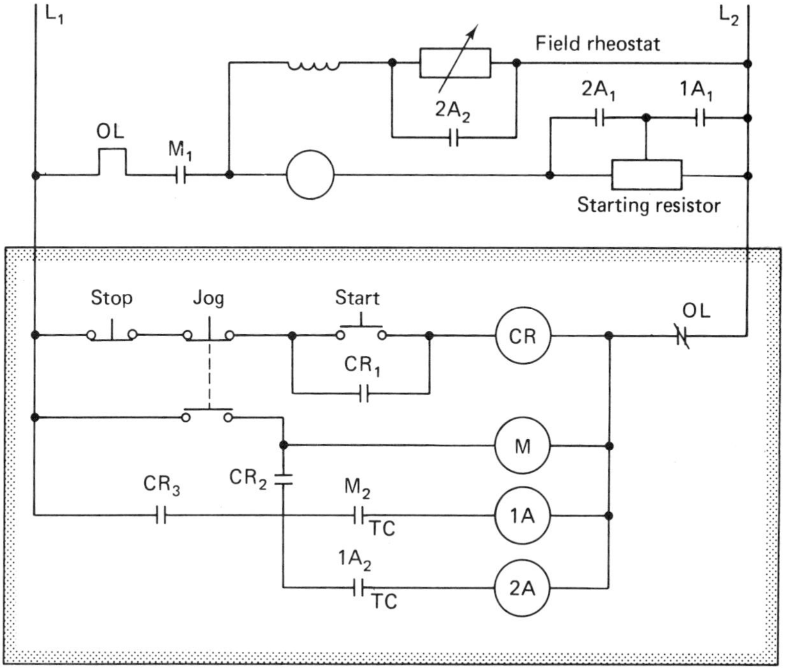

Figure 2-64 Time Acceleration Start and Jog

and a thermal overload element also appear in the power circuit: M_1 is closed during starting and running periods; $2A_2$ short-circuits the field rheostat during the starting period to develop maximum accelerating torque; and $1A_1$ and $2A_1$ permit sequential elimination of starting resistance by shortcircuiting segments of the tapped resistor.

In analyzing the control circuit, it is convenient to first visualize closing of contacts by the START button and then chronologically trace the changes that it triggers. The first significant change is energization of the CR coil; this causes all CR contacts to change from their normal state:

- CR1 closes and seals the CR coil by providing continuity after the START button is released.
- CR2 and CR3 close to provide line-to-line continuity across coil M and the first stage of continuity across coils 1A and 2A.

One coil, M, was energized by the pickup of CR; the next step is to trace circuit changes due to the pickup of M:

- M_1 closes and the motor begins to rotate.
- M_2 (TC) closes after a time delay and energizes coil 1A.

When 1A picks up, $1A_1$ eliminates the first increment of starting resistance and $1A_2$ (TC) energizes coil 2A. After a time delay, 2A picks up and

- $2A_1$ eliminates the remaining starting resistance.
- $2A_2$ opens to place the rheostat in series with the field.

While running, all contacts are in the pickup position. When the STOP button is depressed, coil CR drops out, deenergizing coils M_1, 1A, and 2A. All contacts revert to normal and the motor coasts to a stop. The same stopping action will occur if excessive loading causes the OL contacts to open.

This schematic also incorporates a *jogging* feature. When the motor is stopped, the position of the rotor and the coupled load when rotation ceases is unpredictable, yet possibly critical. For example, the motor may be powering a conveyor belt carrying an item that must be stopped opposite a loading platform. If the system relied upon the judgment of an operator to push the STOP button at the precise moment that results in deceleration to a halt at the required position, the system probably would be more a curse than a boon.

Jogging permits inching the motor to a desired position. Coil M is energized only while the JOG button is depressed. Coil CR is *not* energized and contacts CR_2 and CR_3 isolate coils 1A and 2A. All the starting resistance remains in series with the armature so that the speed will be less than normal running speed even if the JOG button is kept closed until steady state is reached. However, jogging is not intended for sustained rotation—in fact, prolonged depression of the JOG button probably would cause tripping of the OL contacts; the intent is to nudge the motor to the desired position.

A control circuit similar to Fig. 2-64 is termed *timed acceleration*. The amount of delay in the closing of M_2 TC and $1A_2$ TC is adjusted to match the acceleration time in reaching an E_g that allows elimination of starting resistance without a consequent jump in armature current above the permissible level. An alternative to timed acceleration is *current-limit acceleration*.

Current-limit acceleration and bidirectional run are illustrated in Fig. 2-65. The armature current in this form of the control circuit is not sampled directly. Coils 1AX and 2AX are positioned across V_A and are adjusted for differing pickup voltages. $V_A = e_g + i_a R_{ac}$; the pickup voltage is $e_{g1} + I_a R_{ac}$ for 1AX and $e_{g2} + i_a R_{ac}$ for 2AX. The required values of these voltages are determined by the methods used in Example 2-6; as each is attained, contacts close to energize coils 1A and 2A and thereby eliminate starting resistance.

The control circuit has both a FOR (forward) and REV (reverse) pushbutton. Depressing either will energize the related F or R coil. In the motor circuit, the polarity of the field is the same for both F and R; however, the armature current is reversible, depending upon which set of contacts, F or R, is closed, thereby allowing a choice of the direction of rotation. An *interlock* is provided between the F and R coils; when running in the forward direction, the normally closed F TC contacts in series with coil R are open; these contacts are time-delayed closing in

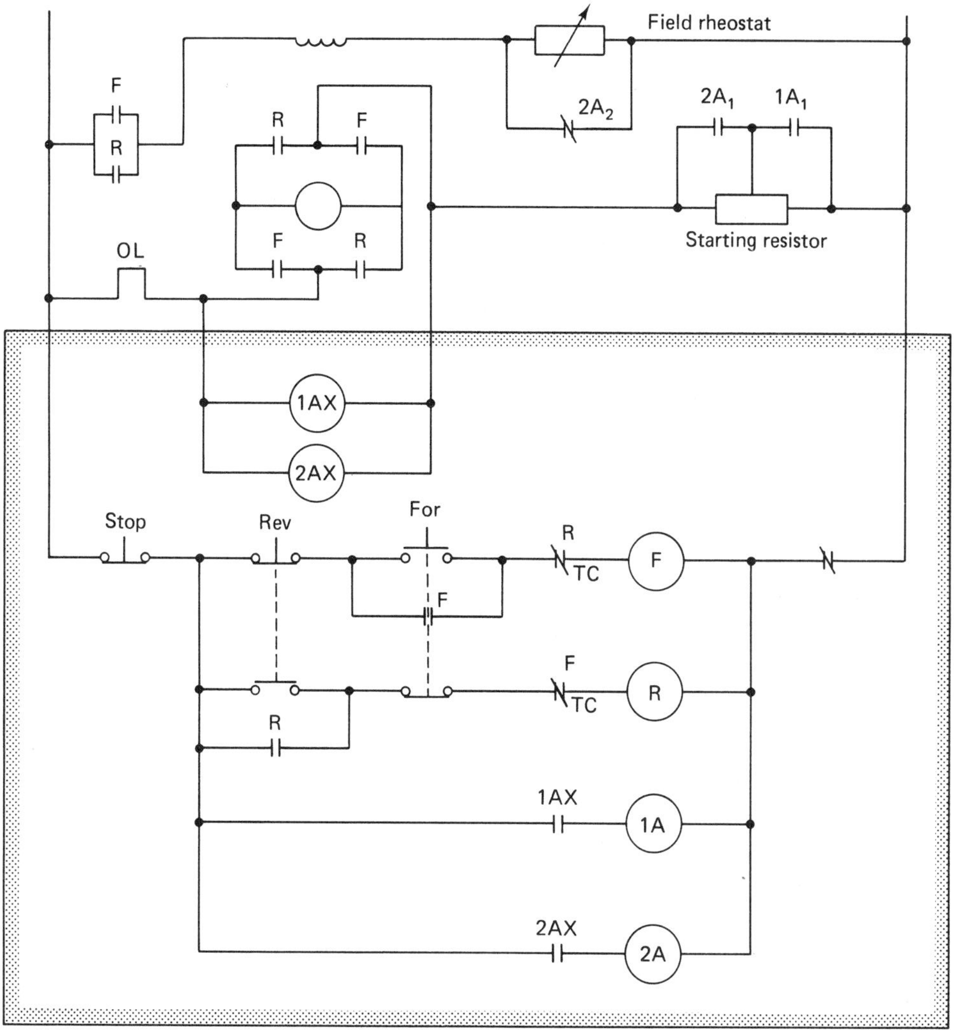

Figure 2-65 Current-Limit Acceleration and Bidirectional Run

order that the motor, stopping after forward rotation, may come to rest or decelerate to an acceptably low speed before coil R can be energized. A similar arrangement provides the required protection during reverse rotation.

The illustrated starters and control circuits lack automatic regulation or control of the speed. Only the starter of Fig. 2-62 has built-in provisions for manually adjusting the running speed. Automatic speed regulation will be considered in Chapter Three. At this point, illustration of a Ward–Leonard speed control system is appropriate.

A Ward–Leonard system has a main generator directly supplying the motor

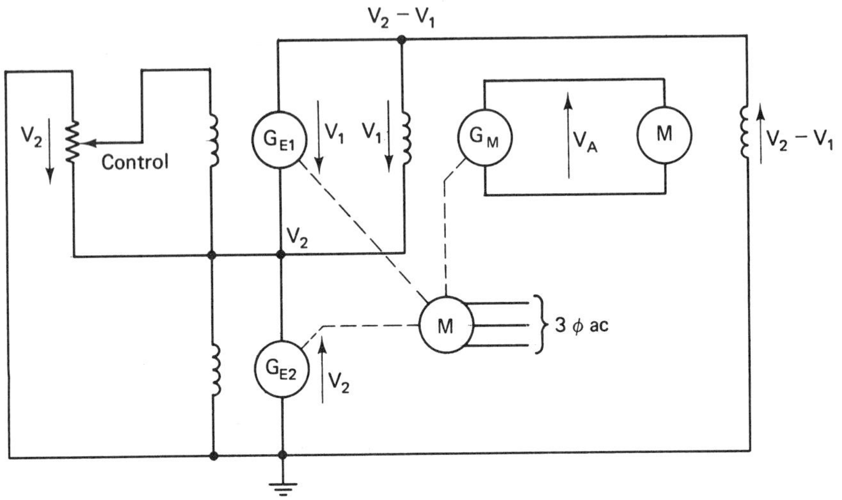

Figure 2-66 Ward–Leonard Speed Control

armature, together with one or more exciter generators to supply field currents. The system of Fig. 2-66 shows one arrangement. Two exciting generators are used. A polyphase ac motor drives all three generators.

For analytical purposes, a ground is assumed on the low side of G_{E2}. The potentials at key nodes with respect to this ground are shown to permit identification of the voltage across each element. The control element is a voltage divider that allows adjustment of V_1. V_2 is constant.

Assume that the motor has been running at a steady speed and the control now is adjusted to increase V_1. This will cause an increase in V_A, and a decrease in the difference $(V_2 - V_1)$; both changes contribute to an increase in the motor speed. The motor is both armature and field controlled, and a change in the system control causes armature and field effects to reinforce one another in decreasing or increasing the speed.

The use of four machines to control the speed of one motor appears to be extravagant, but it is not as serious as it first appears. Use of an ac motor–dc generator set to supply dc for plant motors is routine. Only the two exciters are unusual additions. There are no substantial power demands on either; both are relatively small and have a minimal effect on the system efficiency. The procurement and operating costs are a small price to pay when the adjustable-speed capabilities are critical to the application.

A strong claim can be made that Ward–Leonard systems, at best, are obsolescent because of the advent of solid-state power devices. But replacement of the many operational Ward–Leonard systems is unlikely to occur solely to exploit a new technology. Many of the machines in current Ward–Leonard systems will not perform satisfactorily on thyristor–rectifier drives.

SECTION 2-4. SPECIAL DC GENERATORS

2-17. SPECIAL DC GENERATORS A number of machines depart from the conventional form and characteristics discussed in preceding sections. Several are presented in this section.

Three-Wire Generator. A three-wire dc system is the dc equivalent of three-wire single-phase ac residential service, with two differing line voltages available for use. Although there are alternative methods of providing this dc service, only the three-wire generator is considered here.

The armature winding of Fig. 2-67 is that of Fig. 2-42. In the latter, two coils —coils 2 and 8—occupy identical slots and the instantaneous voltages induced in the coils must be identical. In Fig. 2-67, instantaneous voltages are identified by a double subscript; $e_{2\text{-}8}$ is the instantaneous voltage in coils 2 and 8 and it appears twice on the circuit diagram, alongside the coils located in the positions of coils 2 and 8 in Fig. 2-42. This pattern is applied to all coils. With an assumed ground at the negative brush, the potential at the positive brush is V.

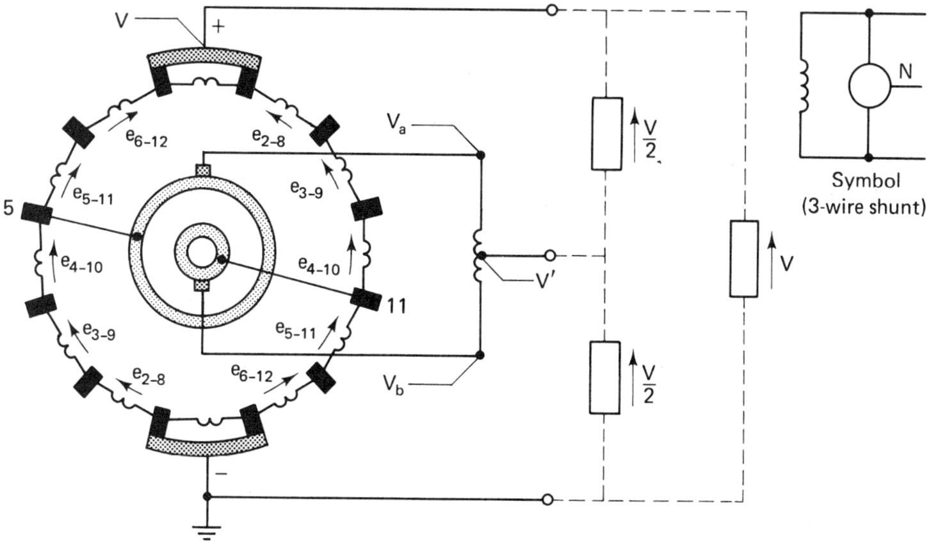

Figure 2-67 Three-Wire Generator

From commutator segment 5 a connection is made to a rotating slip ring; a second slip ring is connected to segment 11. Brushes permit connection of a center-tapped reactor to these rings. The potentials at the terminals of the reactor are $V_a = V - e_{6\text{-}12} - e_{5\text{-}11}$ and $V_b = e_{6\text{-}12} + e_{5\text{-}11}$. At the center tap, the potential is

$$V' = V_b + \tfrac{1}{2}(V_a - V_b)$$
$$= (e_{6\text{-}12} + e_{5\text{-}11}) + \tfrac{1}{2}[(V - e_{6\text{-}12} - e_{5\text{-}11}) - (e_{6\text{-}12} + e_{5\text{-}11})]$$
$$= \frac{V}{2}$$

Rotated and analyzed at any position, V' always is at the potential $V/2$, thereby providing the terminal for the center line of a three-wire system.

As the armature rotates, the voltage across the reactor, $V_a - V_b$, is a sinusoidal voltage with a frequency $\omega P/4\pi$. An ac current will flow through the reactor and the armature winding; however, the inductance of the reactor is large and the ac current is negligibly small. On the other hand, the resistance of the reactor is low; if the loading across the low-voltage lines is unbalanced, a dc current flows into or out of the center tap. When this current divides between the two halves of the reactor coil, the magnetic saturation levels in the two halves will differ; the potential at V' may vary from $V/2$, but the variance is slight.

Dynamotor. An armature having two coil sides per slot is represented in Fig. 2-68. The M coil has one turn but the G coil has two. Two commutators are mounted on the rotor with M coils connected to one and G coils to the other. Not pictured but present is one set of field windings.

If line voltage is applied to the M windings and the field, the machine will run as a motor. The G coils, rotating in the field flux, generate power that can be extracted through the second commutator for application to a load. This is a dynamotor; its function is to convert dc at one voltage level to a second, different level. The same task could be accomplished with a motor coupled to a separate generator, but the dynamotor does it with a single frame, field, and rotor and but one P_{fwc} loss.

The voltage equation, $E_g = PZ\omega\Phi/2\pi a$, applies to both the motor and the generator winding. P, ω, and Φ are common to both. This results in a voltage ratio,

$$\frac{E_G}{E_M} = \frac{Z_G a_M}{Z_M a_G}$$

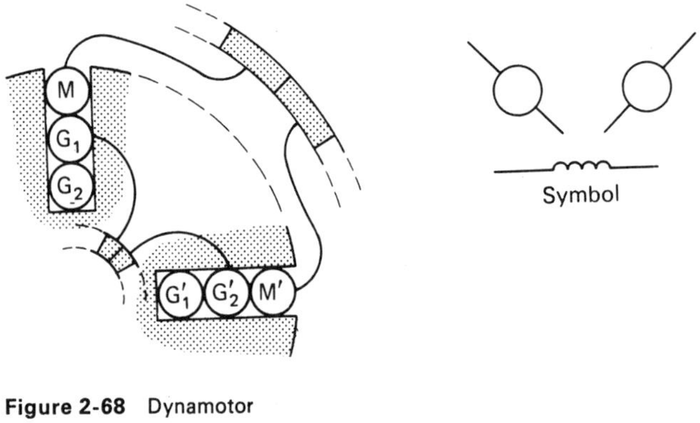

Figure 2-68 Dynamotor

A highly significant point is that the flux, although considered, has no effect on this ratio. If field control were applied to a dynamotor connected to a constant input voltage, the running speed would alter but the output voltage would be unchanged. For this reason, no rheostat is used on a dynamotor; the field often is internally connected and inaccessible to the user.

In the ideal machine, $E_g = V_L$ and the mechanical P_d equals the electrical P_d. If parallel paths in the windings also are equal,

$$V_G I_G = V_M I_M$$

$$\frac{V_G}{V_M} = \frac{I_M}{I_G} = \frac{Z_G}{Z_M}$$

These ratios are counterparts of those in a transformer, with the conductor ratio equated to the transformer turns ratio.

The dynamotor can be constructed with several independent generator windings, each with its own commutator, to permit several output voltages from a single machine.

Rotating Amplifier. The conventional self-excited generator may be viewed as a rotary amplifier with input to the field and output from the armature. The power-amplification factor generally ranges between 20 and 100. In an *amplidyne* this factor can be on the order of 20,000. This level is achieved by exploiting armature reaction.

A conventional separately excited generator is shown in Fig. 2-69(a). The field flux and the armature reaction flux are pictured; armature reaction flux is proportional to the load current.

Because the armature resistance is small, if the brushes are short-circuited, a small voltage will produce a large short-circuit current and a large armature

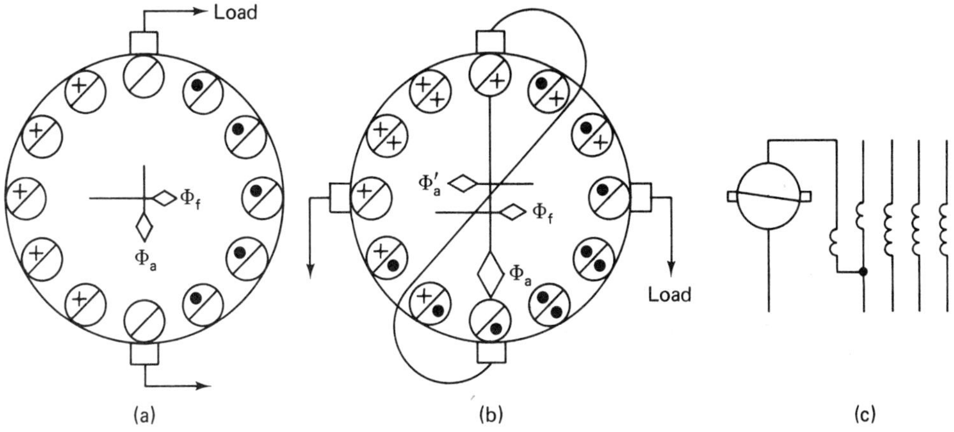

Figure 2-69 Amplidyne

reaction flux. An induced voltage, attributable to Φ_a, is superposed upon the voltage due to field flux. With a second set of brushes on the direct axis, this voltage can be extracted and applied to an external load. This form of the generator commonly is termed *metadyne*.

The field current is the input signal to the metadyne. If the machine is to respond to small changes in I_f, the voltage generated by the field flux must be proportional to the field current—that is, operation is on the linear portion of the saturation curve and the speed is constant. Consequently, the voltage equation, $E = k\Phi\omega$, can be written $E_x = k_x i_x$, in which E_x is the induced voltage attributable to current i_x in any winding x. At no-load (i.e., quadrature-axis brushes short-circuited but direct-axis brushes open), two currents in the system result in

$$E_f = k_f I_f$$

(E_f is the voltage generated in the conventional manner which, with the brushes short-circuited, fixes the armature current.)

$$I_a = \frac{E_f}{R_{ac}}$$

$$E_a = k_a I_a$$

This voltage is the open-circuit voltage between the direct-axis brushes. But if a load is connected across the terminals, the magnitude of the load voltage will drop substantially due not only to the $I_L R_{ac}$ drop but also to a *second* armature reaction flux, Φ'_a, located on the direct axis in opposition to Φ_f. This is illustrated in Fig. 2-69(b). Under load,

$$E_f = k_f I_f - k_L I_L$$

$$I_a = \frac{E_f}{R_{ac}} = \frac{k_f I_f - k_L I_L}{R_{ac}}$$

$$E_a = k_a I_a = \frac{k_a(k_f I_f - k_L I_L)}{R_{ac}}$$

$$I_L = \frac{E_a}{R_{ac} + R_L} = \frac{k_a(k_f I_f - k_L I_L)}{R_{ac}(R_{ac} + R_L)}$$

Solving this last equation for I_L,

$$I_L = \frac{k_a k_f I_f}{R_{ac}(R_{ac} + R_L) + k_a k_L}$$

In this equation, the product $k_a k_L$ is substantially greater than $R_{ac}(R_{ac} + R_L)$; if R_L is varied and I_f is constant, I_L remains fairly constant, a characteristic of the metadyne that may be exploited.

By addition of a well-designed compensating winding, Φ'_a may be eliminated—this machine is the *amplidyne*. In the amplidyne, if I_f is constant, E_f and E_a will be constant and the line voltage, $E_a - I_L R_{ac}$, is essentially constant.

In lieu of a single field winding, several independent windings may be placed on the pole cores. Each can be excited by a proportional electrical signal that is representative of some physical quantity in a system under the control of the amplidyne—quantities such as speed, torque, acceleration, and load current. Properly combined, these signals will cause the system to automatically maintain the controlled characteristics within predetermined limits. This use of the amplidyne is illustrated in Chapter Three. The symbol appears in Fig. 2-69(c).

Homopolar Generator. In Paragraph 1-10, the induction of currents in a disc, a portion of which passes between the poles of a magnet, was illustrated. If brushes were positioned with one at the center and the other at the rim of the disc, dc could be extracted. This is one of the earliest known dc generators and is unique in that the generated power is not ac, which requires commutation, but a true dc. However, for many years it was little more than a laboratory curiosity because of a lack of effective static contacts. More recently, research in the use of liquid-metal contacts promises a useful machine for low-voltage, high-current employment.

Tuned Generator. The buildup of voltage in a self-excited generator was discussed in Paragraph 2-2 and illustrated in Fig. 2-13, but with the omission of one possible condition. In Fig. 2-70(a), the resistance in the shunt-field circuit, R_c, is such that the plot I_fR_c coincides with the linear portion of the saturation curve. R_c is termed the "critical resistance." The terminal voltage can build to any level between V_1 and V_2. If no control is applied, the user cannot predict the operating point. Further, the operation is unstable; for example, if the machine is supplying a load at the level V_A and a momentary disturbance occurs in the system, the operating point can move to any other location within the V_1–V_2 range and, upon removal of the disturbance, either remain at this new point or slide to yet another level. Such performance obviously is undesirable, but the sensitivity to extremely small flux

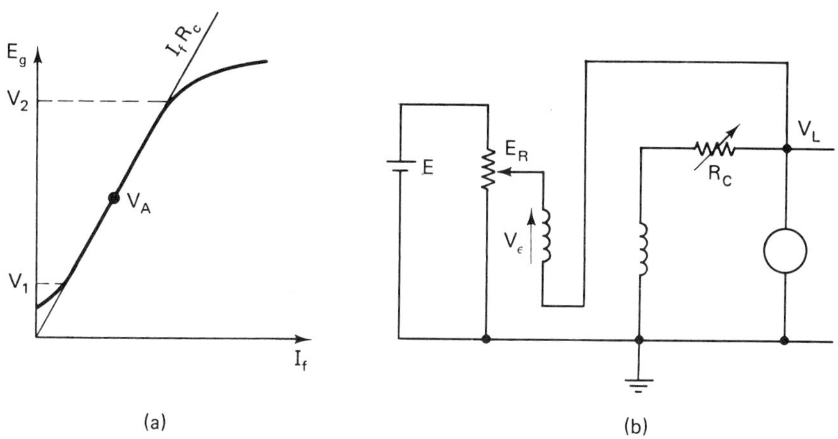

Figure 2-70 Tuned Generator

change is an asset in that the machine is highly susceptible to control. One possibility is shown in Fig. 2-70(b).

A control winding is added to the normal field windings of the generator. An error voltage, V_ϵ, equal to $E_R - V_L$, occurs across this field. E_R is constant after calibration for the desired V_L; V_L is variable by changing E_R. When $V_L = E_R$, the current in the control field is zero. But if V_L is less than E_R, current flows from point A to point B, creating a flux increment that adds to the normal pole flux and thereby causes V_L to rise. Conversely, if V_L rises to a value greater than E_R, the control winding current will flow from B to A, creating a flux that subtracts from the normal pole flux to cause a drop in V_L. Thus, a stabilized operating point is established—one that is easily adjusted to any desired level by means of the voltage divider.

The high sensitivity of the generator to control requires a very small control flux, and the current and the power in the control field are small.

SECTION 2-5. TESTING

A comprehensive test code for dc machines is published in IEEE Std 113. The complete range of tests is broad, including measurement of items such as air-gap length, insulation resistance, and vibration. There are procedural differences when the power supply for the tests is a thyristor rectifier rather than a ripple-free source (battery, dc generator). For most tests, alternative procedural methods are described.

This discussion of testing is consistent with Std 113 but restricted in scope. It assumes a ripple-free source and includes tests to determine performance characteristics discussed in earlier sections of the text only. The selection from alternative methods favors those that can be executed in a modestly equipped laboratory with commonplace instruments and equipment.

2-18. ELECTRICAL MEASUREMENTS

To achieve reasonably discriminating results, several practices are basic:

1. The range of a meter should be chosen so that the needle deflects into the upper portion of the scale in order that the fraction of a major division may be estimated with reasonable precision.

2. Voltage measurements are taken at machine terminals. Applied voltages should be as near as practical to rated voltages.

3. Ammeters should be located at points where only the current through the tested component is recorded. Where this is impractical, the recorded current must be corrected.

4. Power may be computed from voltage and current or measured directly with a wattmeter.

5. Resistance preferably is measured with a bridge but may be computed from measured voltage across and current through an element. The change in ohmic value of a resistance with temperature must be recognized by correcting the measured value if the operating temperature differs from the temperature at the time of measurement. For copper, the correction is

$$R_1 = \frac{R_0(T_1 + 234.5)}{T_0 + 234.5}$$

R_0, T_0: measured resistance R_0 at temperature T_0

R_1, T_1: resistance R_1 corrected to temperature T_1

2-19. TESTS COMMON TO THE MOTOR AND GENERATOR

Resistance. All inherent machine resistances, isolated to the extent permitted by machine construction, are measured. The standard terminal markings for a dc motor or generator are F_1 and F_2 (shunt field), A_1 and A_2 (armature circuit to include compensating and commutating field windings), and S_1 and S_2 (series field).

Shunt- and series-field resistances are measured at the marked terminals, but the armature circuit resistance is measured by components.

1. *Rotor winding*: The rotor must be blocked during this measurement to prevent rotation due to residual flux. A controlled voltage is applied to the armature terminals to produce a constant current through the armature; the duration and magnitude of this current must be limited to prevent damage from excessive heating. Voltmeter leads are applied, in turn, to commutator segments directly under the center of each pair of adjacent brushes; the measured potential drops are averaged and used with the measured current to calculate R_a.

2. *Compensating and commutating windings*: One end of these windings usually is connected to a brush rigging and the other end to the A_1 terminal. The resistance is measured between these points.

3. *Brushes*: It is not practical to obtain a meaningful measurement of the nonlinear brush resistance. The potential drop across the brushes can be measured, but only with a special test harness and readings over an extended period. In lieu of measurement, the code assigns voltage drops by type of brush; for carbon and graphite with an attached flexible lead the value is 2 V.

Magnetic Saturation. The machine is driven at rated speed. With the armature circuit open, a controlled dc voltage is applied to the field in order to permit step increases in the field current from zero to the value producing a generated voltage of approximately 125% V_R. A sequence of simultaneous field current and generated voltage readings are obtained either by starting at zero-field current and increasing to maximum or beginning at maximum and descending to zero. In either case, three readings should be as near as practical to 90, 100, and 110% rated voltage, and the residual voltage must be measured.

To avoid error due to hysteresis effects, step changes cannot be made by overrunning and returning to the desired test point. For ascending readings, if the desired point is overrun, the field current must be reduced to zero, then returned to the desired level; the converse applies to descending readings. Sufficient readings are taken to permit an accurate plot of the saturation curve.

If a similar test is performed with the shunt field unexcited and controlled currents in the series field, N_{se}/N_{sh} can be determined.

Friction, Windage, and Core Losses. Two methods are described for this measurement; one permits identification of component losses while the other yields only the aggregate. For either method, the machine should be at operating temperature.

If the machine, motor or generator, is run as a motor at rated speed, without load, and with a generated voltage equal to the full load E_g, the developed power is P_{fwc} for rated conditions. When run at no-load, the armature current is small and E_g approximately equals V_L. Therefore, for this test the voltage applied to the terminals should be adjusted to a value V_L' which is E_g at rated load: that is,

$$V_L' = V_R \pm (E_B + I_R \sum R_{ac})$$

E_B: value listed in the test code (2 V for flexible lead carbon and graphite)

$\sum R_{ac}$: resistance of the armature winding and all windings in series with it

$\pm$: + for generator, − for motor

The power input is measured directly or computed from V_L' and the measured current. P_{fwc} (P_d under the test conditions) is computed either by $E_g I_a$ or $P_{in} - \sum P_{cu}$. This method is termed *running light*.

If component losses are to be measured, the machine is driven by a motor whose power output can be determined. The output of the motor is input to the machine undergoing test. Power is recorded at rated speed with the following conditions in the tested machine:

1. *Brushes removed from holders*: The input is windage and bearing friction loss.

2. *Brushes restored, no field excitation, armature open-circuited*: Added input is brush friction loss.

3. *Field excited to produce rated E_g, armature open-circuited*: Added input is full-load core loss.

2-20. EFFICIENCY

Efficiency may be computed by the segregated loss method. Using this method, computations can be based exclusively on electrical measurements; that is, for the motor, $P_o = P_{in} - \sum P_{loss}$ and for the generator $P_{in} = P_o + \sum P_{loss}$. Copper losses are computed using measured resistances corrected for temperature. The brush loss is $E_B I_a$. Losses in field rheostats and exciters usually are charged against the plant, not the machine.

In addition to copper losses and P_{fwc}, a stray-load loss will be present in the machine, attributable to small imbalances and irregularities. This loss cannot be calculated or measured with common test equipment. Standard 113 provides that 1% of the output power shall be taken as the stray-load loss.

2-21. GENERATOR VOLTAGE REGULATION The generator should attain a stabilized operating temperature by continuous operation at rated speed and load before this test is performed. The field rheostat setting should remain fixed at the point that produces rated voltage at rated output power.

Readings of voltage at 25% steps in current are made between no-load and the desired overload. Starting at rated load, gradually decrease the load current to zero, then increase to the overload, and return to rated, recording readings at each step. If the initial and final readings at rated load do not agree, repeat the test. The test should be accomplished quickly in order that the temperature of the armature windings remains essentially constant.

2-22. MOTOR SPEED REGULATION The motor should be at a stabilized operating temperature from continuous operation at rated load before performing the test. The field rheostat remains at the setting for rated speed at rated load. Starting at rated load and speed, gradually decrease the armature current in 25% steps to no-load, recording the speed at each step. Return to rated load and continue readings to the desired overload, then return to rated load. If the initial and final speed at rated load differ, repeat the test.

SECTION 2-6. RATINGS

2-23. RATINGS A broad range of standardized machine ratings appears in NEMA Pub MG 1. For continuous-duty, general-purpose machines, generator voltages are 125, 250, 500, and 700 V, with speeds from 200 to 3450 r/min and power ratings up to 6400 kW. Motor voltages are comparable to generator voltages with the addition of several lesser voltages for machines designed for operation on rectified power; speeds range from 50 to 3500 r/min with power ratings up to 8000 hp.

All machines have a nameplate that includes data on the rated-load power, voltage, and speed. In general, machines can sustain overspeeds to 125% rated in short-duration emergencies and overloads of 150% rated power for 1 minute without significant damage. There is no simple factor to apply in determining permissible motor speeds by armature or field control; neither should be attempted without first determining the acceptable range for the specific motor. Motors will operate satisfactorily at voltages up to 110% rated.

The importance of temperature in the rating of equipment cannot be overemphasized. The permissible operating temperatures primarily are limited by the thermal characteristics of the insulation. Excessive temperatures reduce the dielectric strength of the insulation, accelerate aging, and render it brittle and vulnerable to mechanical disintegration. Generally, the ratings assume an ambient temperature (temperature of the general area in which the machine is located) of 10 to 40°C and no external obstructions to ventilation. Insulation is classified according to thermal endurance. The temperature data on the nameplate will provide either the rated temperature rise or the maximum ambient temperature and insulation class, from which the permissible rise can be determined by referring to NEMA Pub MG 1. For example, at 40° ambient with class A insulation the rated rise in armature windings is tabulated as 50°C for open enclosures.

PROBLEMS AND QUESTIONS

2-1. An expression $e = 2\omega\Phi/\pi$ for the average voltage generated in a single-turn loop appears in Paragraph 2-1. (a) What is the expression for effective (rms) voltage? (b) Express both the average and effective voltage with S, the speed in r/min, used in lieu of ω.

2-2. A conducting loop with the dimensions 25 × 15 cm is rotated at 60π rad/s in a field having a flux density of 0.8 T. (a) What is the average voltage induced in the loop? (b) With 20 identical loops evenly divided between two parallel paths, what is the average voltage across the circuit?

2-3. To determine the flux density of a magnetic field, a rectangular loop with a length of 12 in and a width of 8 in is pivoted at the midpoint of the shorter dimension and rotated in the field at a speed of 1000 r/min. The voltage generated in the coil is measured with an ac voltmeter and found to be 4.6 V. (a) What is the flux density in kilolines/in^2? In tesla? (b) If the pivot is changed to the midpoint of the longer side and the coil is rotated at 1200 r/min, what voltage will register on the voltmeter and what is its frequency?

2-4. At what speed must a five-turn coil be driven in order to generate an average voltage of 5.65 V if the maximum flux through the loop is 3.32×10^2 kilolines?

2-5. The average voltage generated in a coil driven at an angular velocity of 60π is 7 V. What voltage is generated when the coil is driven at a speed of 1200 r/min?

2-6. A six-pole generator has 90 coils, each coil having six turns. The coils are arranged in six parallel paths. The flux in each pole is 0.004 Wb and the generator is driven at 56.7π rad/s. What is the generated voltage?

2-7. A four-pole generator with 5800 armature conductors is driven at a speed of 1200 r/min. The pole flux is 450 kilolines. What is the generated voltage if there are (a) 4 parallel paths? (b) 8 parallel paths? (c) 16 parallel paths?

2-8. The voltage of a dc generator is 250 V. What voltage is generated if: (a) The flux is reduced by 10%? (b) The speed is increased 20%, retaining the original flux? (c) The original flux is reduced 10% while simultaneously increasing the speed 20%?

2-9. When the speed of a dc generator is raised by 15%, by what percentage must the flux be reduced in order to retain the original voltage?

2-10. The field intensity per pole of a dc generator is 2 kA/m. The applicable saturation curve is the cast-iron curve of Fig. 1-10. The generator has an open-circuit terminal voltage of 500 V. (a) If the intensity is raised to 2.5 kA/m and the speed is unchanged, what is the terminal voltage? (b) To obtain 450 V with a concurrent speed reduction of 20%, what intensity is required?

2-11. If the cast iron of the generator in Problem 2-10 were replaced by cast steel with no other changes in the machine, what fraction of the original field current is required to generate 500 V at the original speed?

2-12. With a permissible coil current of 15 A in

the generator of Problem 2-6, what is the rated power of the generator?

2-13. A generator is rated 500 V 1000 kW. It is desired to alter the terminal characteristics by changing the number of parallel paths; after the change, a rated current of 3000 A is required. What will be the new rated voltage?

2-14. If the winding connections of the generator of Problem 2-12 are changed to provide two parallel paths, what are the new voltage, current, and power ratings?

2-15. The armature current of a compound generator at rated load is 130 A and the field current is 5 A. The mmf of the shunt field of each pole is 4000 A and the series field has seven turns per pole. (a) At rated load, what is the mmf per pole? (b) What is the ratio of N_{se} to N_f? (c) What is the difference in mmf between rated and open-circuit conditions?

2-16. A separately excited generator is rated 125 V 3 kW and has armature resistance 0.317 Ω, $P_{fwc} = 300$ W, and $P_f = 90$ W. What is the full-load efficiency?

2-17. A separately excited generator has a field resistance of 36 Ω, armature resistance 0.13 Ω, and constant brush drop of 2 V. The no-load generated voltage is 263 V and the rated current is 84 A. Field excitation voltage is 125 V and P_{fwc} is 1200 W. What is (a) the rated voltage and power, and (b) the efficiency?

2-18. A 25-kW 250-V shunt generator has the following machine constants: $R_{ac} = 0.1059$ Ω, $R_f = 75$ Ω, and $P_{fwc} = 1100$ W. What is the (a) full-load efficiency, (b) no-load voltage, and (c) % voltage regulation? (Assume that ϕ remains constant.)

2-19. To what value will the voltage of the generator of Fig. 2-71 build if the total shunt field circuit resistance is 40 Ω? 80 Ω?

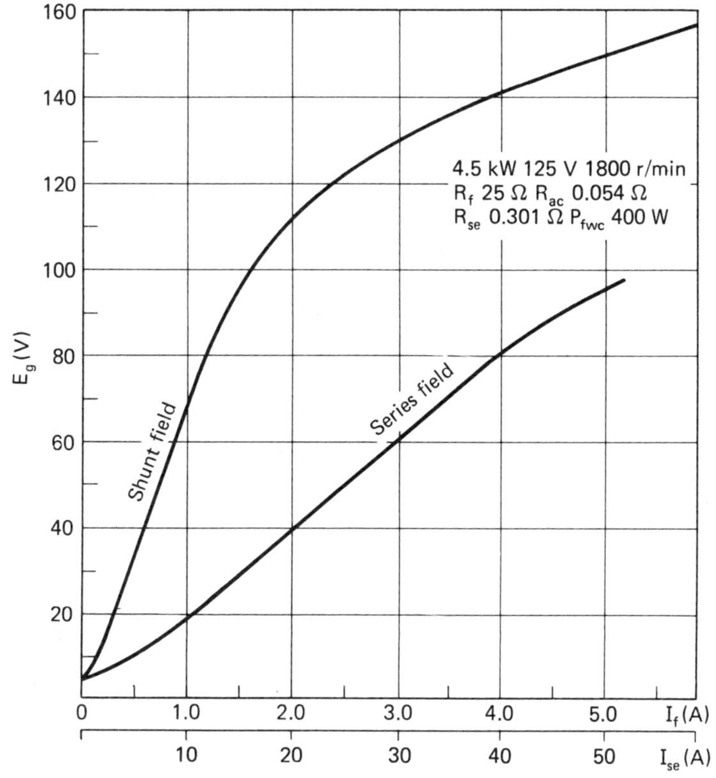

Figure 2-71 Problems 2-19 to 2-22 and 2-57

2-20. For the generator whose characteristics are given in Fig. 2-71, what rheostat resistance in series with the field is required for an open-circuit voltage of 125 V? At this setting, what is the value of $k\Phi$?

2-21. The characteristics of the machine to which this problem applies are shown in Fig. 2-71.

(a) The first step in the analysis of this machine when connected as a compound generator is to calculate the full-load E_g from the electrical characteristics. However, although the line current may be determined, the field current initially is unknown. From inspection of the graph, it is estimated to be 3 A.

(1) Assuming that the 3-A estimate proves accurate within ± 1 A, what is the accuracy of the computed E_g?

(2) Does a ± 1-A tolerance in the estimated field current significantly affect the contribution of the *series* field to the full-load E_g?

(3) Using the 3-A estimate, calculate the full-load E_g.

(b) (1) What is the ratio N_f/N_{se}? (2) When related to the shunt winding, what equivalent field current produces the same mmf as that provided by the series field at rated load?

(c) What is the actual required shunt-field current? To achieve this current, what is the required field rheostat resistance?

(d) With the rheostat setting of the preceding paragraph, what is the no-load terminal voltage? What is the %VR of the generator?

(e) What is the full-load efficiency?

(f) Discuss any tacit assumptions in the foregoing analysis.

2-22. Using Problem 2-21 as a guide, analyze the machine of Fig. 2-71 when employed as a shunt generator.

2-23. The generator in Problem 2-22 is running at rated speed with the field circuit open. When the field circuit is closed, the voltage builds from the low residual value to the open-circuit voltage. Before reaching the steady state, the voltage from the $I_f R_f$ line is less than the corresponding voltage on the saturation curve. Considering Ohm's law, how do you account for this discrepancy?

2-24. A 170-kW 250-V, flat-compounded generator has the following machine constants: $E_b = 0.5$ V, $R_{se} = 0.004$, $R_{ac} = 0.0069$, $R_f = 24\ \Omega$, and $P_{fwc} = 6$ kW. What is the efficiency at 25, 50, 75, 100, and 125% I rated?

2-25. One means of reducing the effects of armature reaction on commutation is to rotate the brushes to the displaced q-axis. Discuss the advantages and/or disadvantages of this method. Will it counteract the drop in generated voltage due to uncompensated armature reaction?

2-26. How many compensating-winding conductors are required in each pole face of a generator in which there are six poles that together cover 68% of the armature surface? There are 1944 armature conductors in six parallel paths with an armature current of 100 A.

2-27. The shunt-field resistance of a generator is 27 Ω and the field requires 2.3 A to deliver rated power at rated load. The rated voltage is 125 V. A slide-wire rheostat rated 5.5 A 52 Ω is available for control of the field current. What is the total power loss at rated load in the rheostat and field circuit if the rheostat is: (a) Inserted in series with the field? (b) Used as a voltage divider to supply the field?

2-28. A 125-V 3-kW laboratory generator has a terminal board with six terminals marked F_1 and F_2 for the ends of the shunt field, S_1 and S_2 for the series field, and A_1 and A_2 for the armature. None of these components are interconnected within the machine.

(a) Sketch the connection diagram for a compound generator, arbitrarily assigning subscripts 1 and 2 to each component terminal. Include a rheostat for field control and identify the load terminals.

(b) The machine is connected according to your diagram and run at rated speed. The open-circuit terminal voltage is 4 V. What is wrong and how will you correct it?

(c) After correction, the open-circuit voltage is 131 V. A graduated load is applied and concurrent voltage–current readings are taken with the following results: 79.8 V–0.8 A, 64 V–2.5 A, 36.3 V–3.7 A, 18.7 V–

2.8 A. Regardless of how the load is changed, the maximum load current that can be obtained is 4.3 A. What is wrong and how will you correct it?
(d) Following the preceding corrections, the terminal voltage is 135 V when the load current is 24 A. Are you satisfied with this combination? If not, what changes will you make?

2-29. A 500-V 100-kW compound generator has an open-circuit voltage of 500 V and a terminal voltage of 525 V when delivering rated current. From an analysis of the saturation curves it is determined that a series-field mmf of 1200 A will provide flat compounding. The series field has eight turns and a resistance of 0.012 Ω. What diverter resistance is required for flat compounding? (The field current is negligibly small.)

2-30. For the generator of Example 2-4, what value of shunt field resistance and series-field diverter resistance would result in flat compounding with a terminal voltage of 240 V? Assume that the speed is constant at 1800 r/min.

2-31. When the generator of Problem 2-30 is delivering rated load, what electromagnetic torque is developed in the generator?

2-32. A dc armature has 72 slots. If coils are inserted with 160 electrical degrees between sides, what is the mechanical degree separation when the stator has: (a) Two poles? (b) Four poles? (c) Eight poles?

2-33. The voltage generated in each coil side of the two-pole machine of Problem 2-32 is 12 V. (a) What is the coil voltage? (b) Does the number of poles affect the coil voltage if all other quantities are unchanged?

2-34. If the voltage per coil side of the preceding two-, four-, and eight-pole machines is the same, the machines must differ in some respect. Discuss the possible differences.

2-35. If the two-, four-, and eight-pole machines are simplex-lap-wound, how many parallel armature paths does each have? How many paths are in each if they are simplex-wave-wound?

2-36. Each of 72 slots holds 12 conductors. The commutator has 144 segments. How many coils are there? How many turns are in each coil? How many coil sides lie in each slot?

2-37. The armature current of a six-pole duplex-lap generator is 600 A. What is the current in each coil? How many brushes are required and what is the current in each?

2-38. A 240-V 40-hp 1750-r/min shunt motor has the following characteristics: rated I_L = 140 A, $R_f = 62\ \Omega$, $R_{ac} = 0.0761\ \Omega$, and $P_{fwc} = 1.4$ kW. Find (a) the full-load output horsepower, torque, and efficiency; and (b) the no-load speed and speed regulation.

2-39. A series field with a resistance of 0.0139 Ω is added to the motor of Problem 2-38. The other characteristics are unchanged. At no load, the flux is 10% less than at rated load. Repeat the requirements of Problem 2-38.

2-40. The power developed in a motor armature is 100 kW. Find the torque in both US and SI units for speeds of (a) 1800 r/min, and (b) 38π rad/s.

2-41. A motor has six poles, 300 conductors, and six parallel paths. The flux per pole is 1.5×10^{-2} Wb and the armature current is 100 A. (a) At a velocity of 57π rad/s, what is the developed torque and power in SI units? (b) At 1200 r/min, what is the developed torque and power in US units?

2-42. With the constants combined in equation 2-3 [US], $T_d = 0.1173PZ\Phi I_a/a$. This equation appears in some texts. Another equation you may encounter is $F_m = BlI_cpZ/1.13 \times 10^7$, in which F_m is the total force developed in the armature, B the flux density per pole in lines/in^2, l the conductor length in inches, I_c the coil current, and p is defined as the percentage of the armature surface directly opposite pole faces or, alternatively and with the same sense, the percentage of the conductors opposite pole faces at any instance. Show that the product $F_m r$, with r the radius of the armature in feet, is equal to equation 2-3.

2-43. The armature of a rotor has a diameter of 10 in and a length of 12 in. Pole faces cover 76% of the armature surface. Flux density is 0.9 T. There are 1200 conductors, with a coil current of 15 A. (a) What is the magnetic force and torque per conductor? (b) What is the net armature torque?

2-44. What is the net torque in Problem 2-43 if:
(a) Φ decreases 10% and I_c is unchanged?
(b) Φ is unchanged and I_c increases 15%?
(c) Φ decreases 10% and I_c increases 15%?

2-45. A 240-V compound motor has a series-field resistance of 0.14 Ω and an armature resistance of 0.15 Ω. The brush drop is 3 V. When the armature current is 90 A, what is the counter emf generated in the motor armature? What is the developed power?

2-46. A 120-V shunt motor has a field resistance of 30 Ω and an armature circuit resistance, including brushes, of 0.19 Ω. When the line current is 42 A, what is the counter emf and developed power?

2-47. A 240-V compound motor has a brush drop of 2 V and the following resistances: $R_f = 30\ \Omega$, $R_{se} = 0.009\ \Omega$, and $R_a = 0.0105\ \Omega$. When the counter emf is 226 V, what are the field, armature, and line currents? What is the armature current if the counter emf is zero?

2-48. Two shunt motors are identical except that one has a stabilizing field winding while the other does not. Which one has the better speed regulation? The better rated torque?

2-49. A coil having eight turns carries a 15-A current and rotates at 1800 r/min in a two-pole stator having a flux of 0.005 Wb per pole. (a) What is the generated voltage and the electrical power of the coil? (b) What is the torque and mechanical power of the coil in both SI and US units?

2-50. A 15-hp 240-V 3450-r/min motor is compared with a 10-hp 240-V 1750-r/min motor. (a) Which motor develops the larger torque? The larger armature current? (b) Assuming that the counter emfs are equal, which has the larger field flux if the poles, conductors, and parallel paths are the same in both machines? (c) One machine is physically larger than the other; which one and why?

2-51. When driven at 1800 r/min, the open-circuit voltage of a separately excited generator is 240 V. As a motor with 240 V applied to the terminals, the field excitation is unchanged. R_{ac} is 0.2 Ω. When the armature current is 50 A, what is the speed of the motor?

2-52. A 3-hp 120-V motor is driving a constant-torque load. The armature current is 24 A, armature circuit resistance is 0.2 Ω, and speed 1700 r/min. The field circuit accidentally is opened, resulting in the flux dropping to a residual of 6% of the excited value. Theoretically, neglecting time constants and practical limitations, there is a rise in current immediately after the flux reduction and the motor accelerates to a new steady speed where the load torque is restored. In theory, (a) what is the armature current immediately after the drop in flux? (b) What is the armature current with the original torque restored? (c) What is the final running speed? (d) Considering these results, what alternatives do you perceive for protecting the motor against this circumstance?

2-53. A 240-V series motor is operating on the linear portion of the saturation curve. The armature circuit resistance is 0.25 Ω. At a speed of 1000 r/min, the armature current is 50 A. (a) At 2000 r/min, what is the armature current? (b) By what percentage has the torque increased or decreased?

2-54. The sketch of a proposal for a small dc motor is shown in Fig. 2-72. Will it work? Justify your answer.

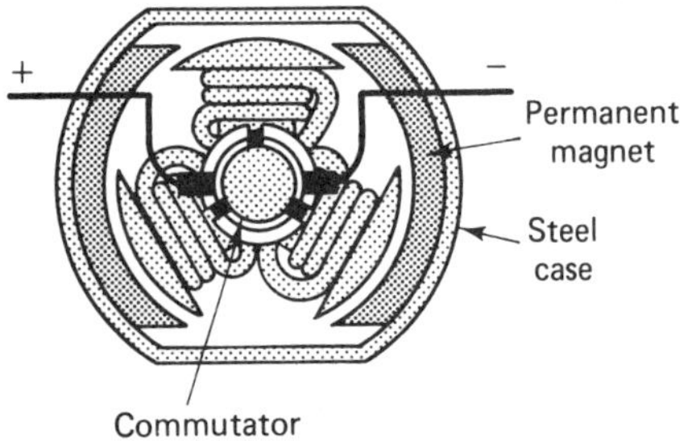

Figure 2-72 Problem 2-54

2-55. A dynamometer permits a direct reading of the shaft torque of a motor. One reference gives the equation $P = kTS$ for output power in watts from measured torque and speed, the latter in r/min. The value of k depends upon the units of the measured torque as follows: N·m, $k = 0.1047$; lb-ft, $k = 0.1420$;

and oz-in, $k = 0.000740$. Verify the correctness of the given constants.

2-56. Two identical shunt machines are mechanically coupled and connected to a common supply (Fig. 2-73). One operates as a motor and the other as a generator. This arrangement may be used to test for efficiency of large machines. With $V_L = 240$ V, the following currents are measured: $I_{fm} = 3.8$ A, $I_{am} = 136.2$ A, $I_{fg} = 4.1$ A, and $I_{ag} = 113.7$ A. For each machine, $R_{ac} = 0.0761\ \Omega$ and the P_{fwc} losses are identical.

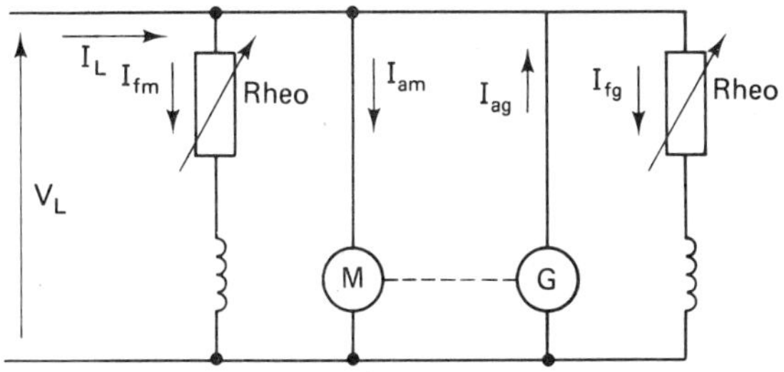

Figure 2-73 Problem 2-56

(a) The power output of the motor ($P_{dm} - P_{fwc}$) is the power input to the generator ($P_{dg} + P_{fwc}$). Find P_{fwc}.
(b) Find the efficiency of each machine for loads corresponding to the given armature currents.
(c) Show that the line power input, $V_L I_L$, represents the total losses in the two machines.

2-57. The characteristics of a machine when operated as a generator are given in Fig. 2-71. This machine is to be analyzed, first, as a shunt motor and, second, as a compound motor. For both, the line voltage is 125 V and the rated conditions are those when the armature current is 36 A and the speed is 1750 r/min. At no load the developed power equals P_{fwc} and the speed is that which occurs with the field rheostat set for *rated* conditions. For each form, shunt and compound, determine the following:
(a) Counter emf at rated load.
(b) Rheostat resistance for rated conditions.
(c) No-load E_g and I_a.
(d) No-load speed with (1) no rheostat resistance in the field circuit, and (2) the rheostat resistance of part (b).
(e) Load torque at rated conditions.
(f) Percent speed regulation.
(g) Percent efficiency.
(h) Rated horsepower.
(i) Starting torque with the armature current limited to 75 A.

2-58. During starting, the armature current of the shunt motor in Problem 2-57 is to be limited to 75 A, with successive increments of starting resistance eliminated when the current falls to 36 A. Find the number and ohmic values of required starting resistances.

2-59. The resistance of copper increases when the temperature rises. Suppose that a motor is started at ambient temperature and immediately adjusted to its base speed. Discuss the nature of possible speed changes when the motor reaches operating temperature if the machine is (a) shunt and (b) compound.

2-60. With a reduction in speed, the heat-dissipating ability of a self-ventilated motor is reduced. In general, a motor is capable of operation at 50% base speed at 67% rated torque.

A 15-hp 240-V motor has a base speed of 1800 r/min. At rated load and speed, $I_A = 54$ A. $R_{ac} = 0.23\ \Omega$. The speed is to be adjustable within the foregoing limitations. The control arrangements of Fig. 2-57 are under consideration.
(a) For control by series resistance [Fig. 2-57(a)], what is the required ohmic value of the rheostat at 50% S_R and 67% T_R? At this

speed and torque, what is the power loss in the rheostat?

(b) Can the power loss be reduced through use of a voltage divider [Fig. 2-57(b)]? If so, suggest appropriate values for R_1 and R_2 at 50% base speed.

(c) The field current remains constant at 2.2 A. P_{fwc} losses at base speed are 1.0 kW; this loss is proportional to speed. With the series resistance from part (a), what is the efficiency of the motor at 50% base speed and 67% torque?

(d) With the resistance of part (a), the no-load armature current is 2.5 A. What is the speed regulation? (*Note:* The 50% base speed and 67% rated torque are the applicable full load.)

(e) Using the Ward–Leonard system of Fig. 2-57(c) with the generator field adjusted for 50% base speed at 67% rated torque, what is the %SR?

(f) If the motor will be operated at the reduced speed for a substantial portion of its duty cycle, do you consider the Ward–Leonard method to be economically competitive with the methods of parts (a) and (b)? Discuss.

2-61. A motor will operate satisfactorily at 110% V_R and can sustain 25% overspeed in emergencies. The motor of Problem 2-60 has the field adjusted for rated conditions. With no change in the field setting, at no load and 110% V_R, is the speed likely to exceed 125% S_r? (*Note:* Estimate unknown data.)

2-62. When running at rated speed, the motor of Problem 2-60 is to be stopped by electrical braking. The maximum permissible armature current is 150% rated. What is the required braking resistance for: (a) Dynamic braking? (b) Plugging?

2-63. Analyze the control circuit of Fig. 2-74 and prepare the schematic of a compatible power circuit for a shunt motor.

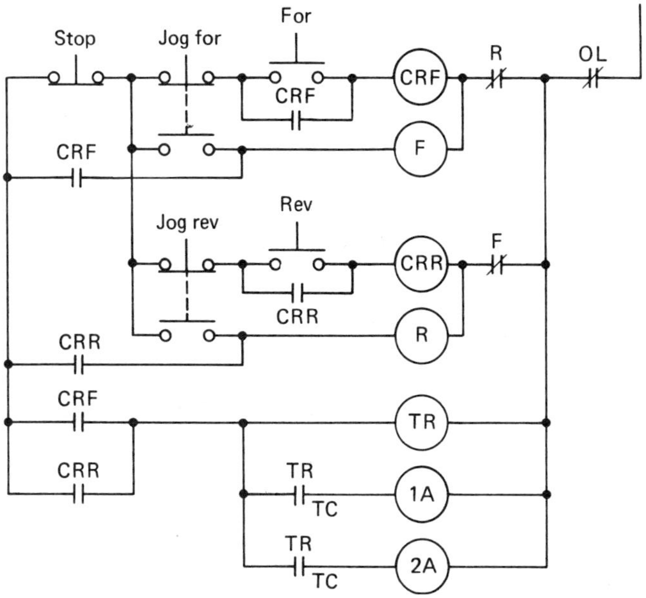

Figure 2-74 Problem 2-63

2-64. The motor of Problem 2-60 is to be controlled by the circuit of Fig. 2-74. The starting current shall not exceed $2I_R$; this may require more than the two accelerating resistors now included in the diagram. All the capabilities incorporated in the circuit of Fig. 2-74 are to be retained, but the following are to be added: (a) the option to run at 50%

base speed in the forward direction only, and (b) dynamic braking. Amend the schematic and determine the ohmic value of the non-machine resistances. (Figure 2-75 shows one method sometimes used for dynamic braking.)

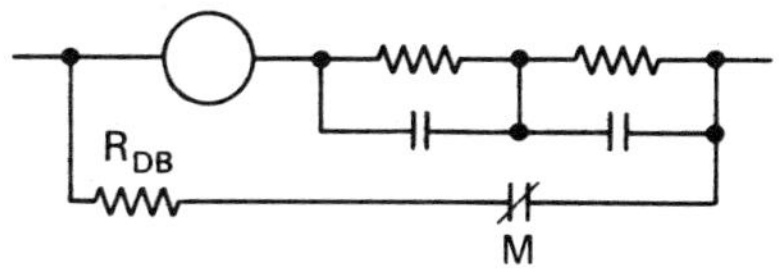

Figure 2-75 Problem 2-64

2-65. In a dynamotor there is no need for compensating windings to correct for armature reaction. Why?

2-66. A dynamotor has an input voltage of 28 V and two output windings, one rated 350 V 115 W and the other 500 V 100 W. If the efficiency is 91%, what is the rated input current?

3

Introduction to Feedback Regulation and Control

This chapter introduces regulation and control of a system output, often involving the application of feedback. The term *control* refers to methods whereby the performance of a machine or system is governed—for example, field rheostat control of a shunt motor in order to adjust the speed. *Regulation* refers to methods employed to stabilize a characteristic of the output—for example, the voltage of the critically tuned generator in Fig. 2-70(b); note that this generator also can be controlled by changing the magnitude of E_R.

Feedback was employed to regulate the line voltage of the critically tuned generator by returning part of the output to the input in order to stabilize the performance.

3-1. BLOCK DIAGRAMS

The performance of a simple machine or system can be interpreted from a schematic diagram, but as systems become more complex it is convenient to represent the cumulating contributions of components in a block diagram. The diagram also provides a ready means for establishing the mathematical relationship between the system output and input and determination of the system *transfer function*, the ratio of the output to the input. The components, development, and use of a block diagram will be illustrated with a separately excited generator.

Multiplying Element (Forward path). A multiplier is represented by a rectangle. In the forward transmission path, a signal enters from the left and is multiplied by the expression in the block; the product exits to the right and continues through the transmission path.

EXAMPLE **3-1:** The current in a winding is related by Ohm's law to the voltage across the resistance of the winding. In Fig. 3-1, $I_f = E_R/R_f$. E_R is the entering signal, $1/R_f$ the multiplier, and I_f the exiting signal.

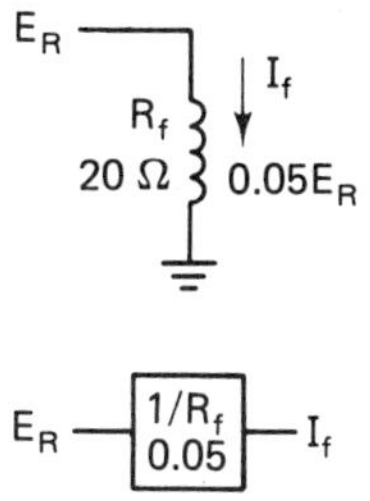

Figure 3-1 Multiplying Element

Multiplying Elements May Be Combined If No Diagrammatic Symbols Appear Between Them. This is the first step to synthesize a system. The combination is simply the product of the multipliers.

EXAMPLE **3-2:**

Assumptions:

1. The excitation of the separately excited generator is confined to the linear range of the saturation curve. Consequently, $\Phi = k_\Phi I_f$. (k_Φ is the slope of the saturation curve.)

2. The driven speed of the generator is constant. From this and the first assumption,

$$E_g = k\Phi S = k(k_\Phi I_f)S = k_v I_f \qquad (k_v = kk_\Phi S)$$

In this expression, k_v is the slope of the linear E_g–I_f curve at constant speed S [Fig. 3-2(b)].

The winding from Example 3-1 is combined with the armature of a separately excited generator in Fig. 3-2(a). Within the shaded areas, the circuit schematic and block representation of each component is shown. Beneath these symbols is the single multiplying element, k_v/R_f, that fully represents the generator for open-circuit conditions.

Transfer Function. The transfer function of the system is the ratio of the output to the input. For Fig. 3-2, $V_o = (k_v/R_f)E_R$ and the transfer function is V_o/E_R

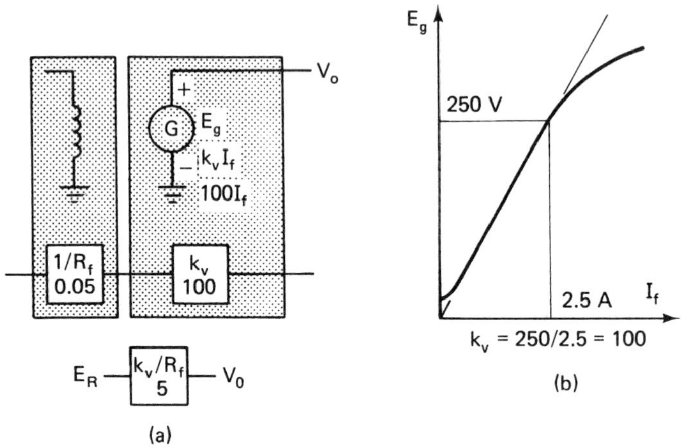

Figure 3-2 Combined Multiplying Elements

$= k_v/R_f$ with a numerical value $100/20 = 5$. Thus, for $E_R = 30$ V, $V_o = (30)(5) = 150$ V; if $E_R = 48$ V, V_o is 240 V.

Summing Point. At a summing point, two or more signals combine and continue in the forward path as the algebraic sum. The entering signals may be positive or negative.

EXAMPLE 3-3: Although Fig. 3-2 provides the open-circuit voltage of the generator, it is invalid for the line voltage with the generator loaded. The voltage drop across the armature circuit can be introduced at a summing point as illustrated in Fig. 3-3. The line voltage is $E_R(k_v/R_f) - I_L R_g$, where R_g is the armature circuit resistance. For $E_R = 30$ V, $R_g = 0.15\ \Omega$ and $I_L = 100$ A, $V_L = (30)(5) - (0.15)(100) = 135$ V.

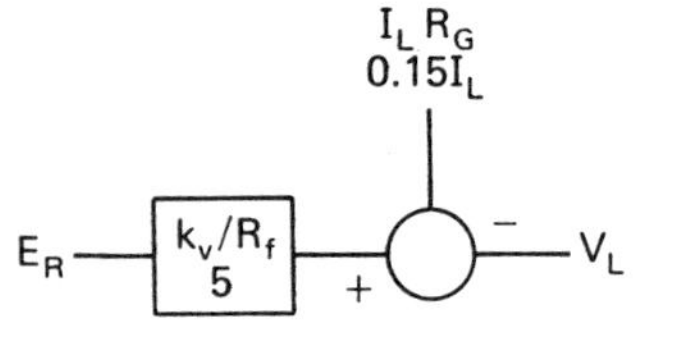

Figure 3-3 Summing Point

Feedback Element. At a pick-off point anywhere within the forward transmission path, a characteristic signal may be extracted and returned, directly or through a multiplier, to a preceding summing point.

EXAMPLE 3-4: Figure 3-3 provides the line voltage as a function of line current. Line current as a function of the load resistance may be represented by Fig. 3-4.

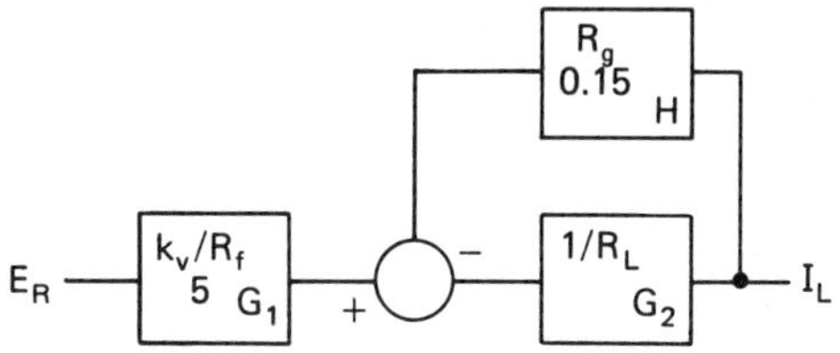

Figure 3-4 Feedback Element

The signal leaving the summing junction, as in the preceding example, is V_L. Multiplying by the reciprocal of R_L, the load resistance, produces I_L. In turn, this is picked off and returned to the summing point through the feedback multiplier R_g in order to subtract the armature resistance voltage drop from the generated voltage.

Letting $G_1 = k_v/R_f$, $G_2 = 1/R_L$, and $H = R_g$, the transfer function can be determined in general terms and the actual quantities inserted thereafter.

$$I_L = G_2(G_1E_R - HI_L)$$

$$I_L(1 + G_2H) = G_1G_2E_R$$

$$I_L = \frac{G_1G_2E_R}{1 + G_2H}$$

$$\frac{I_L}{E_R} = \frac{G_1G_2}{1 + G_2H} = \frac{5/R_L}{1 + 0.15/R_L} = \frac{5}{R_L + 0.15}$$

For $E_R = 30$ V and $R_L = 1.35\ \Omega$, $I_L = (30)(5)/(1.35 + 0.15) = 100$ A.

3-2. FEEDBACK REGULATION

In Fig. 3-4, feedback degrades the performance of the generator. On the other hand, feedback may be deliberately introduced as a means of improving system performance. If two signals, X_1 and X_2, enter a summing point and depart in the forward transmission path as the sum, $X_1 + X_2$, the feedback is positive; conversely, when they combine to continue as $X_1 - X_2$, it is negative feedback. Generally, positive feedback can cause system instability; negative feedback is more common and the one to which this discussion is directed.

Since the purpose of a regulating system is to stabilize an output characteristic, the system must monitor that characteristic and feed back a proportional electrical signal. For voltage or current, sampling can be done with nothing more than a voltage divider or resistor. Other output characteristics, regardless of their nature, may be converted to proportional electrical signals by *transducers*; for example, speed may be sampled by a tachometer, a generator with a permanent-magnet field and a terminal voltage directly proportional to speed. A number of transducers are introduced at various points in the text; suffice it to say that there probably is no physical characteristic for which there is no existing transducer.

To maintain an output at a desired level, the system must have a reference against which the output can be compared. This usually is a reference voltage, E_R. The difference between the reference and feedback voltages, $\epsilon = E_R - E_f$, enters the transmission path and determines the output. The system may be designed to have a correct output for either $\epsilon = 0$ or an absolute magnitude. For a system in which ϵ should equal zero, a low output causes a positive and a high output a negative ϵ; similarly, for an active signal a low output increases and a high output reduces ϵ.

The use and effectiveness of feedback in regulatıng an output will be illustrated with an armature-controlled motor supplied by the separately excited generator of the preceding examples. Without regulation, the system is shown in Fig. 3-5. R_m is the armature circuit resistance and E_m the counter emf of the motor.

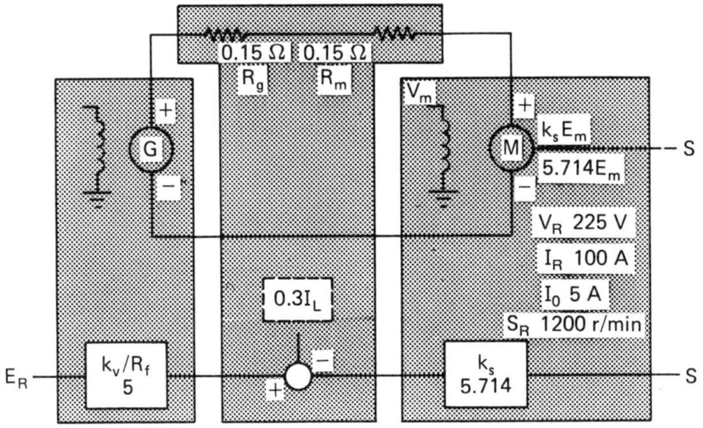

Figure 3-5 Open-Loop System

The field of the motor is excited f.om a separate source; $k\Phi_m$ is constant. From the equation $S = E_m/k\Phi_m$, a speed constant can be determined:

$$k_s = \frac{1}{k\Phi_m} = \frac{S}{E_m}$$

At rated conditions, $E_m = 225 - (100)(0.15) = 210$ V. Therefore, $k_s = 1200/210 = 5.714$ r/min/V and $S = 5.714E_m$ r/min.

The block diagram of the system is also shown in Fig. 3-5. This system is open loop—that is, the output is permitted to vary in accordance with the inherent system characteristics; this contrasts with a closed-loop system to be illustrated in the succeeding discussion in which a proportional output signal is introduced at a summing point in order to automatically control the output according to a designed plan.

From the block diagram, $S = 5.714(5E_R - 0.3I_L)$. Since $I_L = 100$ A at 1200 r/min,

$$E_R = \frac{S + (5.714)(0.3)I_L}{(5.714)(5)} = \frac{1200 + (5.714)(0.3)(100)}{(5.714)(5)} = 48 \text{ V}$$

and the speed equation is

$$S = (5.714)(5)(48) - (5.714)(0.3)I_L = 1371 - 1.714I_L$$

At no load, $I_L = 5$ A, $S = 1371 - (1.714)(5) = 1362$ r/min, and $\%SR = (100)(1362 - 1200)/1200 = 13.5\%$.

In order to improve the speed regulation, an amplidyne and a tachometer are added to the system (Fig. 3-6). The generator and motor of the open-loop system are intact, but the voltage across the generator field is no longer constant and will automatically change as the load is varied. Speed is monitored by the tachometer, whose generated voltage, E_t, is connected to one end of the amplidyne control winding; the other end of the winding is connected to a constant reference voltage. $E_t = k_tS$, where k_t is the tachometer constant. The electrical signal to which the system responds, $\epsilon = E_R - k_tS$, therefore changes in response to speed variations. When the speed drops, ϵ increases; if speed rises, ϵ decreases.

For the illustration it is assumed that the armature circuit resistance of the amplidyne, including the compensating and series field, is negligible. The voltage

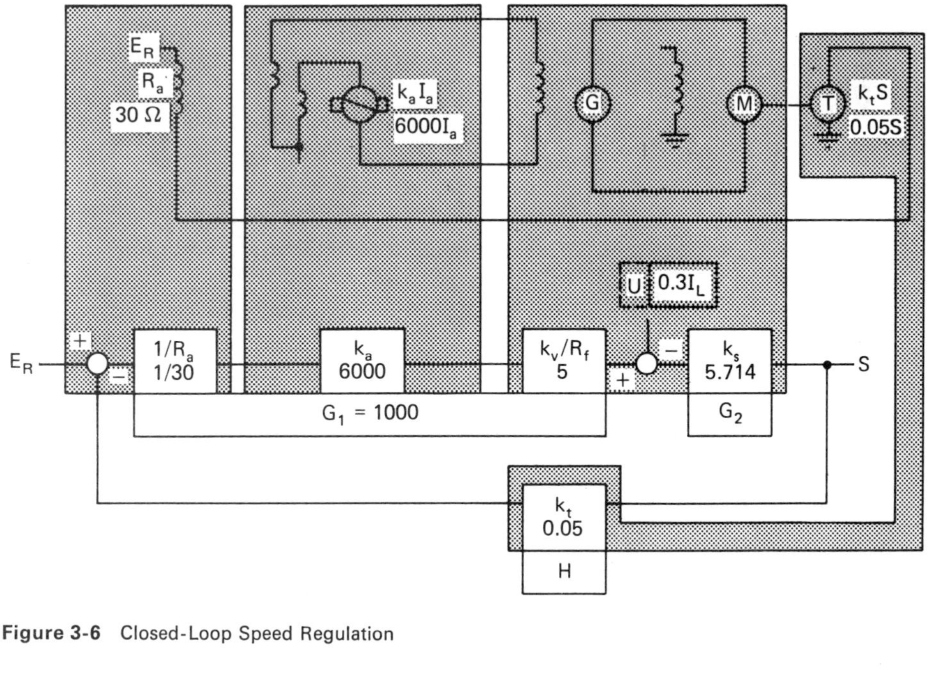

Figure 3-6 Closed-Loop Speed Regulation

generated by the amplidyne, $k_a I_a$, is applied directly to the field of the generator.

From the block diagram,

$$\begin{aligned} S &= G_2[G_1(E_R - HS) - U] \\ &= \frac{G_1 G_2 E_R - G_2 U}{1 + G_1 G_2 H} \\ &= \frac{(1000)(5.714)E_R - (5.714)(0.3)I_L}{1 + (1000)(5.714)(0.05)} \\ &= 19.930E_R - 5.979 \times 10^{-3} I_L \end{aligned}$$

At rated load, $I_R = 100$ A and $S_R = 1200$ r/min. Solving for E_R,

$$E_R = \frac{1200 + (5.979 \times 10^{-3})(100)}{19.93} = 60.24 \text{ V}$$

$$\begin{aligned} S &= (19.93)(60.24) - 5.979 \times 10^{-3} I_L \\ &= 1200.6 - 5.979 \times 10^{-3} I_L \text{ r/min} \end{aligned}$$

At no load, with $I_0 = 5$ A, the computed speed is 1200.57 r/min and the speed regulation is reduced from the 13.5% in the open-loop system to essentially zero with the closed loop.

Regulation through feedback is not as simple as is suggested by this illustration. The analysis was based exclusively on steady-state conditions. However, following each load change there is a transient period before steady state is achieved. The duration and nature of the transient is governed by the time constants in the system.

There are three electrical circuits in the system: (1) tachometer, reference voltage, and control winding; (2) amplidyne armature and generator field winding; and (3) generator and motor armatures. Each of these circuits is inductive, with a delay in the rise or fall of current when the net circuit voltage is changed. Since the circuits are interrelated, the individual circuit delays are cumulative in the system.

Mechanically, the significant time constant results from inertia of the motor and the attached load. Inertia opposes acceleration and deceleration and delays the system response.

The nature of the response to changed loading can better be understood if a numerical analysis first is performed for a hypothetical system. The system of Fig. 3-6 is operating unloaded at rated speed when a load equal to the rated torque suddenly is thrown on the motor. In the torque equation, $T = 7.04k\Phi I_A$, $k\Phi$ is constant; loading will impose an immediate demand for the rated current of 100 A. The analysis assumes that (1) the response of each circuit component to a change in its parameters is instantaneous, but (2) there is a microscopic delay between the adjustment to change by one component and the receipt of its new output signal by the next component in the transmission path.

Immediately prior to imposing the load, E_m is 210 V; $E_g = 210 + (5)(0.3) = 211.5$ V. Under the assumptions, I_L immediately jumps to 100 A, but E_g will not be affected until the consequence of this current change travels through the system and arrives as a new voltage across the generator field.

The equations that define the successive changes, listed in sequence, are

$$E_m = E_g - (0.3)(100) = E_g - 30$$
$$S = 5.714E_m$$
$$E_t = 0.05S$$
$$\epsilon = 60.24 - E_t$$
$$E_g^* = 1000\epsilon$$

The equation for E_g^* neglects magnetic saturation in the amplidyne and the generator. Assuming saturation at 300 V, the actual E_g will be greater than 300 V for any computed value in excess of 300 V. However, for this analysis, a conservative estimate of 300 V will be used for continued calculations whenever the computed E_g exceeds 300.

$$I_L = \frac{E_g - E_m}{0.3}$$

If the computed I_L is anything other than 100 A, the assumptions require the motor to instantaneously accelerate or decelerate to a speed where 100 A is restored.

The computed results for three successive transmission cycles are as follows:

	First	*Second*	*Third*
E_m (V)	181.5	270	−330
S (r/min)	1,037	1,543	−1,885
E_t (V)	51.85	77.14	−94.28
ϵ	8.39	−16.90	154.52
E_g^*	8,385	−16,900	15,452
$E_g^\dagger$	300	−300	300
I_L (A)	395	−1,900	2,100

Because of the assumption limiting the magnitude of $E_g^\dagger$ to 300 V, succeeding cycles will repeat the second and third, with the motor oscillating between −1885 and +1543 r/min. Without this limitation, the computed performance would alternate between wildly increasing positive and negative speeds.

When transient periods are considered, the tendencies in the hypothetical example remain but the response differs. Mathematical analysis of the transient in a realistic system is beyond the scope of this text, but the nature of the transient can be rationalized.

When the load is thrust on the motor, rotational inertia opposes deceleration. The speed drops—not instantaneously as assumed in the earlier analysis, but gradually. Concurrent with the declining speed, the voltage across the amplidyne increases. This voltage is processed through the system and causes the line current to increase, but, because of time delays in the system, the instantaneous current reflects the speed at a slightly earlier time. As the current gradually rises, it first reduces the deceleration, then accelerates the motor. When the speed rebuilds to 1200 r/min, the delayed current still is providing an accelerating torque; the speed continues to rise but feeds back a signal that, after a delay, returns a decelerating torque. This torque drops the speed below 1200 r/min and triggers a new accelerating torque. Throughout the transient period, the motor will alternately rise above and fall below the correct speed, but the speed at each succeeding reversal will be closer to and eventually equal the steady-state speed, resulting in a transient period as illustrated in Fig. 3-7.

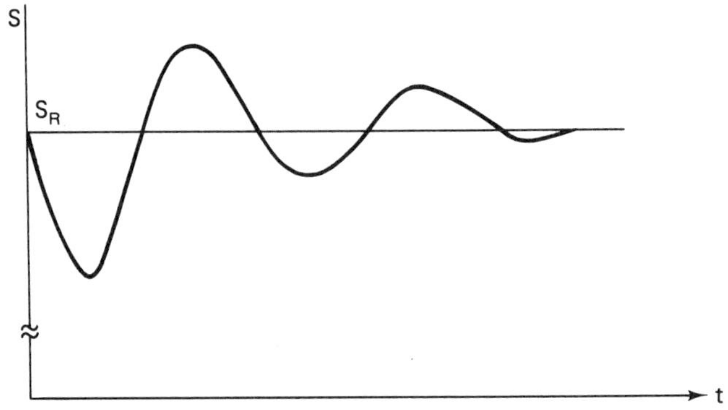

Figure 3-7 Transient Period

The peak deviation and the duration of the transient period can be reduced by careful design with an intricate balance between mechanical and electrical characteristics in the system; with proper modifications, overshoot can be eliminated.

3-3. POSITION CONTROL

In automated systems there are innumerable motor requirements for intermittent operation—not continuous rotation—in order to relocate some element in the system (e.g., moving an elevator cage to a new floor, reorienting an antenna, positioning a machine tool at an X–Y coordinate, or simply rotating a needle indicator to an index mark on a calibrated dial). The dc motor is well suited to this purpose; either field or armature control may be used, but armature control is more common.

Feedback is an essential element of positional control. There are available many transducers to sense position with varying degrees of accuracy, including potentiometers. Both a circular and a linear potentiometer are illustrated in Fig.

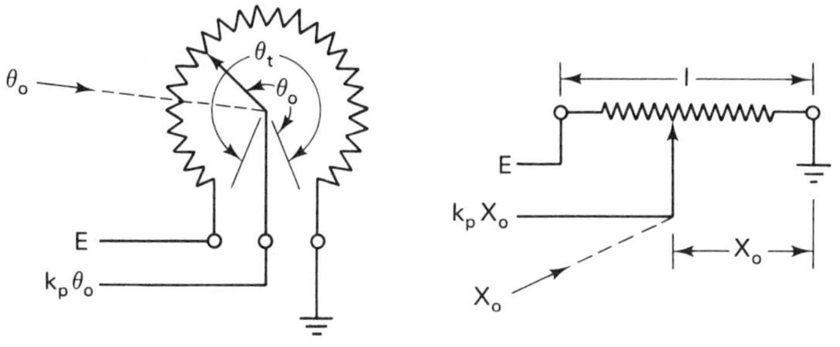

Figure 3-8 Position Transducers

3-8. In the circular potentiometer, $k_p = E/\theta_t$ V/rad is the potentiometer constant; with the pick-off arm coupled to a controlled device, an angle of θ_o radians results in $E_o = k_p\theta_o$ V. Similarly, in the linear potentiometer $k_p = E/l$ and an input position X_o results in $E_o = k_pX_o$. A drive motor may be coupled to the pick-off arm directly or through gearing that requires a number of motor revolutions in order to move the arm through a relatively small angle or distance; similarly, gearing can be used in coupling a motor to a load. Consequently, although there is a fixed ratio between movements of the controlled element, rotor, and pick-off arm, this ratio need not be unity. Also, while the movement of the rotor is rotary, by suitable couplings—for example, winch and pulleys—the movement of the controlled device and/or transducer pickoff may be linear.

An elementary positioning system is illustrated in Fig. 3-9. Two identical potentiometers, one input and the other output, are connected to a common source and therefore have identical k_p. The input command, θ_i, and the output, θ_o, determine the input voltage to the amplifier; the latter can have any of several forms such as rotary (e.g., amplidyne) or electronic. The error voltage, $\epsilon = k_p(\theta_i - \theta_o)$ is amplified and applied to the armature of the dc motor. The field of the motor may be a separately excited winding or, in low-power applications, a permanent magnet; in either case, it provides a field flux that is constant in magnitude and direction. The output of the amplifier can be positive or negative; the motor will revolve in opposite directions for opposite-polarity error voltages. When properly

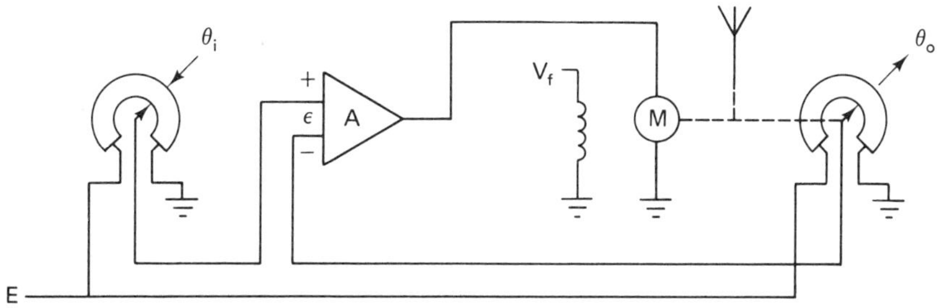

Figure 3-9 Position Control

connected, the motor is null seeking—that is, it rotates in a direction to reduce ϵ to zero.

The controlled element, in this case an antenna, and the arm of the output potentiometer are mechanically coupled to the rotor. When the system is dormant, the azimuth of the antenna corresponds to θ_i, $\theta_o = \theta_i$, and $\epsilon = 0$.

For an armature-controlled motor, the relationship between torque, speed, and applied voltage may be developed:

$$T = 7.04k\Phi I_a$$

$$= 7.04k\Phi\frac{V_A - E_m}{R_{ac}}$$

$$= \frac{7.04\,k\Phi V_A}{R_{ac}} - \frac{7.04(k\Phi)^2 S}{R_{ac}}$$

Since $k\Phi$ is constant,

$$\boldsymbol{T = k_1 V_A - k_2 S} \tag{3-1}$$

Stall torque (T_s) is the torque developed when $S = 0$; in the formula, it is evident that stall torque is directly proportional to the applied voltage. The torque decreases linearly as the speed increases, and the slope of the *T–S* curve, $-k_2$, is independent of V_A; hence, the curves for differing V_A are parallel lines (Fig. 3-10).

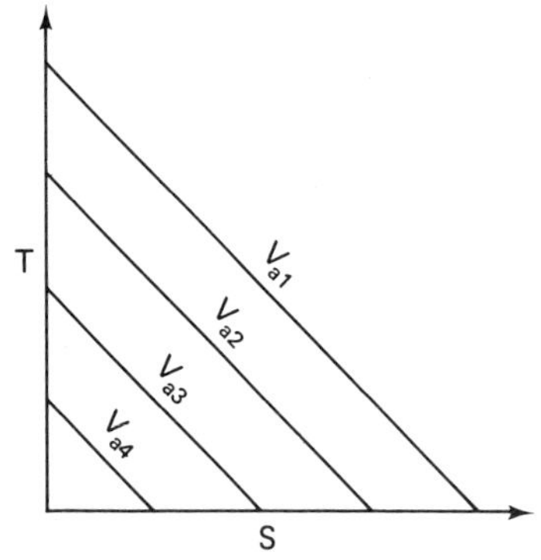

Figure 3-10 Armature Control

The magnitude of V_A in the position control system is $A\epsilon$, where A is the amplification. The error signal is small; A dominates the magnitude of V_A and is definitive in governing system performance while resisting external disturbances or responding to new input commands.

A disturbance such as a high wind may dislocate the antenna from its proper azimuth. If amplification is large, stall torque is impressive and a slight dislocation is heavily opposed, but as the amplification is decreased, the disturbance causes greater displacement before an equalizing countertorque is developed. From this viewpoint, a large amplifier gain is desirable.

The system response is identical for a new input angle or recovery from a comparable displacement angle caused by disturbance. If the amplification results in a torque greater than that of static friction, the rotor and antenna accelerate toward the null position. Movement is opposed by inertia and viscous friction. The acceleration is $\alpha = T/J$; for a massive antenna, the moment of inertia is large and a large torque is required if the system is to be responsive. With a low amplifier gain, the ability of the system to correct small displacements declines and the accuracy of the system is reduced. This is a second argument for a high gain.

A third consideration favoring high gain is the response time. As amplification and the resulting torque increases, acceleration increases and the elapsed time in moving to the null position is reduced.

But there are arguments against high amplification. A large torque can accelerate the motor and load to a relatively high speed; as the motor approaches the null position and V_A declines, the electromagnetic torque falls below that necessary to sustain constant speed, but inertia now opposes deceleration. Also, the time constant of the electrical circuit aggravates the circumstance; when ϵ declines to zero as the rotor arrives at the correct position, the line current, which ideally would be zero, continues due to the inductive characteristics of the circuit. Thus, both the inertia and the current lag contribute to overshoot of the null position. Upon passing the null point, an error voltage of reversed polarity occurs across the input terminals of the amplifier and a delayed current reversal occurs in the motor. This current builds, decelerates the motor, brings it to a momentary halt, and then accelerates it in the opposite direction. As a result, the system will exhibit one of several possible characteristics depending upon the magnitude of the gain and the system contants.

1. *Unstable*: If the amplification is sufficiently large, the angle of overshoot will be greater than the initial displacement; the motor reverses, then overshoots in the opposite direction by an even larger displacement angle. Usually, the output potentiometer can revolve through 360°; in this case, the overshoot can build to the point where the pick-off arm crosses the gap between the positive and negative terminals. Before jumping the gap, current was building to oppose the overshoot; after crossing, the current quickly reverses and *accelerates* the motor in the overshoot direction, causing sustained rotation that might build the speed to destructive levels.

2. *Sustained oscillation*: For a unique value of A, the angular overshoot will equal the initial displacement. Recovery from the overshoot will drive the motor to the initial displacement angle before it again reverses. The system will continuously swing back and forth, with the angular overshoot always equal to the initial displacement.

3. *Damped oscillation*: If A is less than that required for sustained oscillation but large enough to cause an overshoot, the angular overshoot will be less than the initial displacement angle; the motor will undergo one or more declining, reversed overshoots before coming to rest. This condition is labeled "underdamped" in Fig. 3-11.

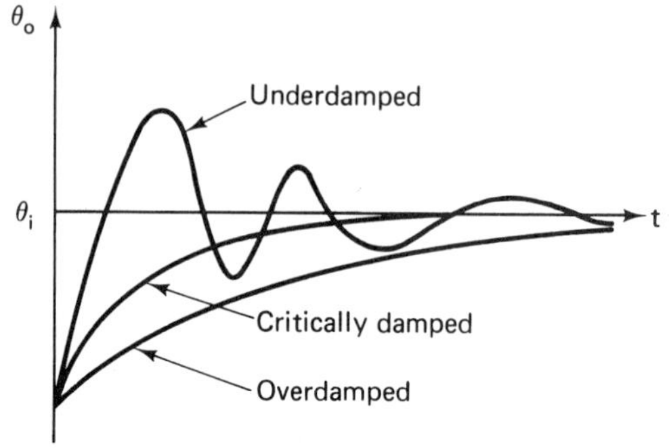

Figure 3-11 Damping

4. *Overdamped*: The amplification can be reduced to the point where the acceleration is low, the current drops below that necessary to sustain rotation before arrival at the null position, and the motor halts with an error in the antenna orientation.

5. *Critical damping*: For a unique value of A and a unique initial displacement angle, the system responds without oscillation and with a delicate balance between acceleration and deceleration that halts the rotor at the null position. Critical damping is the desired circumstance. However, for the unsophisticated system of Fig. 3-9, the response would be slow even for the unique angle; at other displacements, response would be slow and the positional accuracy poor.

The performance of the illustrated system seldom would be acceptable without modification. Ideally, the response to an input should be rapid, without overshoot, and with no error when the response is complete.

If the mechanical torque is equated to the electrical torque of equation 3-1,

$$J\alpha = k_1 V_A - k_2 S \tag{3-2}$$

In this equation, $\alpha = \Delta\theta/\Delta t^2$, $S = \Delta\theta/\Delta t$, and $V_A = Ak_p(\theta_i - \theta_o)$. Thus, the angular position of the rotor can be defined as a function of time. But an algebraic equation is meaningless because acceleration and velocity are not constant. With calculus and several techniques specifically developed for feedback analysis, equation 3-2 provides the basis for analyzing the system in order to identify modifications to ensure performance compatible with the application needs. Typically, one required modification is the addition of velocity feedback; speed cannot be allowed to build indiscriminately, and a tachometer, coupled to the motor, supplies a speed voltage that subtracts from the error voltage supplied by the potentiometers.

The illustrated system is a positional servomechanism. A *servomechanism* is a feedback control system in which at least one of the system signals represents mechanical motion. Other servomechanisms may control variables such as speed and acceleration.

Historically, the development of servomechanisms began in the mid-1930s and was greatly accelerated to meet requirements during World War II. Today, servomechanisms are widely used in manufacturing, navigation and guidance, and power generation. In addition to dc components, ac servos, discussed in Chapter Seven, are employed.

An excellent illustration of the potential of servomechanisms is the automatic anti-aircraft system developed during World War II:

1. A conical radar antenna emits a signal along a beam axis, receives an echo from the target, and senses both the difference between the target azimuth and beam axis and the range to the target. These sensings are transmitted to a dc servo that rotates and elevates the antenna to, and maintains, coincidence of the beam axis and target azimuth.
2. The target azimuth and range sensings are continuously transmitted to a computer. The computer calculates the speed and bearing of the target and the ballistic trajectory that will cause a projectile to arrive at a point in space at the same time as the target.
3. The computer feeds the proper intercept azimuth and elevation to a servo that correctly positions the antiaircraft guns.

Replacing the radar with a sensor to lock on to the light from a star and the gun with the steering mechanism of a spaceship, one can conceive a space navigation system. Similarly, coupling an input command bearing with a gyroscope suggests a possible automatic pilot for an aircraft. The potential for automated control is limitless.

PROBLEMS AND QUESTIONS

3-1. An amplidyne may be represented as two cascaded generators, as shown in Fig. 3-12. Find the expression for V_L in terms of V_f, I_L, and the machine constants.

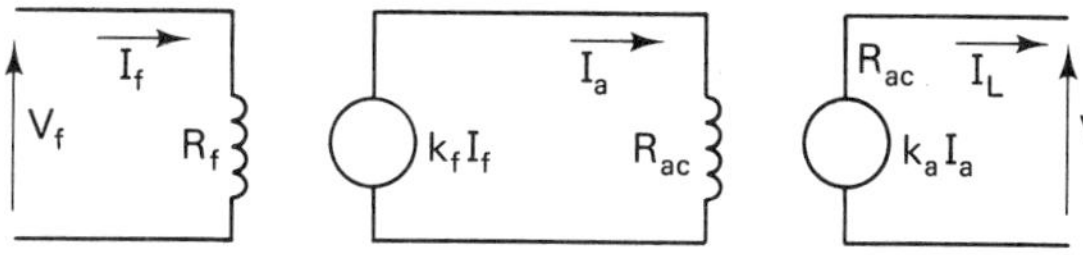

Figure 3-12 Problem 3-1

3-2. The machine constants of the amplidyne of Problem 3-1 are as follows: $k_f = 100$ V/A, $k_a = 60$ V/A, $R_f = 20\ \Omega$, and $R_{ac} = 5\ \Omega$. The amplidyne is rated 5 kW 250 V.

(a) For rated conditions, find V_f and I_f.

(b) Find the power amplification, $P_R/V_f I_f$.

(c) $P_{fwc} = 5\% P_R$. Find the efficiency at rated conditions.

3-3. Find the transfer function, O/I, for Fig. 3-13.

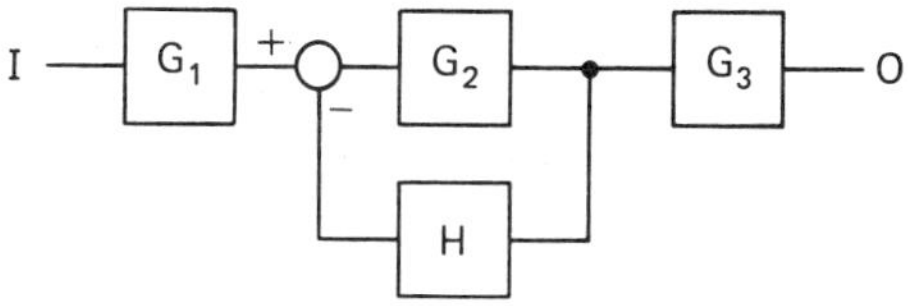

Figure 3-13 Problems 3-3 and 3-4

3-4. The constants in Fig. 3-13 are G1 = 0.05, G2 = 350, G3 = 10, and H = 0.01. Find O when $I = 2$.

3-5. The text analysis of Fig. 3-6 assumed that the armature circuit resistance of the amplidyne was negligible. The resistance is 5 Ω; repeat the analysis and determine the speed regulation.

3-6. The line voltage of the generator in Fig. 3-14 is regulated. The generator is rated 125 V 12.5 kW.

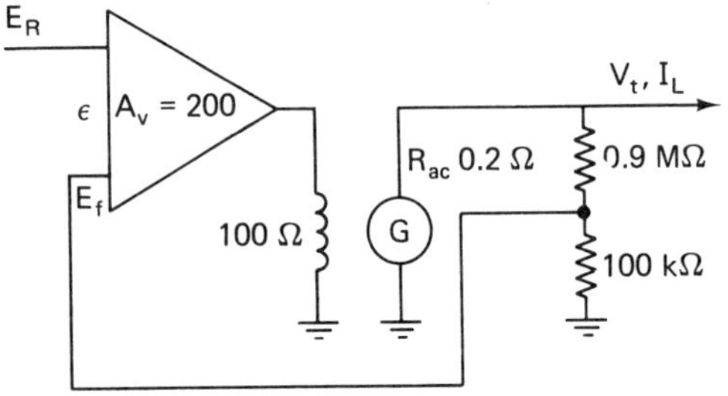

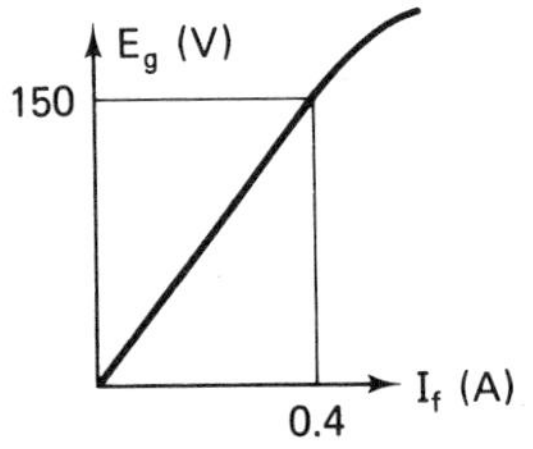

Figure 3-14 Problem 3-6

(a) The feedback voltage is obtained from a voltage divider. Is the current in the divider a significant consideration in a steady-state voltage regulation analysis?

(b) Convert the schematic to a block diagram.

(c) Determine the transfer function.

(d) To obtain rated power at rated voltage, what is the value of E_R?

(e) With the E_R of part (d), what is the % voltage regulation?

3-7. The system of Fig. 3-15 regulates load current.

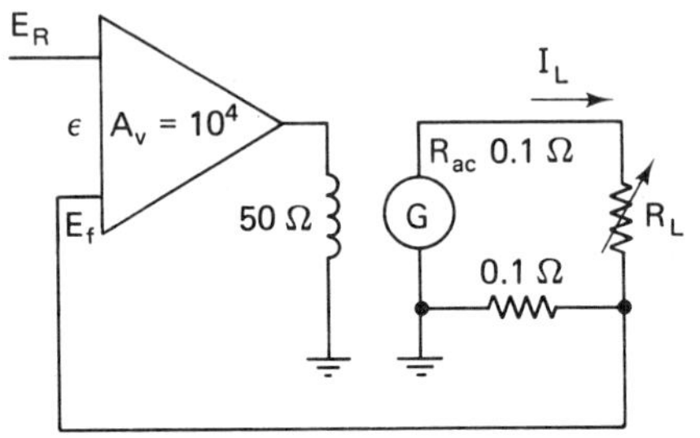

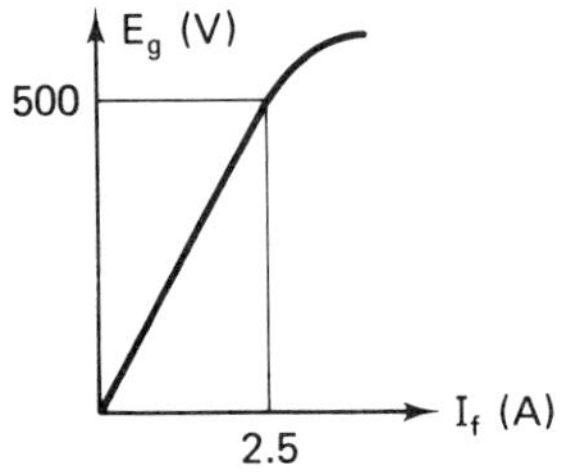

Figure 3-15 Problem 3-7

(a) Find the transfer function, I_L/E_R.

(b) For $R_L = 25\ \Omega$ and $I_L = 20$ A, what is E_R?

(c) With the E_R of part (b), what is I_L if $R_L = 10\ \Omega$? 0 Ω (short circuit)?

(d) From the transfer function and the E_R of part (b), what is I_L for an R_L of 50 Ω? Do you accept this as realistic? If not, explain and roughly estimate the actual current.

3-8. A circular rheostat has an effective arc of 320°. With 120 V applied, what is the potentiometer constant, k_p? For an input angle of 80°, what is the output voltage?

3-9. In Fig. 3-10, with $V_{A1} = 240$ V, the stall torque is 280 lb-ft and the theoretical no-load speed (i.e., x-axis intercept) is 1800 r/min.

(a) What is the stall torque for $V_A = 180$ V?

(b) With $V_A = 120$ V and $T = 60$ lb-ft, what is the developed power?

3-10. A family of T–S curves for an armature-controlled motor is shown in Fig. 3-10. In Fig. 2-59, torque is shown as constant for the range $0.1\omega_b - 1.0\omega_b$, also with armature control. Are these figures consistent? Explain.

4

Transformers

4-1. INTRODUCTION One of the most remarkable devices ever developed is the transformer. The construction is simple; there are no moving parts. It can be hung on a utility pole, exposed year-round to the elements, yet perform its assigned task around-the-clock while unattended and seldom maintained. Its efficiency is unusually high, and although costly to procure, its long useful life makes it economical. It can fulfill diverse purposes: impedance matching, electrical circuit isolation, step or continuously variable voltage transformation, phase shifting, and others. Whatever the purpose, the operational principles are unchanged although emphasis on functional characteristics may differ. This chapter concentrates upon power and distribution transformers, common elements in every power transmission system.

Several reasons for the incorporation of transformers in power distribution systems are developed in this chapter, but one deserves special attention at this point. Observing the utility system in any community, you can gain an appreciation of the number of transformers employed—the many distribution transformers hung on utility poles, each servicing a handful of residences, and the large substation transformers to supply the feeders. Transformers are costly and introduce power losses in the system, but these aspects are miniscule when compared to the advantages of power transmission at high voltages. An insight can be gained by analyzing a hypothetical situation; the analysis, although based on idealized assumptions and not rigorously correct, is convincing evidence.

SITUATION: A group of six residences are located at a distance of 1 mile from a power station. Each residence requires 120/240-V three-wire single-phase service with a peak requirement of 14.45 kW at unity power factor per home. Two proposals, each meeting the requirement, are analyzed and compared (Fig. 4-1).

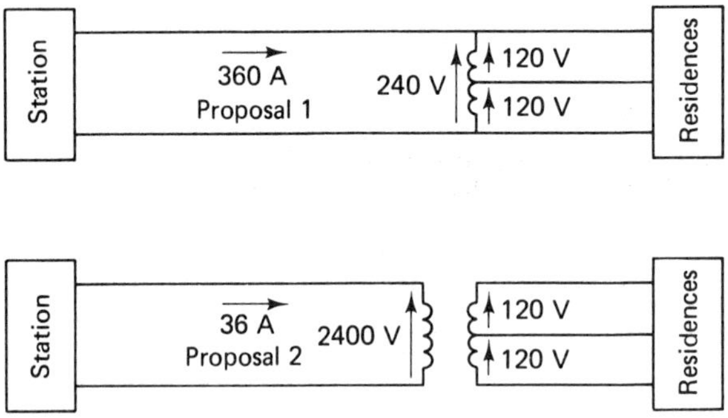

Figure 4-1 Transmission Proposals

PROPOSAL 1: 240 V (0.24 kV) transmission.

PROPOSAL 2: 2400 V (2.4 kV) transmission.
(kV subscripts are used to identify the proposals in the following analysis.)

LOAD POWER. The load power is the same for both proposals:

$$P_{0.24} = \text{kVI} = (0.24)(360) = 86.4 \text{ kW} \qquad P_{2.4} = (2.4)(36) = 86.4 \text{ kW}$$

VOLUME AND WEIGHT OF TRANSMISSION LINE. The current density (I/A) is assumed to be the same for both systems. (A is the area of the cross section of the conductor.)

$$\frac{360}{A_{0.24}} = \frac{36}{A_{2.4}} \qquad \text{therefore, } A_{0.24} = 10A_{2.4}$$

Since the length of the transmission lines is the same for both proposals, the volume and weight of the lines are proportional to the area. At the lower voltage, the weight and volume are 10 times as large as that at the higher voltage, and the supporting structure (e.g., utility poles) also must have 10 times the load capacity, with construction and maintenance tasks proportionately greater. Considering only the copper in the lines, the cost at the lower voltage is 10 times the cost at the higher voltage.

VOLTAGE REGULATION. The resistances of the lines are inversely proportional to the area ($R = \rho L/A$). Thus, $R_{2.4} = 10R_{0.24}$. (For the proposal, $R_{0.24}$ would be approximately 0.25 Ω.) The line-voltage drop is the same for both proposals:

$$IR = 360R_{0.24} = 36(10R_{0.24}) \simeq 90 \text{ V}$$

But the voltage regulation of the transmission line is superior at the higher voltage:

$$\%VR_{0.24} = \frac{90(100)}{240} = 37.5\%$$

$$\%VR_{2.4} = \frac{90(100)}{2400} = 3.75\%$$

EFFICIENCY. The power loss in the lines and the line efficiencies are

$$P_{0.24} = I^2R \times 10^{-3} = (360)^2(0.25) \times 10^{-3} = 32.4 \text{ kW}$$

$$\%\eta_{0.24} = \frac{(86.4)(100)}{86.4 + 32.4} = 72.7\%$$

$$P_{2.4} = (36)^2(2.5) \times 10^{-3} = 3.24 \text{ kW}$$

$$\%\eta_{2.4} = \frac{(86.4)(100)}{86.4 + 3.24} = 96.4\%$$

All these considerations favor transmission at higher voltage; preferably, the higher the power to be transmitted, the higher should be the voltage. Within the United States, classes of voltage and preferred voltage levels have been established as shown in Table 4-1.

Table 4-1. Commercial and Industrial Voltage Classes

Voltage Class	*Inclusive Voltages (rms)*	*Preferred Voltages*
Low	Maximum 1000 V	(3-wire single phase) 120/240 V (3-wire 3-phase) 480 V
Medium	1.0+ to 72.5 kV	(3-wire 3-phase) 4.16; 13.8; 69 kV (4-wire 3-phase) 12.47/7.2; 13.2/7.62; 24.94/14.4; 34.5/19.92 kV
High	72.5+ to 242 kV	(3-wire 3-phase) 115; 138; 230 kV
Extra high	242+ to 800 kV	(3-wire 3-phase) 345; 500; 735–765 kV

Currently, 765 kV is the highest-level operational system in the United States, but research and development is in progress on systems in excess of 1 megavolt. Also under development are dc transmission systems—preferable to ac primarily because of lesser transmission-line losses. A number of factors have spurred interest in dc—increasing costs and politics of oil supply; scarcity, high cost, and competitive demands for low-sulfur-content coal; objections to nuclear plants, particularly near population centers; and so on. The trend is toward locating generator plants in sparsely populated regions, near the mine mouth, where fuel transportation savings also are realized. Necessarily, this means longer transmission lines and a need for greater transmission-line efficiency.

In a dc system, generated ac is converted to dc at the sending end and reconverted to ac at the receiving end of the line, formerly by mercury-arc converters but increasingly with thyristors. The transformer remains an essential component of the system—its use is not curtailed and actually is more complex, owing to the inherent characteristics of a converter system. Transmission levels are on the order of ±500 kV. One such system will transmit 1000 MW over 400 miles at ±400 kV.

4-2. THE IDEAL TRANSFORMER

Figure 4-2 represents the essential construction of a transformer. A magnetic core supports two independent windings of N_p and N_s turns. Idealized, the sequential (but essentially simultaneous) steps in the operation are as follows:

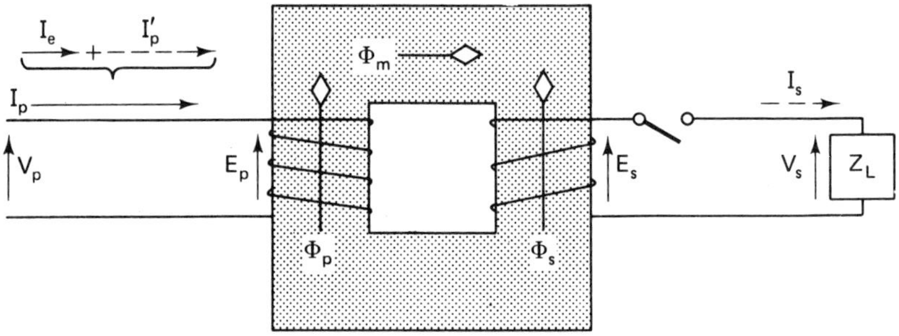

Figure 4-2 Ideal Transformer

- A sinusoidal voltage, V_p, applied to the primary causes current I_e to flow in the primary windings.
- I_e creates a sinusoidal flux, Φ_m, in the ferromagnetic core.
- Φ_m (1) induces a voltage, E_p, in the primary that, by Lenz's law, opposes V_p and thereby limits I_e, and (2) induces E_s in the secondary.
- Closing the switch in the secondary results in a load current, I_s, through the secondary winding.
- I_s creates Φ_s in opposition to Φ_m.
- The reduced core flux, $\Phi_m - \Phi_s$, causes a drop in E_p. This increases the net voltage across the winding impedance, $V_p - E_p$, and results in an added current increment, I'_p, in the primary winding.
- The flux, Φ_p, created by I'_p, opposes and neutralizes Φ_s, leaving Φ_m as the significant flux to maintain a constant E_p and E_s.

The value of the power transformer lies in its ability to accept an apparent power input at one combination of voltage and current and deliver it to a load at a differing combination. This is possible because Φ_m is common to both the primary and the secondary windings; consequently, the voltage induced in a single winding turn, $e_t = 4.44f\Phi_m$ (equation 1-2) is identical in both the primary and the secondary, resulting in the following idealized relationships:

$$E_p = N_p e_t \qquad E_s = N_s e_t$$

(N_p and N_s are the number of turns in the primary and secondary windings, respectively.)

$$\frac{E_p}{E_s} = \frac{N_p e_t}{N_s e_t} = \frac{N_p}{N_s}$$

The subscripts p and s distinguish between the primary and secondary; the primary is the input winding and the secondary is the output. If the input is less than the output (i.e., $V_p < V_s$) the transformer is step-up; conversely, in a step-down transformer $V_p > V_s$. The *turns ratio* is defined as the ratio of the number of turns in the high-voltage winding to the number of turns of the low-voltage winding; the turns ratio always is greater than (or sometimes equal to) unity. Generally, either side of a transformer can be used as the primary. The symbol n will be used to represent the turns ratio; that is, $n = N_H/N_L$ ($= E_H/E_L$), where H and L are the high- and low-voltage windings, respectively.

The *voltage ratio*, represented by the letter a, is the ratio of primary voltage to secondary voltage—that is, $a = V_p/V_s$. The voltage ratio will be less than unity for a step-up transformer. In an ideal transformer, $V_p = E_p$, $V_s = E_s$, and there are no power losses. This leads to the relationship

$$V_p I_p = V_s I_s$$

$$\frac{V_p}{V_s} = \frac{I_s}{I_p} = a \tag{4-1}$$

Particularly note that (1) currents are inversely proportional to voltages, and (2) $a = n$ in a step-down transformer but $a = 1/n$ in a step-up transformer.

Transformers used in the commercial power distribution system are highly efficient. Equation 4-1, though based on the ideal, is used in much of the engineering analysis of the practical transformer. However, the idealized transformer concept neglects three elements that must be considered in assessing performance characteristics: winding resistance, leakage reactance, and core loss (hysteresis and eddy currents).

4-3. THE REALISTIC TRANSFORMER

With the secondary winding open-circuited and V_p applied to the primary, the exciting current (I_e) must (1) create the flux in the iron circuit, (2) supply the core power losses (hysteresis and eddy current), and (3) supply the copper loss in the primary winding. But the exciting current and primary resistance are small enough for the copper loss to be negligible; essentially, the exciting current has two components, a true power component, I_c (core loss), and a reactive power component, I_m (create Φ_m).

I_c, I_m, and E_p all are functions of the frequency and amplitude of the mutual flux, Φ_m. They can be interrelated by the circuit representation of Fig. 4-3(a) with R_c and jX_m a resistance and an inductive reactance of proper magnitude to satisfy the Ohm's law relationship between I_e and E_p.

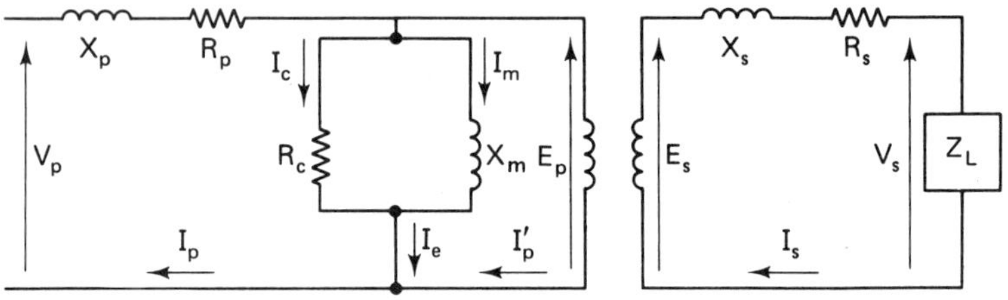

(a) Equivalent circuit

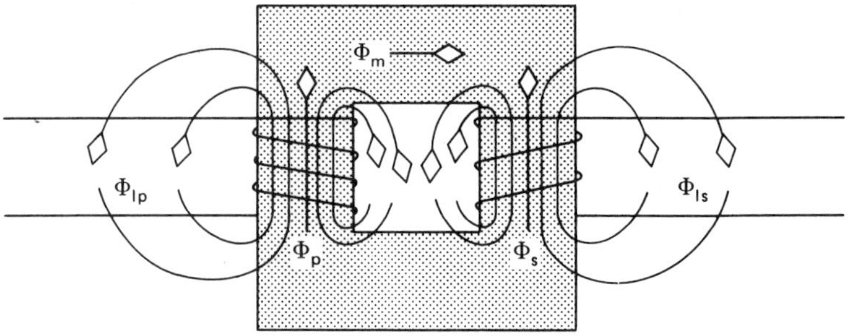

(b) Leakage flux

Figure 4-3 Realistic Transformer

When the secondary is loaded, voltage drops and copper losses in R_p and R_s, the primary and secondary winding resistances, become significant in determining the voltage regulation and efficiency of the transformer. Additionally, voltage regulation is affected by leakage flux in the primary and the secondary.

Figure 4-3(b) shows the flux pattern in the operational transformer. Although the coils are tightly wound and fully occupy the central space of the iron core, not all of the flux created by I_p and I_s will be confined to the magnetic circuit. Φ_{lp} and Φ_{ls} are leakage fluxes following an iron path through turns of the primary and secondary, respectively, but closing their loops by return paths as shown. Leakage fluxes induce voltage in the turns they link but are noncontributory to transfer of power from the primary to the secondary. Their effect is that of inductive voltage drops, represented by jX_p and jX_s.

For application analysis, the schematic of Fig. 4-3(a) can be simplified with negligible degradation of the computed performance. Being small, the exciting current is significant only from the viewpoint of core losses; these losses are constant throughout the practical loading range of the transformer if the frequency and amplitude of V_p remain constant. Since variations in frequency and amplitude are small in a commercial power distribution system, R_c and X_m may be deleted with the understanding that a constant core loss, P_c, not represented in the schematic, must be included in efficiency analysis.

Further simplification of the schematic is possible by eliminating the electrical isolation of the primary and secondary and referring all quantities to one side. Transfer of voltage and current is accomplished through application of equation 4-1. An additional requirement is a means for transferring impedance.

In Fig. 4-4(a), reflection of the load impedance must be consistent with Ohm's law—that is, $Z_s' = V_p/I_p$. From the voltage ratio, $Z_s' = V_p/I_p = (aV_s)/(I_s/a = a^2(V_s/I_s) = a^2Z_s$. Any impedance on the secondary side, load or winding, can be transferred to the primary by multiplying it by a^2. Similary, multiplying a primary impedance by $(1/a^2)$ reflects it to the secondary side.

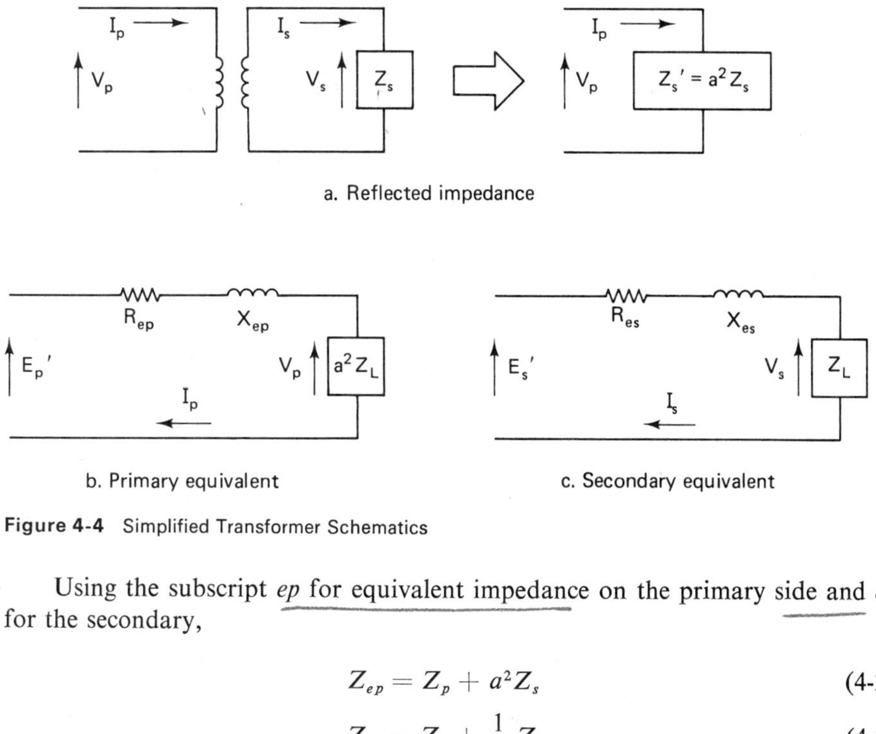

Figure 4-4 Simplified Transformer Schematics

Using the subscript *ep* for equivalent impedance on the primary side and *es* for the secondary,

$$Z_{ep} = Z_p + a^2 Z_s \tag{4-2}$$

$$Z_{es} = Z_s + \frac{1}{a^2} Z_{ep} \tag{4-3}$$

Applying equations 4-1, 4-2, and 4-3 and eliminating exciting current components results in the simplified circuits of Fig. 4-4(b) and (c), either of which is valid for computation. E_p' and E_s' are artificial quantities that do not actually occur in the transformer; however, their circuit relationship to V_p and V_s allows realistic analysis of voltage regulation.

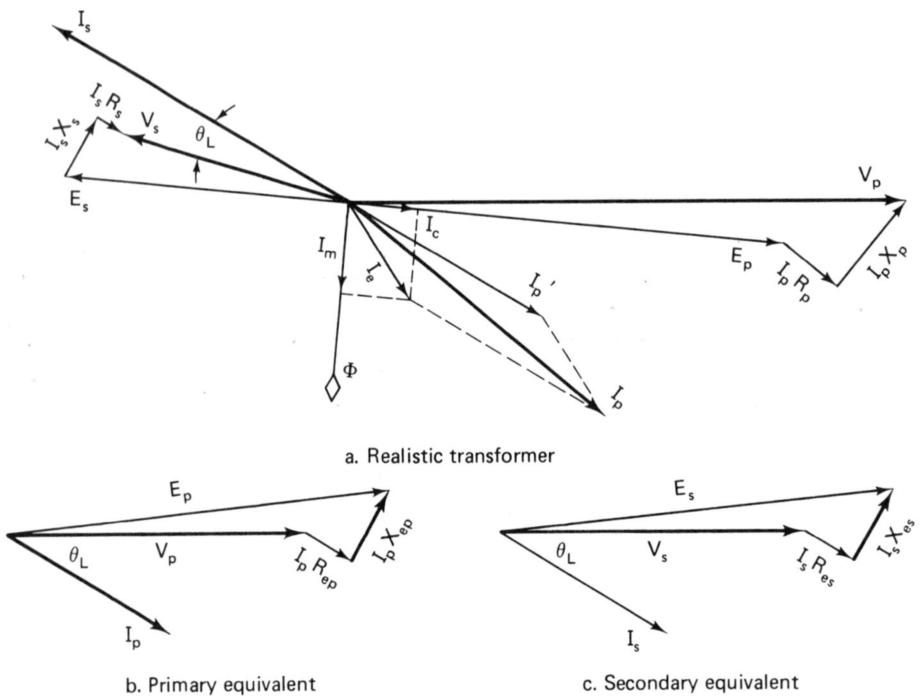

Figure 4-5 Vector Diagrams

Figure 4-5 shows vector diagrams for the realistic transformer and for the simplified schematics. A convenient starting point for sequential analysis of the realistic transformer [Fig. 4-5(a)] is E_p. Since this voltage is across both R_c and X_m, I_e is the vector sum of I_c (in phase) and I_m (lagging by 90°). I_e will lag E_p by a substantial angle.

E_p and E_s both are induced by Φ_m. Since they are electrically isolated, they may be represented in phase opposition. The load current I_s is related to E_s by Ohm's law: $I_s = E_s/(Z_s + Z_L)$. At realistic load levels, Z_L is very much greater than Z_s and will predominate in fixing the angular separation between E_s and I_s. The schematic shows the most common circumstance of lagging power factor loading, but a leading power factor is equally possible.

I_s causes a voltage drop across the secondary impedance Z_s. Rather than using this drop in polar form, it is more convenient to work with the in-phase and quadrature components—that is, I_sR_s in phase with the current and I_sX_s leading the current by 90°. Subtracting these components from E_s yields V_s, the secondary terminal voltage. Note that the angle between V_s and I_s is the load impedance angle, θ_L.

Any current in the secondary creates a flux that must be neutralized by a primary current component I'_p, related in magnitude to I_s by the turns ratio and at the same phase angle. This current appears on the diagram in phase opposition to I_s; vectorially, it adds to I_e to produce I_p, the primary current.

Current flow through the primary impedance causes a voltage drop I_pZ_p ($= I_pR_p + jI_pX_p$). If this drop is added to E_p, the sum is V_p, the primary (applied) line voltage.

In the vector diagram for the realistic transformer, the effect of winding impedances and the exciting current on the angle between V_p and I_p is small. In the simplified schematics, it may be assumed that the angle is fixed by the load alone. Therefore, the simplified circuits are straightforward and require identification of only the load power factor angle in determining the phase difference between V and I. This angle, combined with equivalent impedances referred to either the primary or secondary and the corresponding voltages and currents, leads to the vector diagrams of Fig. 4-4(b) and (c).

4-4. VOLTAGE REGULATION AND EFFICIENCY

Voltage regulation may be computed on either the primary or secondary side, using quantities referred to the side on which the analysis is conducted. Voltage regulation always is based upon full-load and no-load voltages, but the magnitude of the load voltages is affected by the load power factor. A given voltage regulation must specify the power factor upon which it is based. The reasons are demonstrated in Fig. 4-6.

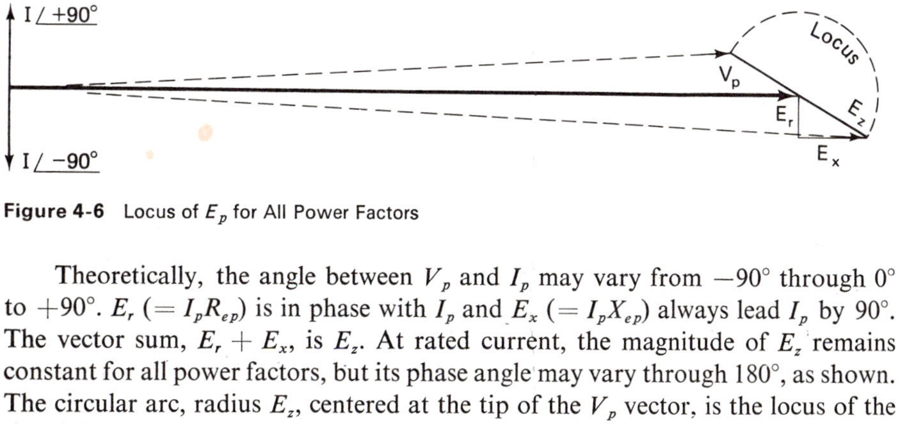

Figure 4-6 Locus of E_p for All Power Factors

Theoretically, the angle between V_p and I_p may vary from $-90°$ through $0°$ to $+90°$. E_r ($= I_pR_{ep}$) is in phase with I_p and E_x ($= I_pX_{ep}$) always lead I_p by $90°$. The vector sum, $E_r + E_x$, is E_z. At rated current, the magnitude of E_z remains constant for all power factors, but its phase angle may vary through $180°$, as shown. The circular arc, radius E_z, centered at the tip of the V_p vector, is the locus of the tip of the E_p vector for all possible power factors.

Although the open-circuit voltage, E_p, and the full-load terminal voltage, V_p, differ in both magnitude and phase angle, the phase-angle variation is of no consequence to the load; only the difference in magnitudes of the voltages is significant, and the determination of this difference is a scalar—not a vector—subtraction. (Note that E_z is the vector difference for all power factors.) The computation of E_p must be done in vector form, but in the voltage regulation equation, $(E_p - V_p)/V_p$, the numerator is an arithmetic difference.

To compute voltage regulation using primary values, the preliminary steps are to identify the rated primary voltage (V_p), rated primary current (I_p), equivalent primary resistance (R_{ep}), equivalent primary reactance (X_{ep}), and the angle of the load impedance (θ_L) for the specified power factor. Typical computations develop as follows:

$$I_p = \frac{\text{kVA}_R \times 10^3}{V_p}$$

$$\theta_L = \text{arc cos (power factor)}$$

$$E_r = I_p R_{ep}\underline{/\theta_L}$$

$$E_x = I_p R_{ep}\underline{/90 + \theta_L}$$

Note: θ_L is the angle from the reference voltage vector to the current vector. It is *positive for leading* and *negative for lagging* power factors.)

$$E_p = V_p\underline{/0} + E_r\underline{/\theta_L} + E_x\underline{/90 + \theta_L}$$

$$= [V_p + E_r \cos \theta_L + E_x \cos (90 + \theta_L)] + j[E_r \sin \theta_L + E_x \sin (90 + \theta_L)]$$

$$\%\text{VR} = \frac{100(|E_p| - |V_p|)}{|V_p|}$$

For efficiency calculations, the power input is determined by adding the transformer losses to the power output. The copper loss is identifiable on the simplified schematics, and the core loss, although not portrayed, is understood to be present. As with voltage regulation, the power factor is significant and must be stated, but whereas voltage regulation always is based on full-load values, efficiency may be determined for any load. A typical development using primary values is as follows:

$$P_o = \text{kVA}_L \times \text{pf} \qquad \text{(pf is the power factor)}$$

The core loss, P_c, is constant.

$$I_p = \frac{\text{kVA}_L \times 10^3}{V_p}$$

$$P_{\text{cu}} = I_p^2 R_{ep}$$

$$P_{\text{in}} = P_o + P_{\text{cu}} + P_c$$

Computation of the efficiency by the preceding method inherently assumes constant loading over sustained periods. Constant loading is a realistic condition when equipment is attended; units may be brought on or removed from the line in response to increased or decreased load demands, and operation of on-line equipment can be maintained at or near rated values. But distribution transformers largely are unattended and operate around-the-clock with substantial variations in the load. In this case, interest focuses on *all-day efficiency*. This efficiency is computed on the basis of energy rather than power. Experience permits development of the typical loading patterns for a 24-hour period, and from this the typical energy output can be determined. Energy losses also can be computed for the typical loading. The all-day efficiency is determined from the input and output energy over the 24-hour period—that is, $\%\eta_{\text{all-day}} = 100W_o/(W_o + W_{\text{losses}})$. The energy unit in these calculations is the kilowatt-hour. $W_{\text{kW-hr}} = P_{\text{kW}} \times t_{\text{hr}}$.

The task of computing copper losses for the 24-hour period can be reduced through use of the ratio of load kVA to rated kVA. In testing a transformer, a short-circuit power input (P_{sc}) is measured. Because of imposed test conditions, P_{sc} is the copper loss at rated load. Using ratios,

$$\frac{P_{\text{cu(any load)}}}{P_{sc}} = \frac{I_L^2 R_{ep}}{I_R^2 R_{ep}} = \left(\frac{V_p}{V_p} \times \frac{I_L}{I_R}\right)^2 = \left(\frac{\text{kVA}_L}{\text{kVA}_R}\right)^2$$

(*Note:* $V_p \neq I_L R_{ep}$. Mathematically, R_{ep} cancels and the current ratio is multiplied by $V_p/V_p = 1$.)

The following example illustrates voltage regulation, efficiency, and all-day efficiency computations.

Example 4-1: A 3-kVA 230/115-V transformer has a primary resistance of 0.3 Ω, secondary resistance 0.05 Ω, primary reactance 0.8 Ω, and secondary reactance 0.1 Ω. The core loss, P_c, is 45 W. At rated load, the copper loss is 85 W. (a) Find the percent voltage regulation for 0.9 lagging pf.

SOLUTION:

$$R_{ep} = R_p + a^2 R_s = 0.3 + (2)^2(0.05) = 0.5\ \Omega$$

$$X_{ep} = X_p + a^2 X_s = 0.8 + (2)^2(0.1) = 1.2\ \Omega$$

$$I_p = \frac{\text{kVA}_R \times 10^3}{V_p} = \frac{3 \times 10^3}{230} = 13\ \text{A}$$

$$\theta = -\cos^{-1}(\text{pf}) = -\cos^{-1} 0.9 = -25.8^\circ$$

(*Note*: The minus sign applies to the lagging pf.)

$$E_r = I_p R_{ep}\underline{/\theta} = (13)(0.5)\underline{/-25.8^\circ} = 6.5\ \text{V}\underline{/-25.8^\circ}$$

$$E_x = I_p X_{ep}\underline{/90 + \theta} = (13)(1.2)\underline{/90^\circ - 25.8^\circ} = 15.6\ \text{V}\underline{/64.2^\circ}$$

$$E_p = V_p + E_r + E_x = 230\underline{/0^\circ} + 6.5\underline{/-25.8^\circ} + 15.6\underline{/64.2^\circ}$$

$$= (230 + j0) + (5.85 - j2.83) + (6.79 + j14.0)$$

$$= 242.6 + j11.17 = 242.9\underline{/2.6^\circ}\ \text{V}$$

$$\%\text{VR} = \frac{|E_p - V_p|\,100}{|V_p|} = \frac{242.9 - 230}{230} \times 100 = 5.6\%$$

(b) Find the efficiency when delivering 3.04 kW at 0.95 pf.

SOLUTION:

$$\text{kVA}_L = \frac{\text{kW}}{\text{pf}} = \frac{3.04}{0.95} = 3.2\ \text{kVA}$$

$$I_p = \frac{\text{kVA}_L \times 10^3}{V_p} = \frac{3.2 \times 10^3}{230} = 13.9\ \text{A}$$

$$P_{cu} = I_p^2 R_{ep} = (13.9)^2(0.5) = 97 \text{ W} = 0.097 \text{ kW}$$

$$\%\eta = 100 \times \frac{P_o}{P_o + P_{cu} + P_c}$$

$$= 100 \times \frac{3.04}{3.04 + 0.097 + 0.045} = 95.5\%$$

(c) Find the all-day efficiency for the following typical loading:

3 hours at 3.5 kVA, 0.95 pf
7 hours at 3.0 kVA, 0.9 pf
1 hour at 3.2 kVA, 0.85 pf
5 hours at 2.4 kVA, 0.8 pf
8 hours at 1.0 kVA, 1.0 pf

SOLUTION:

			W (kW-hr)		
kVA_L	t (hr)	pf	Output [1]	Copper Loss [2]	Core Loss [3]
3.5	3	0.95	9.98	0.35	
3.0	7	0.9	18.90	0.60	
3.2	1	0.85	2.72	0.10	
2.4	5	0.80	9.60	0.27	
1.0	8	1.0	8.00	0.08	
	Total		49.20	1.40	1.08

[1] kVA $\times$ t $\times$ pf: for example, (3.5)(3)(0.95) = 9.98 kW-hr

[2] $\left(\frac{\text{kVA}_L}{\text{kVA}_R}\right)^2 \times P_{sc} \times t$: for example, $(3.5/3.0)^2(0.085)(3) = 0.35$ kW-hr

[3] $P_c \times 24$ hr: for example, (0.045)(24) = 1.08 kW-hr

$$\%\eta_{\text{all-day}} = \frac{100 W_{out}}{W_{out} + W_{cu} + W_c} = \frac{100(49.2)}{49.2 + 1.40 + 1.08}$$

$$= 95.2\%$$

4-5. TRANSFORMER TESTS

To permit voltage regulation and efficiency analysis, the transformer impedance and the core loss must be determined by testing. Two tests, open-circuit and short-circuit, yield the desired data.

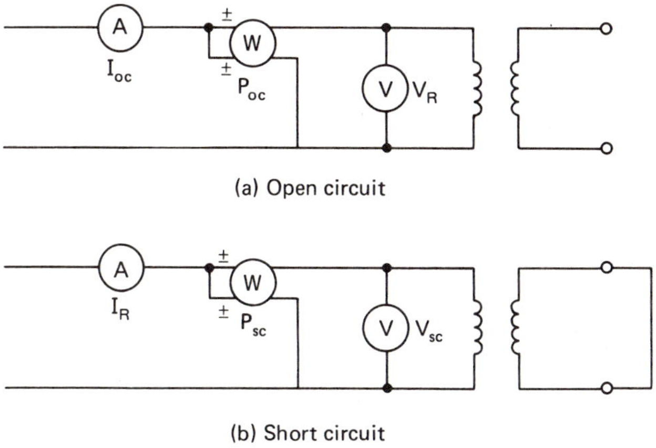

Figure 4-7 Transformer Tests

Open-circuit Test [Fig. 4-7(a)]. Referring to the figure, if rated voltage is applied to the primary and the secondary is open-circuited, the power input (P_{oc}) will consist of the core loss, P_c, and a small, negligible copper loss. Therefore, $P_{oc} = P_c$. The test may be performed on either the primary or the secondary and will yield the same results; the choice of side may be arbitrary, or it may hinge on available voltages or a preference for working at a particular voltage level. (Note that when performed on the low side, the open terminals will be at the higher voltage.) The voltage must be at rated frequency. Power input is measured with a wattmeter; meter selection can be based upon a probable current of less than 10% of the rated value and a power input on the order of 1 to 1.5% rated kVA.

Short-circuit Test [Fig. 4-7(b)]. If a short circuit is placed across the secondary terminals and a controlled voltage at rated frequency is impressed across the primary, a short-circuit voltage, V_{sc}, in the range of 3 to 15% of V_R will cause rated current to flow. Under these circumstances, the core loss is negligibly small—for example, assuming $V_{sc} = 10\%\ V_R$, and recalling that $P_c \propto E_p^2$, the core loss at the reduced voltage is $< (0.1)^2 P_c$ or $< 1\%\ P_c$. Practically, the short-circuit power input (P_{sc}) is $I_R^2 R_{ep}$.

Since E_p and E_s are essentially zero, Kirchhoff's voltage law dictates that V_{sc} is the voltage drop across the equivalent primary impedance, $I_R Z_{ep}$.

The test is conducted with a voltmeter, ammeter, and wattmeter inserted on the primary side and an ammeter short circuiting the secondary. Starting at zero, the input voltage is raised until rated I_s appears on the secondary. Primary parameters V_{sc}, I_p, and P_{sc} are recorded. From these data, $R_{ep} = P_{sc}/I_p^2$, $Z_{ep} = V_{sc}/I_p$, and $X_{ep} = \sqrt{Z_{ep}^2 - R_{ep}^2}$. Note that since rated currents flow in both windings, P_{sc} is the full-load copper loss of the transformer.

It is equally valid to perform the test with the primary short-circuited and the readings taken on the secondary side; the data would then produce R_{es}, X_{es}, and

Z_{es}. Initial meter selection can be based on rated currents, less than 15% rated voltage, and power of less than 5% rated kVA.

Example 4-2: A 3-kVA 230/115-V transformer is tested and yields the following results:

$$\text{Open-circuit test:}\quad V_{oc} = 230\ \text{V} \quad I_{oc} = 1.6\ \text{A} \quad P_{oc} = 50\ \text{W}$$

$$\text{Short-circuit test:}\quad V_{sc} = 4.5\ \text{V} \quad I_{sc} = 13.04\ \text{A} \quad P_{sc} = 30\ \text{W}$$

(a) Find R_{ep}, X_{ep}, R_{es}, and X_{es}.

(b) Find the efficiency when delivering a load of 2 kVA at 0.85 lagging pf.

SOLUTION:

$$\text{(a)}\ Z_{ep} = \frac{V_{sc}}{I_{sc}} = \frac{4.5}{13} = 0.346\ \Omega$$

$$R_{ep} = \frac{P_{sc}}{I_{sc}^2} = \frac{30}{13.04^2} = 0.1764\ \Omega$$

$$X_{ep} = \sqrt{Z_{ep}^2 - R_{ep}^2} = \sqrt{0.346^2 - 0.1764^2} = 0.298\ \Omega$$

$$R_{es} = \left(\frac{1}{a^2}\right)R_{ep} = \left(\frac{1}{2^2}\right)(0.1764) = 0.0441\ \Omega$$

$$X_{es} = \left(\frac{1}{a^2}\right)X_{ep} = \left(\frac{1}{2^2}\right)(0.346) = 0.0865\ \Omega$$

$$\text{(b)}\ \%\eta = \frac{100 \times \text{kVA} \times \text{pf}}{\text{kVA} \times \text{pf} + P_{oc} + (\text{kVA}_L/\text{kVA}_R)^2 P_{sc}}$$

$$= \frac{(100)(2)(0.85)}{(2)(0.85) + 0.05 + (\frac{2}{3})^2(0.03)}$$

$$= 96.4\%$$

4-6. TAPPED TRANSFORMERS

If accessible leads are tapped into a transformer coil at several points, greater versatility is achieved. Consider the transformer of Fig. 4-8. The primary is tapped at 2.5 and 5% of N_p and the secondary has a 50% tap (50% N_s).

Recalling that the voltage per turn, e_t, is identical in the primary and the sec-

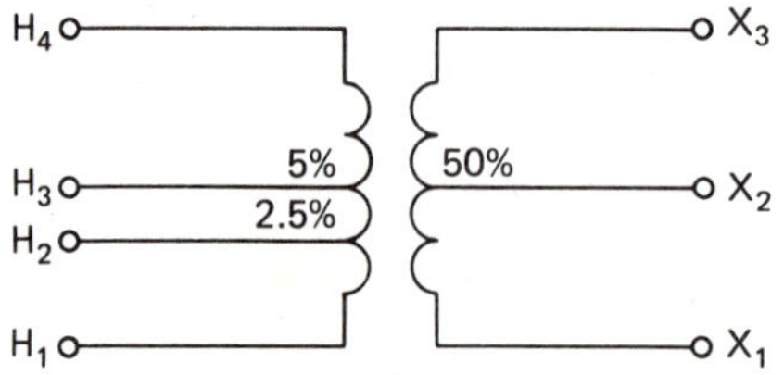

Figure 4-8 Tapped Transformer

ondary, any input arrangements that maintain a constant e_t in the primary turns will maintain a constant V_s across the full secondary. On the primary side, $V_p/N_p = 0.975V_p/0.975N_p = 0.95V_p/0.95N_p = e_t$, a constant. Assuming a 2400/240-V transformer, 2400 V applied to H1–H4, 2340 V applied to H2–H4, or 2280 V applied to H3–H4 all will produce 240 V across the full secondary. The three secondary terminals can be used to provide three-wire, single-phase, 120/240-V service.

A common application of taps occurs in distribution transformers—the relatively small transformers mounted on power poles to which the service entries of several neighboring residences are connected. The primary of the transformer is connected to transmission lines; voltage drops reduce the line-to-line voltages as the distance from the power station increases. The residential consumer, whose lighting and appliances are designed for nominal 120 V or 240 V, expects and has the right to receive power near the nominal voltage level (see Problem 4-30).

Without taps, the illustrated transformer would provide 117/234 V when the line is 2340 and 114/228 V if the line is 2280. Using the 2.5% tap where the line voltage is 2340 V and the 5% tap at 2280 V, the secondaries are 120/240 V.

One caution is in order when using taps—the ratings of the transformer always must be respected. These ratings are based upon the permissible voltage across a single turn and the allowable current through the coil. Application of a voltage across a portion of a coil such that the resulting e_t is greater than V_R/N_p causes excessive insulation stress. A more common abuse is drawing currents that are larger than rated, resulting in excessive heat rise, accelerated deterioration of the insulation, and, probably, premature equipment failure. The permissible kVA must be downgraded when using taps—for example, the continuous-duty rating using the 5% tap is $(0.95V_p)(I_p) = 0.95$ kVA rated. (In practice, ANSI C57.12.00-1973 requires that all taps be full-capacity taps unless the nameplate shows otherwise.) The need for derating explains why some tap combinations, although workable in voltage terms, are impractical. In the foregoing illustration, 60 V applied between terminals H1 and H2 would provide 240 V across the full secondary, but the kVA limit (2.5% of the rated kVA) would prove inefficient and uneconomical.

4-7. TRANSFORMER POLARITY

Figure 4-9 shows two transformers that are identical except for the direction in which the secondary coil is wound around the core. Lenz's law would identify differing polarities of the secondary voltages as shown on the sketches. Polarity is significant when two or more transformer coils are interconnected for purposes to be described in following paragraphs. It can be determined by a simple test.

In each sketch, a connection is placed between one side of the primary and one side of the secondary. For analysis only, a ground is assumed at this connection. Rated voltage is applied to the primary, resulting in the induced V_s in the secondary. In both sketches, the potential of the upper primary terminal with respect to the assumed ground is $+V_p$; however, the potential of the upper secon-

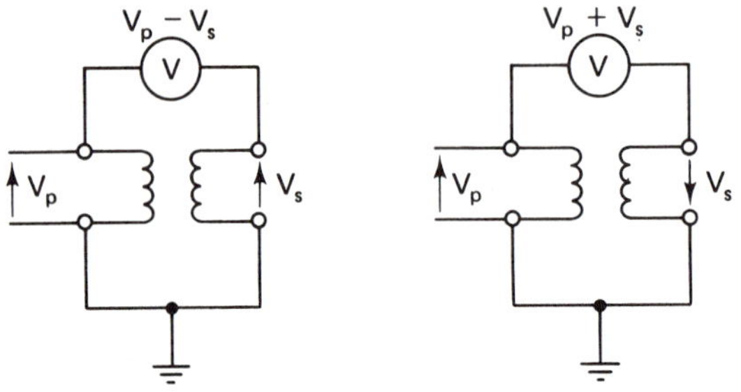

(a) Subtractive (b) Additive

Figure 4-9 Transformer Polarities

dary can be either $+V_s$ or $-V_s$. A voltmeter connected between the upper terminals measures the potential difference. As shown, the terms "SUBTRACTIVE" and "ADDITIVE" are used to identify polarity.

In schematic representations, any convenient polarity may be assumed, but actual connections must be consistent with the desired polarity.

Numbering of transformer terminals has been standardized by ANSI. Each high-voltage terminal is marked with a subscripted *H* and each low-voltage terminal with a subscripted *X*. Consecutive numbered subscripts indicate both the coil terminals and the polarity. A transformer with a split secondary would be marked H1, H2, X1, X2, X3, and X4; it has one high-voltage winding (H1–H2) and two low-voltage windings (X1–X2 and X3–X4). At any instant when H2 is positive with respect to H1, X2 will be positive with respect to X1, and X4 will be positive with respect to X3. Although this is a current standard, many older transformers are unmarked and, in any event, the safe procedure is to check before connecting any circuit in which polarity is a significant consideration.

4-8. AUTOTRANSFORMER

Figure 4-10(a) shows a conventional transformer. The primary is a load on the available power source; the secondary is the power source for the actual load. Polarities of the voltages and currents, consistent with this interpretation, are those shown on the sketch. For the discussion of autotransformers and polyphase transformers that follows, these assumed polarities always will be applied to segments of the configuration where fundamental transformer relationships apply. Summarizing these relationships,

- $a = V_p/V_s = I_s/I_p$.
- $V_pI_p = V_sI_s$ = transformed kVA (kVA_T).
- The phase angles of currents I_p and I_s are identical but are determined by the phase angle of the load. They differ from the voltage angles by the load impedance angle θ_L.

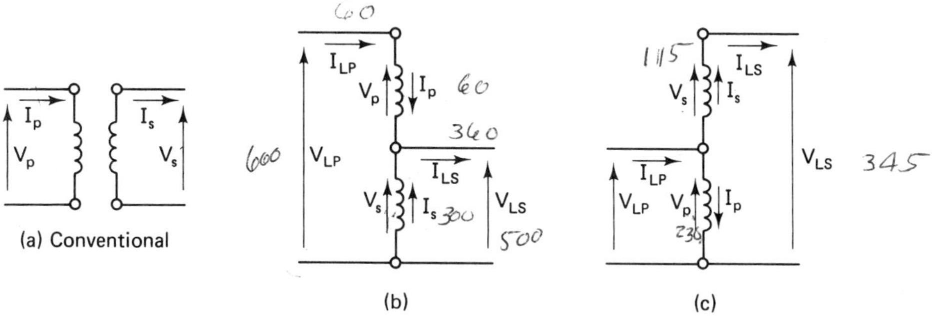

Figure 4-10 Autotransformer

In Fig. 4-10(b), the primary and secondary coils are connected in series and selected input and output terminals are identified. The subscripts *LP* and *LS* designate voltage, current, and kVA at the line primary (input) and line secondary (output) terminals, respectively. Kirchhoff's voltage and current laws interrelate the line and transformer quantities. Thus,

$$V_{LP} = V_p + V_s \quad V_{LS} = V_s \quad I_{LP} = I_p \quad I_{LS} = I_s + I_p$$

Solving for line input and output kVA,

$$\begin{aligned}
\text{kVA}_{LP} &= (V_{LP}I_{LP}) \times 10^{-3} = (V_p + V_s)I_p \times 10^{-3} = (V_pI_p + V_sI_p) \times 10^{-3} \\
&= \text{kVA}_T + V_sI_p \times 10^{-3} \\
\text{kVA}_{LS} &= V_{LS}I_{LS} \times 10^{-3} = V_s(I_s + I_p) \times 10^{-3} = (V_sI_s + V_sI_p) \times 10^{-3} \\
&= \text{kVA}_T + V_sI_p \times 10^{-3}
\end{aligned}$$

Comparing these expressions, $\text{kVA}_{LP} = \text{kVA}_{LS}$: therefore, $V_{LP}I_{LP} = V_{LS}I_{LS}$ and a line-to-line transformation ratio, a_L, can be defined:

$$a_L = \frac{V_{LP}}{V_{LS}} = \frac{I_{LS}}{I_{LP}}$$

Writing line voltages in terms of V_p and V_s,

$$a_L = \frac{V_p + V_s}{V_s} = a + 1 \qquad (a \text{ is the conventional transformer voltage ratio})$$

The line kVA has two components, kVA_T and $V_sI_p \times 10^{-3}$. The former is transformed kVA—it exists simultaneously in both the conventional primary (V_pI_p) and secondary (V_sI_s) coils. The remainder of the line kVA, $V_sI_p \times 10^{-3}$, is conducted kVA—it has been transmitted from the line input to the line output by direct conduction. The principal advantage of the autotransformer is traceable to the conducted kVA; specifically, a conventional transformer can accommodate a load kVA greater than the transformer rating when it is connected as an autotransformer, resulting in greater operational efficiency and reduced equipment cost. However, this advantage is at the expense of electrical isolation of the input and output; if conductive coupling is unacceptable, an autotransformer cannot be used.

Figure 4-10(b) represents but one of several possible selections of input and output terminals. Figure 4-10(c) shows a second possibility—note that the series connection of the conventional primary and secondary coils continues, but the selected line terminals differ. In this arrangement,

$$V_{LP} = V_p \quad I_{LP} = (I_p + I_s) \quad V_{LS} = V_p + V_s \quad I_{LS} = I_s$$

Duplicating the analytical steps applied to Fig. 4-10(b),

$$a_L = \frac{V_{LP}}{V_{LS}} = \frac{I_{LS}}{I_{LP}} = \frac{V_p}{V_s + V_p} = \frac{V_p/V_s}{1 + V_p/V_s} = \frac{a}{1 + a}$$

$$\text{kVA}_{LP} = \text{kVA}_{LS} = \text{kVA}_T + V_p I_s \times 10^{-3} \quad \text{where } V_p I_s \times 10^{-3} \text{ is conducted kVA}$$

Note that the basic relationships applicable to the conventional transformer can be applied not only between the conventional primary and secondary coils but also between line input and output terminals; however, in no practical arrangement will $a_L = a$, and the relationship between a_L and a differs with changes of input and output terminals.

EXAMPLE 4-3: A conventional 3-kVA 230/115-V transformer is connected in autotransformer configuration with a line input of 230 V and output of 345 V. Identify coil and line voltages, currents, and voltage ratios when operating at rated capacity.

SOLUTION:

The configuration of Fig. 4-10(c) applies since $V_{LS} > V_{LP}$. V_p is 230 V and $V_s = V_{LS} - V_p = 345 - 230 = 115$ V.

$$I_p = \frac{\text{kVA}_R \times 10^3}{V_p} = \frac{3 \times 10^3}{230} = 13 \text{ A} \qquad I_s = \frac{\text{kVA}_R \times 10^3}{V_s} = \frac{3 \times 10^3}{115} = 26 \text{ A}$$

$$a = \frac{V_p}{V_s} = \frac{230}{115} = 2$$

$$V_{LP} = 230 \text{ V} \qquad V_{LS} = 345 \text{ V}$$

$$I_{LP} = I_p + I_s = 13 + 26 = 39 \text{ A} \qquad I_{LS} = I_s = 26 \text{ A}$$

$$\text{kVA}_L = V_{LP} I_{LP} \times 10^{-3} = (230)(39)(10^{-3}) = 9 \text{ kVA}$$

$$a_L = \frac{V_{LP}}{V_{LS}} = \frac{230}{345} = 0.67$$

EXAMPLE 4-4: The conventional transformer of the preceding example transforms a line input voltage of 345 V to a line output of 115 V and delivers a load of 2 kW at a lagging power factor of 0.8. Find the magnitudes and phase angles of all voltages and currents.

SOLUTION:

Figure 4-10(b) applies. A reference phase angle must be assigned to one parameter and all others related to it. The voltages are assigned an angle of 0° in this solution.

$$V_{LP} = 345\ V\underline{/0} \qquad V_{LS} = V_s = 115\ \text{V}\underline{/0^\circ}$$

$$V_p = V_{LP} - V_s = 345\underline{/0} - 115\underline{/0} = 230\ \text{V}\underline{/0^\circ}$$

$$\text{kVA}_L = \frac{\text{kW}}{\text{pf}} = 2/0.8 = 2.5\ \text{kVA}$$

$$I_{LS} = \frac{\text{kVA}_L}{V_{LS}} = \frac{2.5 \times 10^3}{115} = 21.74\ \text{A}$$

$$a_L = \frac{V_{LP}}{V_{LS}} = \frac{345}{115} = 3$$

$$I_{LP} = \frac{I_{LS}}{a_L} = \frac{21.74}{3} = 7.25\ \text{A}$$

$$\theta_L = \cos^{-1} \text{pf} = \cos^{-1} 0.8 = 36.9^\circ$$

$$I_{LP} = 7.25\ \text{A}\underline{/-36.9} \qquad I_{LS} = 21.74\ \text{A}\underline{/-36.9}$$

$$I_p = I_{LP} = 7.25\ \text{A}\underline{/-36.9}$$

$$I_s = I_{LS} - I_p = 21.74\underline{/-36.9} - 7.25\underline{/-36.9} = 14.49\ \text{A}\underline{/-36.9}$$

Check: $I_s/I_p = 14.49/7.25 = 2 = a$.

The autotransformer principle is applied in a very useful device, the variac (variable ac), for use in laboratory and other situations that require a continuously variable voltage over broad ranges. In Fig. 4-11, the input and one output terminals are fixed, but the second output terminal is connected to a carbon brush that slides throughout the length of the coil. At any position the coil may be viewed as two segments comparable to the coils of a fixed-terminal autotransformer. The variac performance may be compared to a rheostat voltage divider; however, the variac

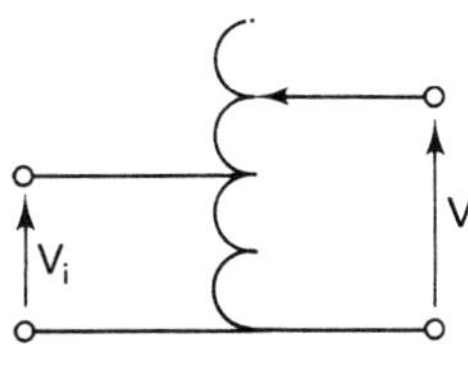

Figure 4-11 Variac

can step up the source voltage but a rheostat cannot, and the power loss in a variac is substantially less than that of a comparable rheostat. With either device, one reminder is appropriate—momentary carelessness can lead to excessively large currents that quickly destroy unprotected equipment.

4-9. POLYPHASE TRANSFORMATION

The primaries of three identical transformers may be connected delta or wye to a three-phase source. Similarly, the secondaries may be connected delta or wye to a three-phase load. Since there is no restriction that the primaries and secondaries have identical connections, there are four possible combinations.

Δ–Δ (Delta Primary–Delta Secondary) [Fig. 4-12(a)]. In the delta–delta configuration the corresponding line and transformer voltages are identical in magnitude on both the primary and secondary sides. All corresponding voltages are at the same phase angles. However, with balanced loading both the primary and the secondary line currents are greater than the transformer currents by the factor $\sqrt{3}$ and differ

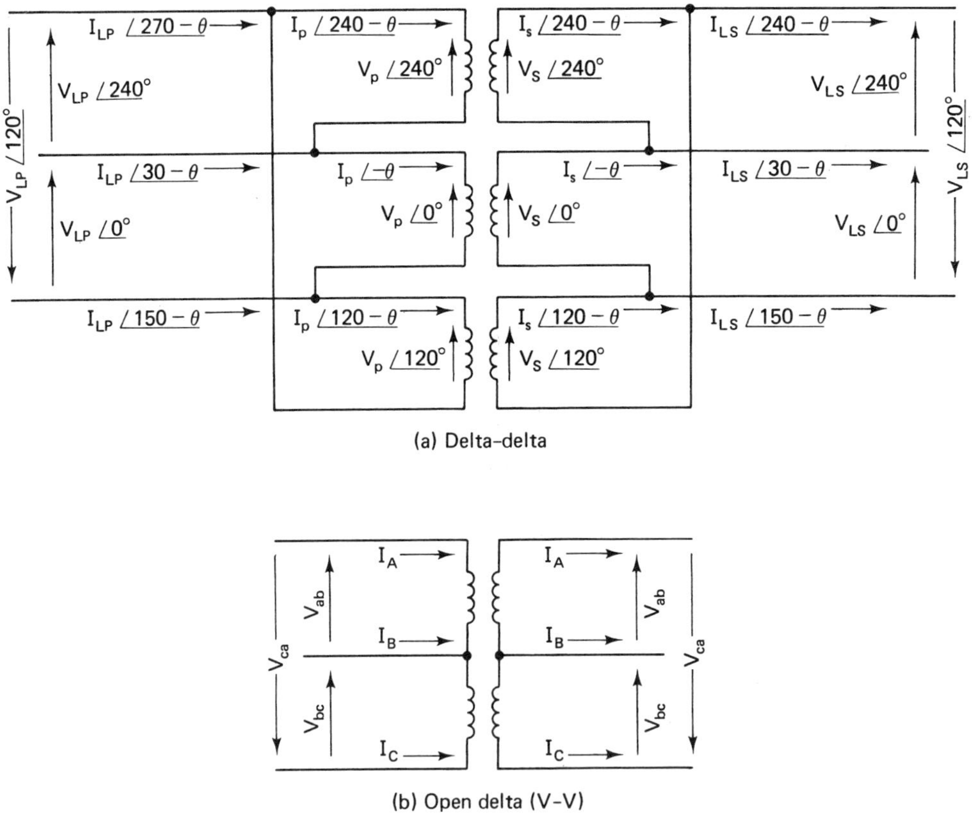

Figure 4-12 Delta Connections

from transformer currents by a phase angle of 30°. The total kVA transformed by the three-transformer bank is $\sqrt{3}\,V_{LP}I_{LP} = \sqrt{3}\,V_p(\sqrt{3}\,I_p) = 3V_pI_p = 3\text{kVA}_T$. Thus the full kVA capacity of each transformer is productively employed in line-to-line transformation; this is true for all three-transformer banks but not for two-transformer banks discussed in later paragraphs. The line-voltage ratio equals the transformer-voltage ratio.

The delta–delta connection is satisfactory for both balanced and unbalanced loading. However, it has the disadvantage of no inherent means for providing a neutral ground.

Since the sum of two phase voltages equals the third voltage in a three-phase system, one of the three transformers in a delta–delta bank may be removed and the two remaining transformers will continue to transform three-phase power. This is the *open-delta* (or V–V) connection [Fig. 4-12(b)]. Note that in open delta the line currents and transformer currents are equal; consequently, the transformed kVA is not 2kVA_T as intuition suggests but rather $\sqrt{3}\,\text{kVA}_T$. Removing one of three transformers and operating open-delta reduces the permissible load kVA to 58% [$\sqrt{3}\,\text{kVA}_T/3\text{kVA}_T$] of the original capacity. Within this restriction, open delta is advantageous in allowing removal of one transformer for repair or maintenance without interrupting service. Open delta may be employed to provide service in a new development area with the expectation of adding a third transformer at a future date to accommodate growth in the power demand; it also is used in rural and suburban areas to provide the usual single-phase residential service and also three-phase for applications such as water pumps and air conditioning.

Y–Y (Wye Primary–Wye Secondary) **(Fig. 4-13).** In wye–wye, the voltage across the transformer is $V_L/\sqrt{3}$. Thus, at a given line voltage level the insulation stress in the wye is less than that for the delta. However, at the same line kVA, the

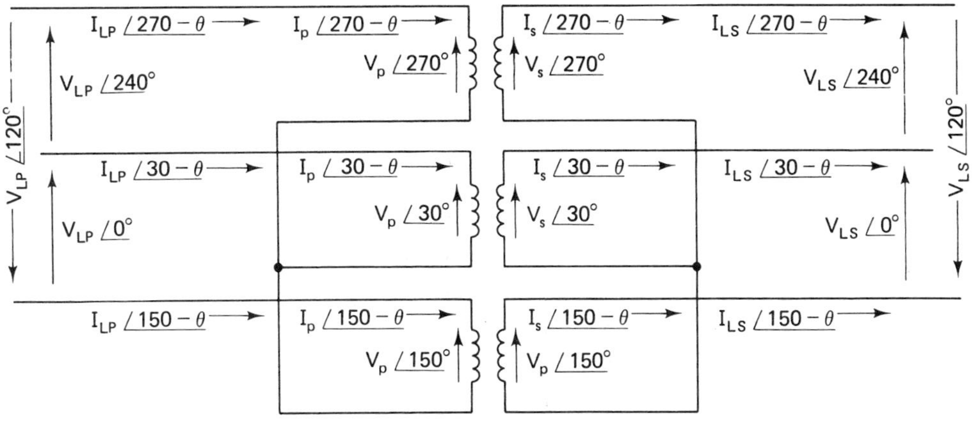

Figure 4-13 Wye–Wye

transformer currents for the wye are larger than those of the delta by a factor of $\sqrt{3}$. The total kVA capacity is 3kVA_T.

Like the delta–delta, the line-voltage ratio is the same as the transformer voltage ratio. There is a phase separation of 30° between line and transformer voltages, but line and transformer currents are in phase.

The Y–Y connection is not satisfactory for unbalanced loading in the absence of a neutral connection. Severe imbalance could cause displacement of the neutral to the extent that there would be no voltage across one transformer and line voltage would occur across the other two. This is readily appreciated by visualizing a short circuit across one phase of the Y and comparing the result with the open-delta schematic.

An additional problem in the Y–Y is traceable to the waveform of the exciting current. With a sinusoidal voltage applied, the flux normally is forcibly sinusoidal, but the exciting current is not. As illustrated in Fig. 4-14, the waveform of the exciting current results from the characteristic hysteresis loop of the core. In the figure, the projection of several points on the flux curve into the hysteresis loop and reflection to the exciting current curve is demonstrated; a complete projection results in the form as plotted.

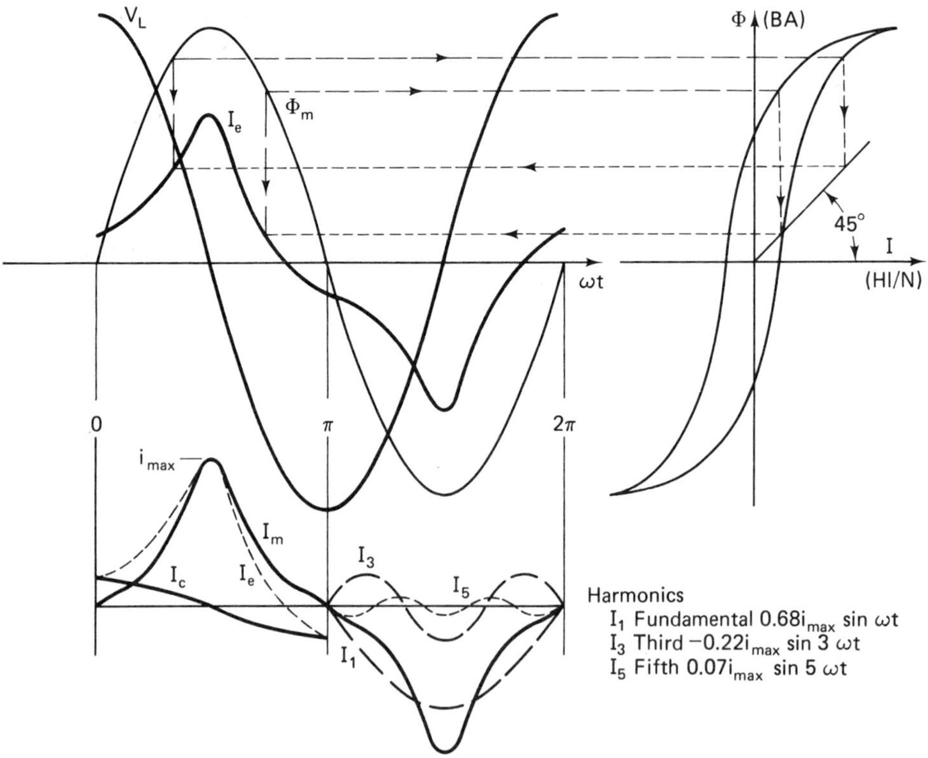

Figure 4-14 Exciting Current

For the interval 0 to π, the exciting current is shown separated into two components; i_c is the core-loss component, in phase with V_L, and the other is I_m, lagging V_L by 90° but not sinusoidal in form. For the interval π to 2π, I_m is further separated into the first three harmonic waves that give I_m its characteristic shape; only odd harmonics are present and the first, third, and fifth almost completely define the wave form.

When three transformers are connected Y–Y, the first harmonics are separated by 120°; at the tie point of the wye, their sum is zero. However, the third harmonic repeats at $\frac{1}{3}$ of the fundamental period—that is, at 120° intervals; consequently, the third harmonics of three-phase currents are in phase, and their sum at the junction of a wye is *not* zero. In a delta, the harmonics have a conducting path through the windings, but there is no path in an ungrounded wye and the result is distortion of both the flux and the induced voltage wave forms. Note, however, that if one side of the transformer is wye-connected and the other delta, the delta provides a discharge path even though the opposing wye is ungrounded.

One method of eliminating the harmonic problem in Y–Y is construction of a three-phase transformer with not two but three windings per phase; the third winding is connected Δ.

The use of Y–Y is less common than other options.

Δ–Y (Delta Primary–Wye Secondary) (Fig. 4-15). The magnitudes and phase angles of voltages and currents are shown in the sketch. Contrasting with Y–Y and Δ–Δ, the line-to-line voltage ratio is no longer a but rather $a/\sqrt{3}$, and corresponding line voltages and currents are not in phase but are separated by 30°. Like Δ–Δ and Y–Y, the kVA capacity is 3kVA_T.

The Δ–Y optimizes step-up of line-to-line voltages and commonly is used at the generator end of a line to transform from medium to high voltage for transmission. The high side windings have less insulation stress than would occur in a delta secondary for the same line voltage. Further, the neutral of the Y can be grounded and the potential of the lines with respect to ground, $V_L/\sqrt{3}$, is less than line voltage.

For medium- to low-voltage transformation, Δ–Y is useful in applications requiring both three-phase and single-phase power. In Fig. 4-16(b), a fourth line is connected to the neutral. As an example, 208-V three-phase service is available from lines A, B, and C, and additionally, 120-V circuits are provided between each of these lines and the neutral line N. The system can accommodate both balanced and unbalanced loading.

Y–Δ (Wye Primary–Delta Secondary) (Fig. 4-16). Much of the comment on Δ–Y applies to Y–Δ. The prominent feature of Y–Δ is the line-to-line transformation ratio, $\sqrt{3}\,a_T$. For stepping down voltages, Y–Δ optimizes the line-to-line ratio and, therefore, is commonly used at the load end of high-voltage-transmission systems.

Although three-phase is the dominant polyphase system, others are employed, including modest use of six-phase. Six-phase is readily obtained by the use of split

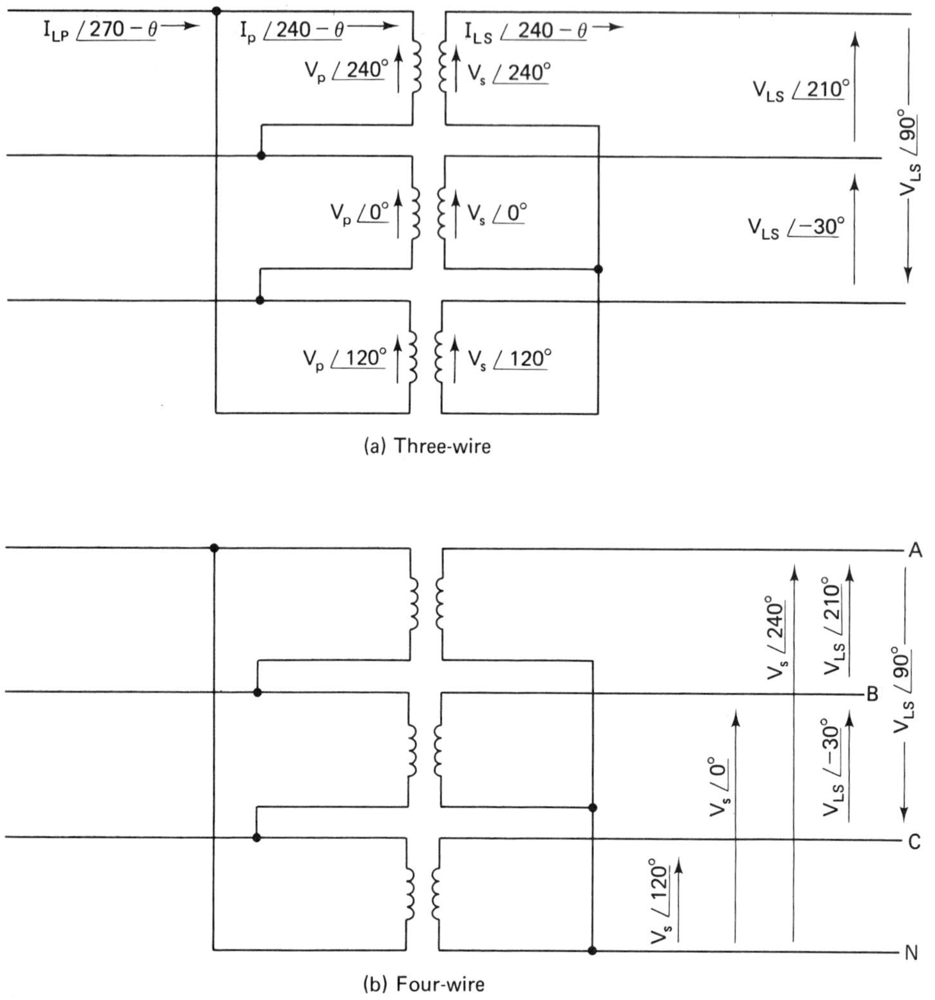

(a) Three-wire

(b) Four-wire

Figure 4-15 Delta–Wye

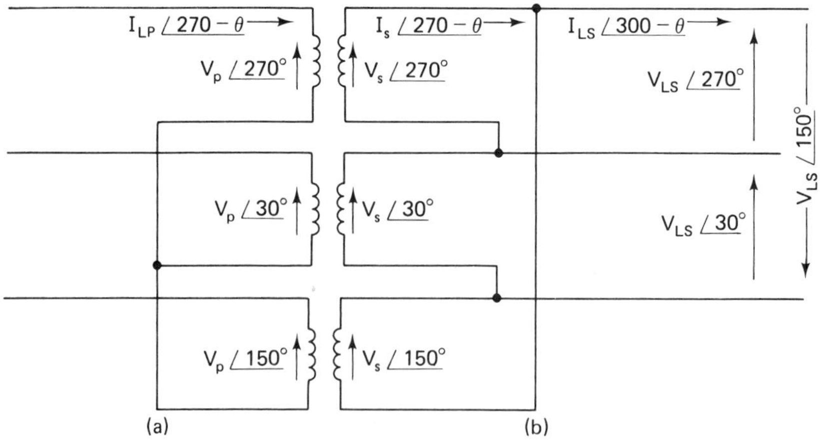

Figure 4-16 Wye–Delta

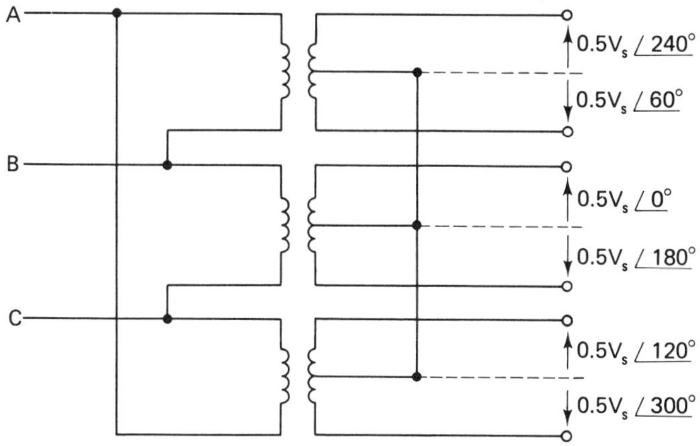

Figure 4-17 Three- to Six-Phase Transformation

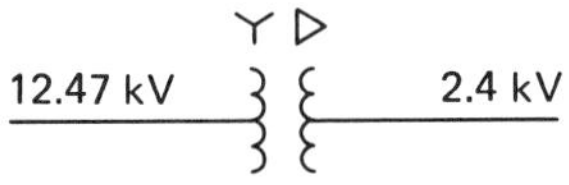

Figure 4-18 One-Line Diagram

coils in each of the secondary phases, (Fig. 4-17). Schematically, three-phase representations may be simplified by one-line diagrams as illustrated in Fig. 4-18.

In analyzing balanced three-phase systems, the problem is less complex if all delta-connected loads are replaced with their equivalent wye. In the wye equivalent of a delta, the phase impedance is one-third of the actual impedance—that is, $Z_Y = Z_\Delta/3$. (There is no change in the impedance angle.) The converted system results in identical phases, each being the equivalent of a simple series–parallel, single-phase circuit. Because of symmetry, analysis of only one phase is required.

EXAMPLE 4-5: For the circuit shown in Fig. 4-19, find a, a_L, V_p, V_s, I_{LP}, I_{LS}, I_p, I_s, and kVA_{LP}.

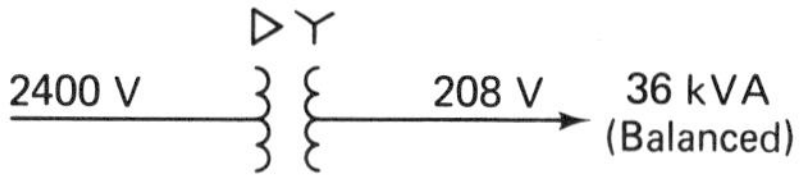

Figure 4-19 Example 5

SOLUTION:

$$V_p = V_{LP} = 2400 \text{ V}$$

$$V_s = \frac{V_{LS}}{\sqrt{3}} = \frac{208}{\sqrt{3}} = 120 \text{ V}$$

$$a = \frac{V_p}{V_s} = \frac{2400}{120} = 20$$

$$I_s = I_{LS} = \frac{\text{kVA} \times 10^3}{\sqrt{3}\,V_{LS}} = \frac{36 \times 10^3}{(\sqrt{3})(208)} = 100 \text{ A}$$

$$I_p = \frac{I_s}{a} = \frac{100}{20} = 5 \text{ A}$$

$$I_{LP} = \sqrt{3}\,I_p = \sqrt{3}(5) = 8.67 \text{ A}$$

$$\text{kVA}_{LP} = \sqrt{3}\,V_{LP}I_{LP} \times 10^{-3} = (\sqrt{3})(2400)(8.67) \times 10^{-3} = 36 \text{ kVA}$$

$$a_L = \frac{V_{LP}}{V_{LS}} = \frac{2400}{208} = 11.54$$

Another method of transforming three-phase power with two transformers is the *Scott-T*, named after its inventor, C. F. Scott. Using two identical transformers having 50% and 86.6% (i.e., $\sqrt{3}/2$) taps in both the primary and the secondary, the low end of one transformer (the teaser) is connected to the center tap of the other (the main) on both the primary and the secondary sides [Fig. 4-20(a)]. The three-phase lines on both primary and secondary sides are connected to the end terminals of the main and the 0.866 tap of the teaser.

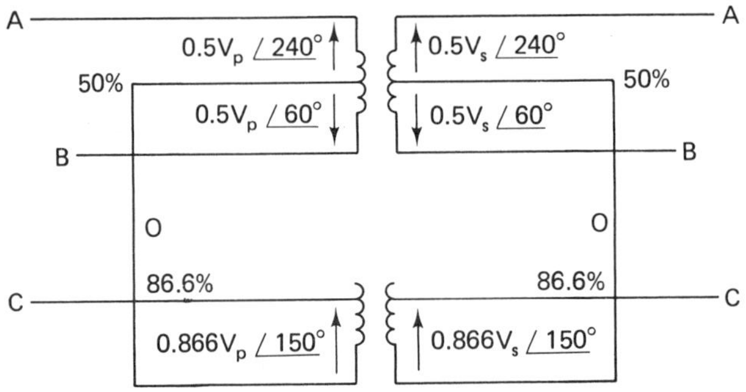

(a) Three-phase transformation

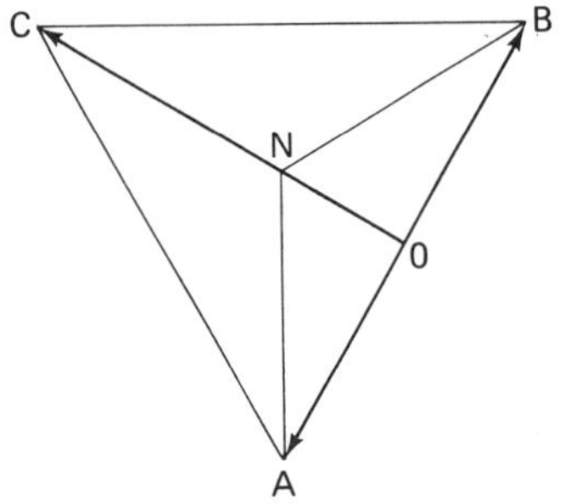

(b) Displaced neutral

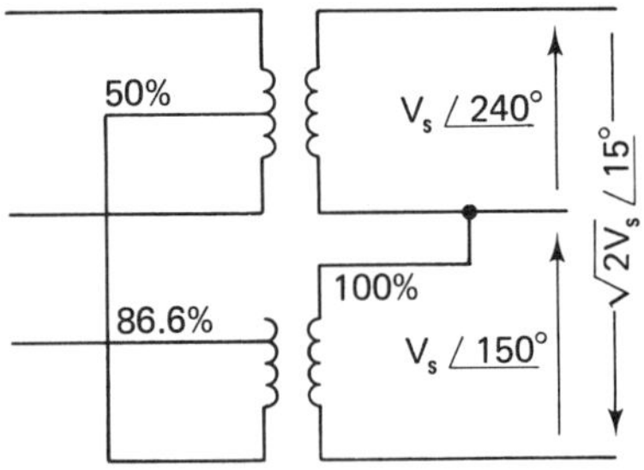

(c) Three- to two-phase transformation

Figure 4-20 Scott-T

Under load, the distribution of the line voltages across the component coils of the transformers is that of an unbalanced wye with a displaced neutral as in Fig. 4-20(b). The result is three-phase transformation with $a_L = a$.

The Scott-T also permits transformation between three-phase and two-phase. Figure 4-20(c) shows the connections. Note two changes in the secondary connections: (1) the connection of the teaser to the main is shifted from the center to an end tap, and (2) the full secondary (not $0.866N_s$) is used. On the teaser, the turns ratio is therefore $(\sqrt{3}\,N_p/2)/N_s = \sqrt{3}\,a_T/2$ and the voltage on the teaser secondary is $(\sqrt{3}\,V_p/2)/(\sqrt{3}\,a/2) = V_p/a$.

4-10. PARALLELING TRANSFORMERS

Transformers often are constructed with split coils that may be connected in either series or parallel to provide several voltage–current combinations. Also, if load requirements exceed the ratings of on-line equipment, additional transformers may be paralleled. But paralleling cannot be done indiscriminately and several conditions must be satisfied.

Figure 4-21 shows two paralleled transformers. One transformer has voltage

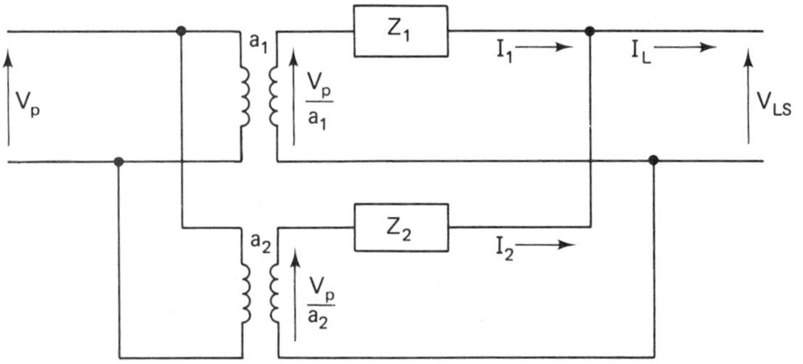

Figure 4-21 Paralleled Transformers

ratio a_1, equivalent secondary impedance Z_1, and coil current I_1; similarly, a_2, Z_2, and I_2 are parameters of the second transformer. I_L and V_{LS} are the line current and line secondary voltage. Applying Kirchhoff's laws,

$$I_2 = I_L - I_1$$

$$V_{LS} = \frac{V_p}{a_1} - I_1 Z_1 = \frac{V_p}{a_2} - (I_L - I_1)Z_2$$

Solving for the currents,

$$I_1 = \frac{Z_2 I_L}{Z_1 + Z_2} + \frac{V_p(a_2 - a_1)}{a_1 a_2 (Z_1 + Z_2)}$$

$$I_2 = \frac{Z_1 I_L}{Z_1 + Z_2} - \frac{V_p(a_2 - a_1)}{a_1 a_2 (Z_1 + Z_2)}$$

Each of these currents has two components; the first component is the transformer's share of the load current, but the second is a circulating current in the windings. Circulating currents are undesirable; they increase the copper loss, overload one transformer, and reduce the permissible load kVA if the transformer ratings are respected. To eliminate them, the voltage ratios must be identical.

When $a_1 = a_2$, the transformer currents are inversely proportional to the transformer impedances; that is, $I_1/I_2 = Z_2/Z_1$. Multiplying each current by V_{LS}, $kVA_1/kVA_2 = Z_2/Z_1$. Therefore, to realize the rated kVA potential of both transformers, the ratio of their kVA ratings should be inversely proportional to their impedance ratio.

The impedances of the transformers are vectors with impedance angles $\theta = \tan^{-1} X/R$. If the angles of both transformers are identical—that is, if $X_1/R_1 = X_2/R_2$—the currents I_1 and I_2 add arithmetically and are in phase with the line current. But if $\theta_1 \neq \theta_2$, the magnitudes of the currents remain inversely proportional to the magnitudes of the impedances, but the phase angles of the currents differ; at rated load, I_L is the vector sum of I_1 and I_2, the magnitude of I_L is less than the arithmetic sum of I_1 and I_2, and the load kVA is less than the sum of the transformer kVAs.

Summarizing, the ideal relationships for paralleling transformers are:

1. Identical primary and secondary voltage ratings.
2. Impedances inversely proportional to the kVA ratings.
3. Identical X/R ratios in the transformer impedances.

Finally, transformer polarities must be observed in making the physical connections.

The conditions above apply equally to the paralleling of two or more three-phase transformer banks but with the added need to ensure that the secondary voltages are at compatible phase angles. A Y–Y and a Δ–Δ may be paralleled if the line voltage ratios are identical, but the phase shifts resulting from Δ–Y or Y–Δ preclude use of either in parallel with a differing configuration.

4-11. CONSTRUCTION AND RATINGS

There are two basic types of single-phase transformer construction, shell and core (Fig. 4-22). In both, the magnetic circuit is built up with laminations generally 0.014 in thick. The common material is silicon steel. In the shell type, the windings are on the central leg and the magnetic circuit surrounds the coils. The core type splits the windings on two legs of the magnetic circuit. To reduce leakage flux, primary and secondary windings are intermixed either by alternate stacking of "doughnut" sections on the core leg or by concentric placement of one winding on top of the other.

Three-phase transformation may employ a single three-phase transformer (Fig. 4-23) or a bank of three single-phase transformers. For a given kVA capacity, the former is smaller, lighter, less costly, and more efficient because of lesser core

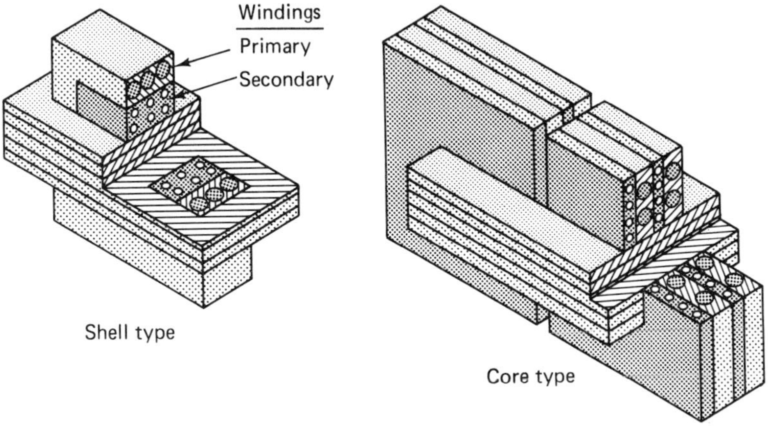

Figure 4-22 Transformer Construction

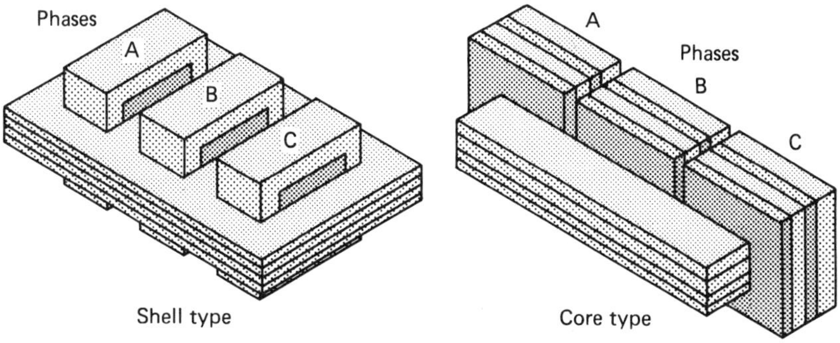

Figure 4-23 Three-Phase Transformer Construction

losses. However, a malfunction requires that the complete transformer be removed for servicing or replacement. With the three-transformer bank, a malfunction usually means replacement of only one of the three transformers and, if connected Δ–Δ, it may be practical to temporarily remove the defective transformer with the other two continuing to supply the load in open-delta configuration. Generally, the modern preference is the single three-phase transformer.

Like dc machinery, the rating of a transformer basically is traceable to the dielectric strength and thermal properties of the winding insulation. The type of insulation determines the permissible operating temperature, and the design of the transformer must provide for heat dissipation at a rate that will keep the temperature below this limit under the operating conditions defined on the nameplate. The dry type of construction relies upon air convection to, and radiation from, the casing to transfer the generated heat to the atmosphere. The air may be static or circulating under pressure.

In the oil-immersed type, the windings and core are submerged in an oil that has both insulating and thermal-conducting properties. Heat is transferred to the casing and radiated into the atmosphere. Forced air may be combined with the oil immersion. Casings may have fins, corrugations, or other special profiles to increase the radiating surface.

The standard ratings for self-cooled transformers range from 3 to 33,333 kVA for single-phase and 15 to 100,000 kVA for three-phase transformers. When force cooled, the ratings increase by as much as 67% for the larger transformers.

From 3 to 500 kVA, transformers are classified as distribution type; above 500 kVA they are power transformers.

4-12. POWER TRANSMISSION AND DISTRIBUTION

Medium voltages are generated by the alternators at a generating site. At a nearby substation these voltages are transformed to high or extra high for transmission to a substation in the locale where the power will be used. At the transmitting end, voltage is regulated only to the extent of allowing power flow at levels compatible with the equipment.

At the receiving substation the voltage is stepped down to medium levels for continued transmission over primary distribution lines; at this point, the voltage is regulated by tap-changing-under-load transformers or induction voltage regulators that raise or lower the voltage in response to power variations. The voltage on the primary feeders will remain within acceptable limits as the line voltage drop varies.

Each branch of the primary distribution system supplies a number of distribution transformers that step down the medium voltage to a low voltage for supply through secondary distribution lines to the consumers. Appropriate taps of the distribution transformers are used to ensure that the secondary voltage remains within permissible limits regardless of variations in the line voltage drop within the primary distribution lines. (See problem 4-30.)

Industrial plants with load requirements up to several hundred kVA are supplied by distribution transformers; for a few thousand kVA the plant is connected to the primary distribution system. Larger plants may be connected directly to the transmission system.

4-13. TRANSFORMER-RELATED DEVICES

Functionally, the key element in the operation of a transformer is the time-variant flux that induces voltages in the coils. The magnitude of the induced voltage is governed by both the magnitude of the change in flux and the time over which the change occurs. If the time is constant (the common circumstance with sinusoidal voltages at standard frequency), the magnitude of the induced voltage can be controlled by varying the magnitude of the flux change. This method is applied in a number of devices, three of which are considered here: saturable reactor, magnetic amplifier, and induction voltage regulator.

In Fig. 4-24 an ac voltage of constant amplitude and frequency is applied to a series branch comprised of a coil and a load resistance. The coil has an open center into which an iron core can be inserted to varying depths. The vector sum

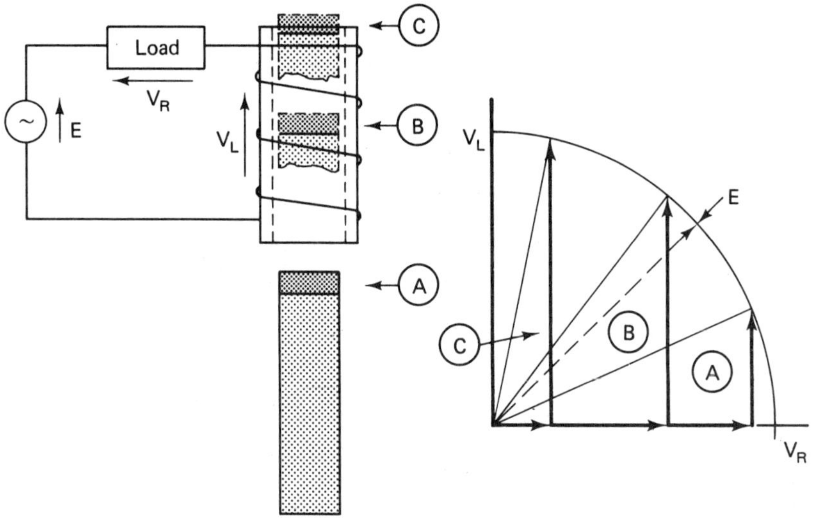

Figure 4-24 Removable-Core Reactor

of the voltage across the coil and the voltage across the resistor always will equal the applied voltage.

With the core removed, the magnetic circuit of the coil is exclusively a high-reluctance air path, resulting in a relatively small flux, a small coil voltage, and a large load voltage. But with the core fully inserted, the flux increases substantially, the coil voltage is large, and the load voltage is small. Continuous control of the load voltage between these extremes is possible by adjusting the depth of insertion of the core. The coil is a variable reactor.

Saturable Reactor. Comparable effects are achieved in the arrangement of Fig. 4-25. Two windings, one dc and one ac, are placed on a closed magnet circuit. The ac winding, load resistance, and ac power source are similar to those of Fig. 4-24, but the core is not removable. With no excitation in the dc winding, the ac flux in the magnetic circuit is large and most of the applied voltage appears across the coil.

The flux in the magnetic circuit is represented on the magnetization curve. With no dc excitation, Φ_m is the maximum flux created when the ac magnetic intensity, H_{ac}, is maximum. During the positive half of the ac cycle, the net change in flux is from zero to Φ_m; thus, the change, which is a measure of the induced voltage, is Φ_m.

In the following discussion it will be assumed that H_{ac} remains constant. Factually, this is not correct. The purpose of the device is to vary the voltage across a constant load resistance; inherently, the load current, which is also the winding current, is proportional to the load voltage. Since the number of turns of the winding is fixed, the magnetic intensity will vary directly with the load current. The simplifying assumption does not invalidate the explanation of effects, but it

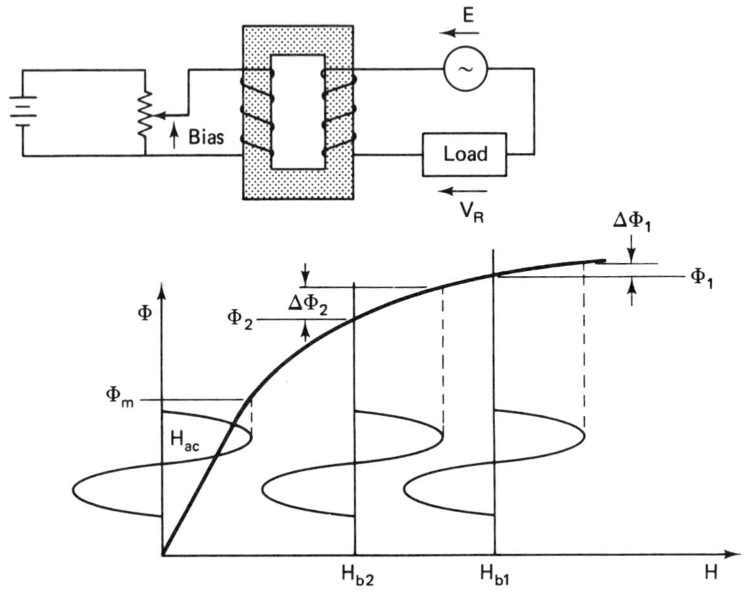

Figure 4-25 Saturable Reactor

does eliminate complexities that add nothing to an understanding of the principles.

Assume now that the dc winding is excited to produce an intensity H_{b1}. This intensity creates a constant bias flux Φ_1. During the positive portion of the ac cycle, H_{ac} adds to H_{b1}, but the resulting change in flux is very small ($\Delta\Phi_1$). The coil voltage will be small and the load voltage will be large.

At an intermediate level corresponding to H_{b2}, the load voltage will be greater than that of the unbiased magnetic circuit but less than that for the saturated condition.

This is a saturable reactor. It permits some variation of the load voltage, but during the negative portion of the cycle the core desaturates and most of the voltage occurs across the reactor. For this reason, the device is termed "desaturating saturable reactor."

Magnetic Amplifier. If a diode is added to the saturable reactor as shown in Fig. 4-26(a), the desaturating portion of the cycle is eliminated and the result is controlled half-wave rectification; this is a magnetic amplifier. In this form it has two principal limitations: (1) with operation limited to half of the cycle, the maximum dc voltage that can be applied to the load is but half that possible from the same source with full-wave rectification, and (2) the polarity of the load voltage is constant—it cannot be reversed. Both of these objections can be eliminated by combinations of multiple amplifier units.

Figure 4-26(b) is an arrangement that permits control of both magnitude and

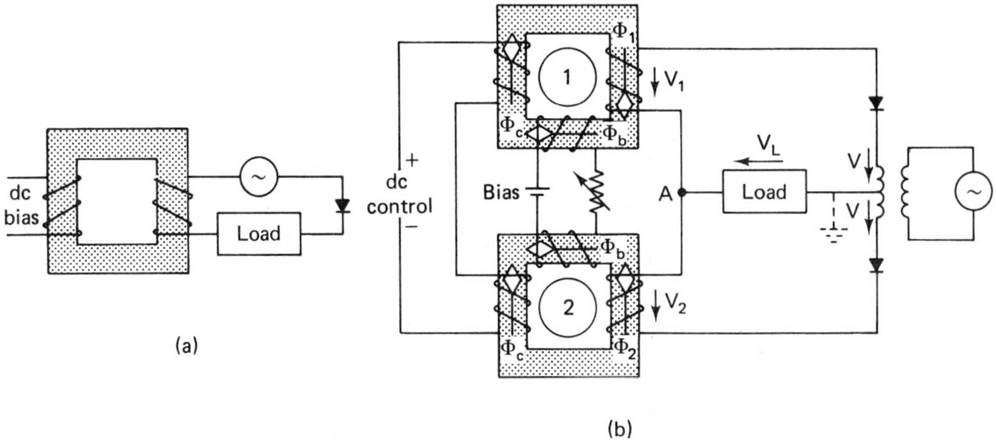

Figure 4-26 Half-Wave Magnetic Amplifier

polarity of the load voltage. Two reactor units are employed. Identical dc biases are applied to both reactors but with opposite polarities in the bias fluxes. Each reactor has a second dc control winding. The control windings, connected in series, are excited by a dc signal; for a positive signal, the control and bias fluxes are additive in reactor 1, driving it toward saturation, but the same positive control flux subtracts from the bias flux in reactor 2, causing it to desaturate. Similarly, a negative dc signal desaturates reactor 1 and saturates reactor 2.

The source voltage is a center-tapped transformer. V is the voltage induced in each half of the transformer winding. For convenience in analysis, a ground is assumed at the centertap. The potential at point A with respect to the assumed ground is the load voltage.

With reactor 1 saturated, the voltage V_1 approaches zero. Applying Kirchhoff's voltage law to the loop comprised of the transformer and the reactor coils, $2V = V_1 + V_2$. If V_1 is zero, $V_2 = 2V$. Tracking from ground through reactor 2 to point A, the potential at this latter point is $+V - V_2 = V - 2V = -V$.

With reactor 2 saturated, $V_2 = 0$, $V_1 = 2$ V, and the potential at A is $+V$. Thus, the load voltage is continuously variable from $+V$ to $-V$. However, this is still half-wave rectification.

Figure 4-27 duplicates the arrangement of Fig. 4-26 with a second pair of reactors, paralleling the original pair but connected to the transformers through diodes that pass the negative half of the transformer cycle. The functioning is unchanged, but the magnitude of the maximum dc load voltage is doubled.

The amplifier is controlled by the dc signal. The power level of the signal is very low, whereas the load power may be as large as 40 kW. Thus, as a power amplifier, the gain is substantial.

Representative uses of magnetic amplifiers are the control of adjustable dc drives, regulation of alternator voltages, and control of lighting and heating.

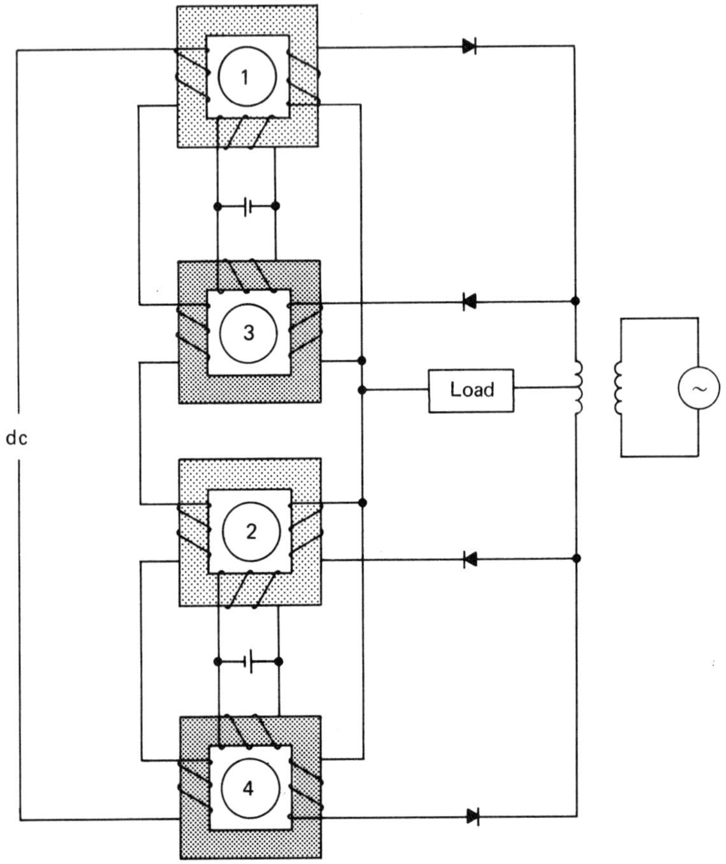

Figure 4-27 Full-Wave Magnetic Amplifier

Induction Voltage Regulator. In Fig. 4-28(a), a loop is positioned with its cross section perpendicular to the axis of a uniform magnetic field with a density $B_m \sin \omega t$; a voltage, $e_t = 4.44 f B_m A$, is induced in the loop.

The loop is static but the field may be rotated. In Fig. 4-28(b), the angle between the plane of the loop and the axis of the field is θ, the projection of the loop cross section into a plane perpendicular to the field axis is $A \sin \theta$, and the loop voltage is $4.44 f B_m A \sin \theta$. Thus, if the field can be rotated through 360°, the voltage in the loop is continuously variable from $+e_t$ to $-e_t$.

In Fig. 4-28(c), a primary winding is placed on a rotor within a stator that contains a secondary winding. With the secondary open-circuited and a voltage V_p applied to the primary, the secondary voltage is $V_s = (N_s V_p / N_p) \sin \theta$. But when a load is attached to the secondary, the load current will create a flux, Φ_s, on the vertical axis. In a conventional transformer, Φ_s is neutralized by Φ_p, the flux

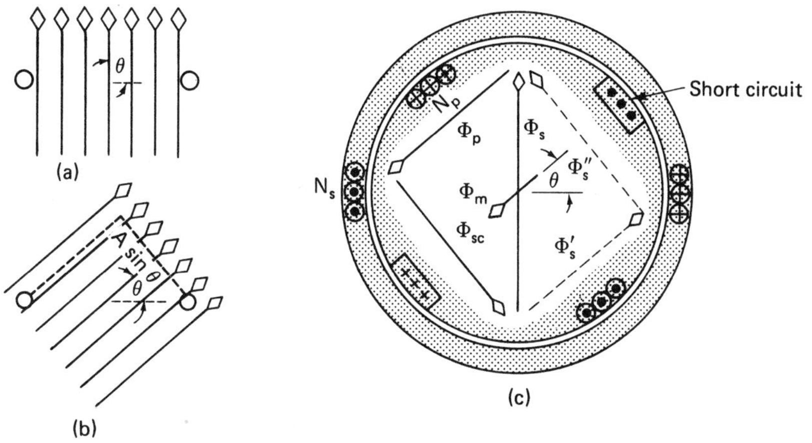

Figure 4-28 Induction Voltage Regulator

attributable to the load component of the primary current. However, in the voltage regulator the secondary flux through the primary coil is not Φ_s but rather $\Phi'_s = \Phi_s \sin\theta$. The vector difference Φ''_s is not neutralized by Φ_p and, therefore, induces a countervoltage in the secondary—that is, its effect is that of a variable inductive reactance in the secondary.

To eliminate Φ''_s, a short-circuited band is placed on the rotor, located in quadrature with the primary winding. Regardless of the rotor position, Φ''_s will be normal to the short-circuiting band, a short-circuit current will flow, and Φ_{sc} will oppose and neutralize Φ''_s.

This device is an induction voltage regulator. Fundamentally, it is a continuously variable transformer and, as such, can be used as a basic transformer or as an autotransformer. In conventional form it commonly is employed in laboratory environments as a controlled voltage source. As an autotransformer, the common employment (and the source of the name) is to compensate for voltage variations caused by line voltage drops as load current varies. With the primary connected across the line and the secondary inserted in the line (Fig. 4-29), the voltage can be boosted. The correction can be automatic through the use of a feedback system that responds to line voltage changes and rotates the primary to the appropriate angle.

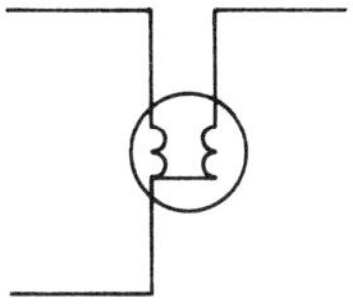

Figure 4.29 Line Voltage Booster

4-1. There are 6.2 V per turn in a 13.2/2.4-kV transformer. How many turns are in the primary? Secondary?

4-2. The rated secondary current of the preceding transformer is 417 A. (a) What is the rated primary current? (b) What is the kVA rating of the transformer?

4-3. The preceding transformer is operating at rated capacity. What is the load current and load power for a power factor of (a) unity, (b) 0.9 leading, and (c) 0.85 lagging?

4-4. The core of a transformer has a cross-sectional area of 10 in^2 and a flux density of 70 kilolines/in^2. The primary has 1280 turns and the frequency of the applied voltage is 60 Hz. (a) What is the primary voltage? (b) How many turns are required for a secondary voltage of 480 V?

4-5. A 2400/240-V 60-Hz transformer has 1600 turns on the high-voltage side. Find (a) e_t, (b) N_s, and (c) Φ. (d) If the flux density is 1.2 T, what is the area of the core? (e) With a current of 83.3 A in the secondary, what is the kVA in the primary coil and the secondary coil?

4-6. The instantaneous flux in the core of a transformer is 516 sin 377t kilolines. The mean length of the core is 17 in and the cross section is 8 in^2. The B–H curve appears in Fig. 4-30. (a) What is the frequency of the applied voltage? (b) Find the rms voltage per turn. (c) The primary has 175 turns. What is V_p? (d) Find the rms exciting current. (e) How many turns are required in the secondary for $V_s = 120$ V? (f) What is the exciting current if 120 V are applied to the primary?

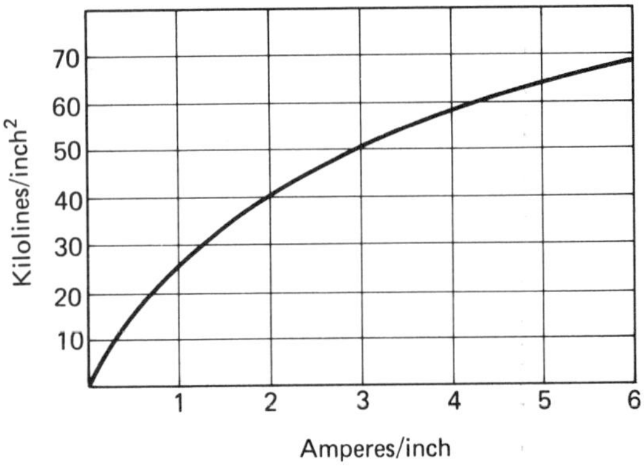

Figure 4-30 Problem 4-6

4-7. (a) What are the rated primary and secondary currents of a 500-kVA 13.8/4.16-kV transformer? (b) What are the primary and secondary currents when the transformer is delivering a load of 320 kW at 0.8 pf?

4-8. The resistance of a primary winding is 1.4 Ω and the secondary is 0.29 Ω. The primary current is 30 A and the secondary current is 60 A. (a) What is the power loss in the primary and the secondary windings? (b) Compute the total I^2R loss by referring quantities to the primary and compare the result with part (a). (c) Repeat with quantities referred to the secondary.

4-9. The primary voltage of a transformer is 440 V. Primary cu rent is 10 A and the impedance, referred to the primary side, is $0.6 + j1.375$ Ω. Find the %VR at power factors of (a) 1.0, (b) 0.8 lagging, and (c) 0.9 leading.

4-10. The preceding transformer is 2:1 step-down. Transfer all quantities to the secondary and repeat the requirements.

4-11. Compute the IZ voltage drop for Problems 4-9 and 4-10. How are they related?

4-12. A 10,000-kVA 13.2/69-kV 60-Hz transformer has the following impedances (ohms): $R_p = 0.0479$, $X_p = 0.628$, $R_s = 1.071$, and $X_s = 20.9$. (a) Prepare the simplified primary and secondary equivalent circuits. (b) Using the primary equivalent, determind the %VR at 0.866 lagging pf. (c) Using the secondary equivalent, determine the full-load efficiency at unity pf. The core loss is 60 kW.

4-13. When the transformer of Problem 4-12 is operating at 80% of rated capacity with a power factor of 0.85 lagging, what is the efficiency?

4-14. The 24-hour load cycle on a 75-kVA 4160/240-V distribution transformer is given below. The core loss is 375 W and the I^2R loss at rated load is 910 W. Find the all-day efficiency:

Hours	*%kVA$_R$*	*Power Factor*
3	150	0.92
5	100	0.85
7	75	0.9
9	25	0.8

4-15. The 20-kVA transformer of Fig. 4-31 has one 2400-V primary winding and two identical 120-V secondary windings. For each subfigure, find the curre t in each winding.

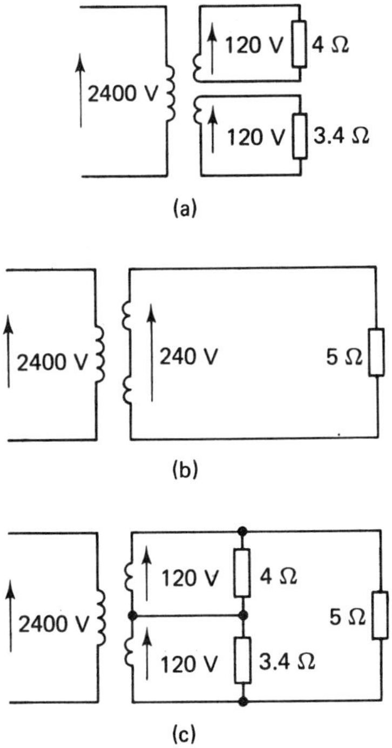

Figure 4-31 Problem 4-15

4-16. Find the current and kVA in each winding of Fig. 4-32.

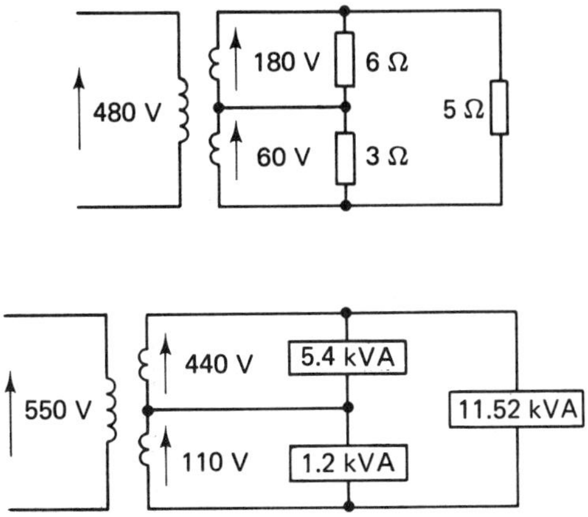

Figure 4-32 Problem 4-16

4-17. The results of open- and short-circuit tests on a 50-kVA 7200/480-V 60-Hz transformer are as follows:

	V	*I*	*P*
(Secondary) open circuit	480	5.2 A	245 W
(Primary) short circuit	157	7 A	615 W

(a) Find R_{ep} and X_{ep}. (b) Calculate the %VR at unity pf. (c) What is the I^2R loss at no load? At full load? (d) What is the core loss at no load? At full load? (e) Find the efficiency for rated load at a pf of 0.90 lagging.

4-18. Tests performed on the low-voltage side of a 250-kVA 13.8/2.4-kV transformer yield the following data:

Open circuit:

$$V = 2.4\text{ kV} \quad I = 4.6\text{ A} \quad P_{oc} = 1.5\text{ kW}$$

Short circuit:

$$V = 230\text{ V} \quad I = 104 \quad P_{sc} = 3.15\text{ kW}$$

(a) What is the voltage between the primary terminals during the open-circuit test? (b) Find the full-load efficiency and voltage regulation at 0.85 lagging pf.

4-19. A 250-μf capacitor is paralleled with the load on the transformer in Problem 4-17(e). What additional unity pf load may be added to the transformer without exceeding its ratings?

4-20. For testing equipment with high voltage and current ratings, instrument transformers normally are used to step down actual quantities and allow readings on 150-V and 5-A meters. For the short-circuit test on a transformer, the connection of instruments and instrument transformers is shown in Fig. 4-33. The turns ratio of the potential transformer (PT) is 10 : 1 and the current transformer (CT) is 150 : 1. The instrument readings during the test conducted on the primary side are 105.8 V, 4.9 A, and 33.3 W. Find R_{ep} and X_{ep}.

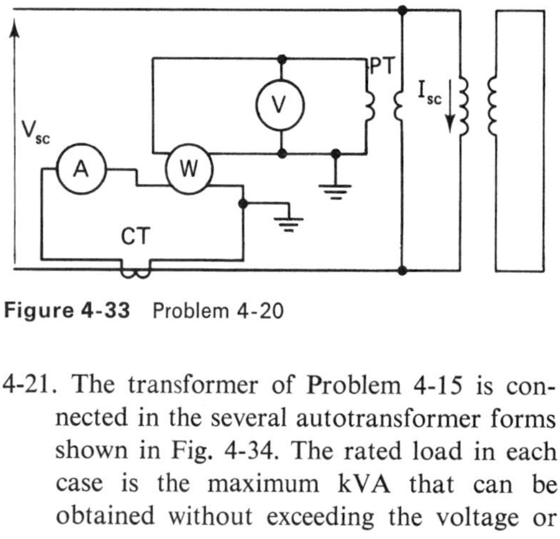

Figure 4-33 Problem 4-20

4-21. The transformer of Problem 4-15 is connected in the several autotransformer forms shown in Fig. 4-34. The rated load in each case is the maximum kVA that can be obtained without exceeding the voltage or current rating of any coil. For each, determine all of the following quantities: voltage across and current in each coil; line input and line output voltage and current; input and output kVA and the kVA in each coil; and the conducted kVA.

4-22. The variac of Fig. 4-35 has a rated line primary voltage of 230 V. The maximum allowable current in any portion of the coil is 10 A. For maximum kVA output at each of the following slider positions, find the line input current and the line output kVA: $0.2N_p$, $0.5N_p$, and $1.15N_p$.

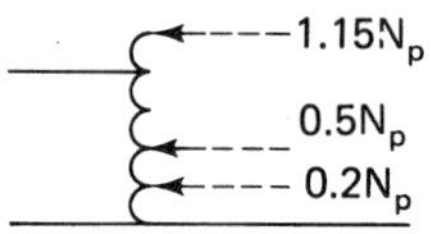

Figure 4-35 Problem 4-22

4-23. Find the line primary and line secondary voltages and currents for balanced loading of three identical 1000-kVA 13.2/69-kV transformers that are connected (a) wye–wye, (b) delta–delta, (c) wye–delta, and (d) delta–wye.

4-24. When two of the preceding transformers are connected open-delta, what are the maximum permissible line voltages and currents with balanced loading?

4-25. A three-phase transformation is required to supply a 60-kW 0.8-pf balanced load at 480 V. The available line is 4160 V. Find the required kVA and voltage ratings of the individual single-phase transformers if the

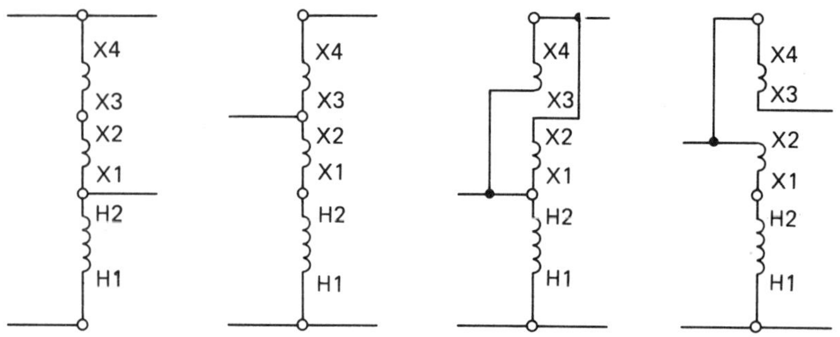

Figure 4-34 Problem 4-21

connection is (a) wye–delta, (b) delta–wye, and (c) open–delta.

4-26. Two variacs, identical to that of Problem 4-22, a e connected open-delta and supply a three-phase balanced load at a line voltage of 120 V and line current 7 A. (a) What is the $\%N_p$ position of the slider? (b) What is the current in each coil segment? (c) Compute the kVA in each segment. (d) What is the load kVA? (e) Is any portion of the load kVA *conducted* rather than transformed; if so, how much?

4-27. Large three-phase motors draw unusually high currents if connected to rated voltage when started. Frequently, open-delta or wye-connected autotransformers are employed during starting to minimize the effects of *IR* drops on the remainder of the distribution system. When started across the line, a 550-V 500-hp motor draws a current of 2900 A. Starting currents are directly proportional to the motor terminal voltage. If the 50% taps on open-delta autotransformers are used during starting, what are the motor and line currents?

4-28. Standard nameplate markings and typical winding diagrams for three-phase transformers are given in IEEE Std 432. Two of these are shown in Fig. 4-36 and defined as follows:

2400/4160 Y/2400: The winding may be Δ-connected in a 2400-V system or Y-connected in a 4160-V four-wire system with 2400 V available between line and neutral.

4160 Y/2400 × 12470 Y/7200: Windings are permanently Y-connected, but each phase has several sections that may be paralleled for 4160Y/2400 or series-connected for 12470/7200.

(a) For each diagram, show the connections for each option. (b) If the rated coil current in each case is 138.8 A, what is the rated line current and kVA for each option?

4-29. A number of three-phase transformer arrangements are used in thyristor converters for dc motor drives. One circuit, designated *delta, zigzag(wye), three-phase, single-way, three pulse* is shown in a common format in Fig. 4-37; this is a vector diagram in which the angular orientation of each heavy line identifies the phase of the winding voltage *E* —for example, the potential at R_1 with respect to N_0 is $-E\angle 0^\circ + E\angle 240^\circ$, where *E* is the rms voltage in each segment. (a) In Fig. 4-37(b), N_0 and R_1 are identified. Complete

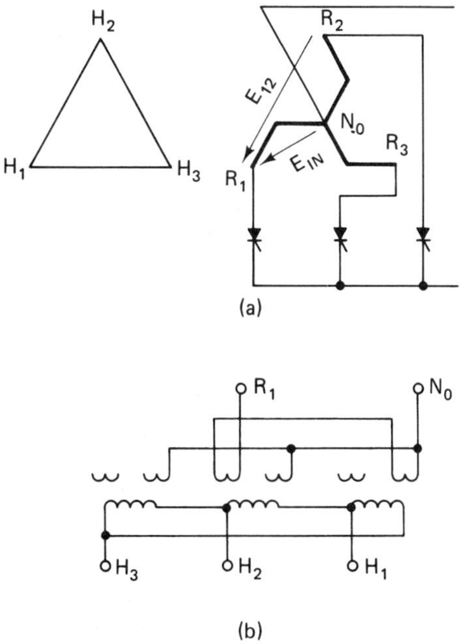

Figure 4-37 Problem 4-29

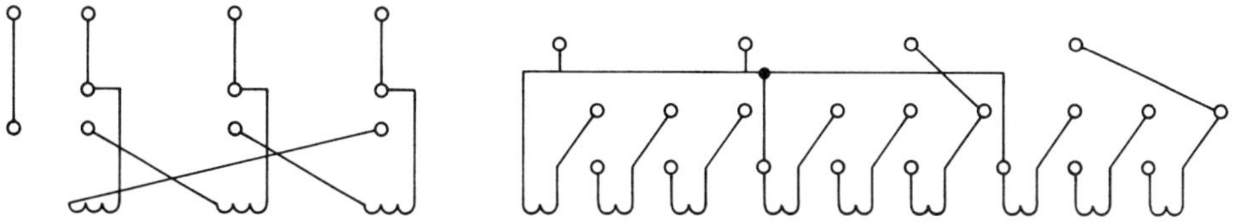

2400/4160Y/2400

4160Y/2400 × 12470Y7200

Figure 4-36 Problem 4-28

the connections and identify R_2 and R_3. (b) Find E_{1N}, E_{2N}, and E_{3N} in terms of E. (c) Find E_{12}, E_{23}, and E_{31}. (d) The turns ratio between each primary winding and one section of the secondary winding is 4 to 1. With a line primary voltage of 240 V, find E_{1N} and E_{12}.

4-30. On a 120-V base, the maximum allowable voltage at the point of connection to functional equipment is 125 V, the limit for incandescent lamps. The minimum is 108 V based upon satisfactory motor performance at 90% rated voltage. The maximum allowable at the secondary of a distribution transformer is 125 V; the minimum is based on an assumed 3-V full-load drop in the transformer and secondary distribution lines and a 6-V drop in building wiring and associated service equipment—it therefore is an open-circuit voltage of 117 V.

(a) A 600-V system is 600/120 = 5 times the 120-V base. For this system the transformer secondary tolerance is 5 × 125 = 625 V max and 5 × 117 = 585 V min. What are the tolerances for 480-V and 240-V systems?

(b) A utility distribution system is shown in Fig. 4-38. By tap-changing-under-load the substation transformer maintains a constant 13.2 kV at its secondary. Transformer A (single phase), located near the substation and only slightly affected by drops in the primary distribution lines, has a primary voltage that varies between 13.2 and 13.09 kV; the transformer is 13.8 kV/480 V with ±2.5 and ±5% taps on the primary side (i.e., the usable options on the primary side are 0.95, 0.975, 1.0, 1.025, and 1.05N_p). Determine all the taps that can be used without exceeding the secondary tolerance. (*Note:* Neglect voltage regulation in the transformer and assume that the rated voltage ratio applies at both no-load and full load.)

(c) Transformer B is one of three Y-connected transformers at a remote distance from the substation. The line voltage at this point varies from 13.0 to 12.2 kV. The transformer is 7200/240 V with ±2.5 and ±5% taps. Determine the usable taps at this location.

Note: The transformers referred to in Problems 4-31 to 4-34 are 37.5 kVA 2400/208 V with equivalent primary impedances as follows:

transformer A: $Z_{ep} = 1.336 + j2.5\ \Omega$

transformer B: $Z_{ep} = 1.229 + j2.3\ \Omega$

transformer C: $Z_{ep} = 0.685 + j2.75\ \Omega$

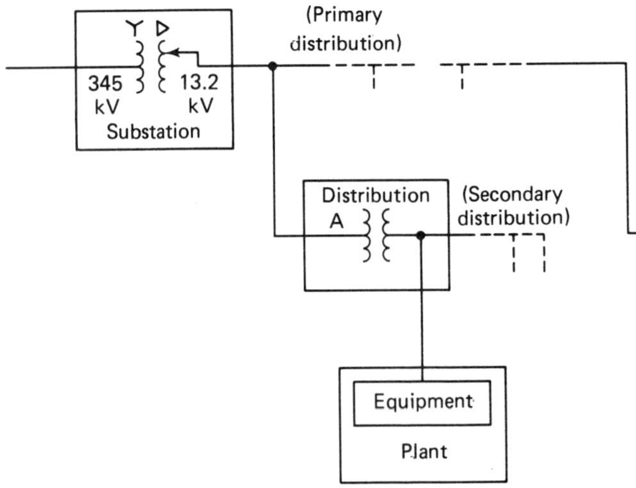

Figure 4-38 Problem 4-30

4-31. Transformers A and B are paralleled. (a) Under load, which one will carry the larger portion of the line current? (b) With rated current in the transformer of part (a), what is the current in the other transformer and what is the line kVA? (Assume identical turns ratios.)

4-32. Transformers A and C are paralleled. Repeat the requirements of Problem 4-31.

4-33. Transformers B and C are paralleled. Repeat the requirements of Problem 4-31.

4-34. The turns ratios of transformers A and B are 11.44 and 11.58, respectively. (a) What is the circulating current at no load? (b) What is the maximum 0.8-pf load that can be delivered without exceeding rated current in either transformer? (c) At this load, what is the kVA in each transformer?

4-35. Transformer quantities frequently are given on a per unit (pu) basis for two reasons: (1) impedances usually lie in the range 0.005 to 0.2 pu, and (2) with several transformed voltages at various points in a system, on a per unit basis the system can be represented by a single-voltage composite circuit for analysis. Base quantities upon which pu values are determined are as follows:

$$\text{kVA}_{\text{BASE}} = \text{kVA}_{\text{rated}}$$

$$V_{\text{BASE}} = V_{\text{rated}}$$

$$I_{\text{BASE}} = \frac{\text{kVA}_{\text{BASE}}}{V_{\text{BASE}}}$$

$$Z_{\text{BASE}} = \frac{V_{\text{BASE}}}{I_{\text{BASE}}}$$

Per unit quantities are the ratio of actual quantities to base quantities:

$$I_{e(\text{pu})} = \frac{I_e}{I_{\text{BASE}}}$$

$$R_{e(\text{pu})} = \frac{R_e}{Z_{\text{BASE}}}$$

(a) Given a 3-kVA 230/115-V transformer, find the base quantities (1) with $V_{\text{BASE}} = V_p$, and (2) $V_{\text{BASE}} = V_s$.
(b) If the per unit transformer impedance is $0.009 + j.015$, find (1) $(R + jX)_{ep}$ using the primary base impedance, and (2) $R_{es} + jX_{es}$ using the secondary base impedance. (3) Are the results of (1) and (2) related by the turns ratio?
(c) Following are the results of open- and short-circuit tests on the transformer:

Open:	230 V	47 W	1.05 A
Short:	5.4 V	40 W	13 A

In per unit quantities, what is (1) I_e, (2) V_{sc}, (3) R_e, (4) X_e, (5) P_{oc} and P_{sc}?
(d) Using the pu quantities determined in part (c), determine the open- and short-circuit quantities that would be obtained if testing was done on the secondary side. Verify the accuracy by applying the turns ratio to primary data.

4-36. Two 3-kVA 230/120-V transformers are connected Scott-T to transform 230 V three-phase to two-phase. (a) Find the two-phase line voltage. (b) Without exceeding rated currents in the transformers, what is the load kVA?

4-37. An induction voltage regulator has one 240-V primary coil and two 60-V secondary coils which may be connected in series or parallel. There are two potential uses for the regulator: (a) continuously variable secondary voltage without conductive coupling between the primary and secondary, and (b) continuously variable secondary voltage when connected in autotransformer configuration. For each, determine the possible ranges of secondary voltages.

5

Polyphase Synchronous Machines

A synchronous machine, generator or motor, is one in which the average speed is directly proportional to the frequency of the electrical system to which it is connected. The many generators supplying commercial power throughout the country are synchronous machines. Synchronous motors are employed in various industries.

A synchronous machine may be single- or polyphase. Single-phase motors are discussed in Chapter Seven. The predominant machine is three-phase; this chapter focuses on three-phase machinery, but the principles apply to any number of phases.

SECTION 5-1. SYNCHRONOUS GENERATOR (ALTERNATOR)

5-1. GENERATED VOLTAGE

In developing the concept of power generation in the dc machine, it was noted that the generated voltage is not dc but ac—the commutating process permitted dc extraction. Except for the number of represented conductors, Fig. 5-1(a) is a reproduction of Fig. 2-15; the armature is rotating clockwise at a uniform angular velocity, the field is static, and the polarity of induced voltages in the conductors changes as each crosses the stationary quadrature axis.

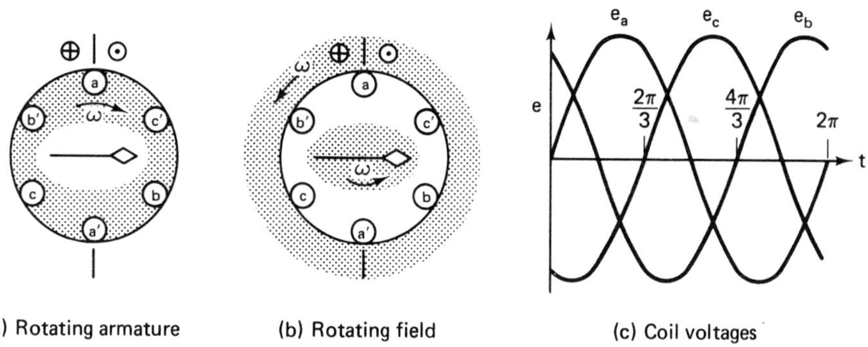

Figure 5-1 Induced Voltages

In Fig. 5-1(b), the rotating and stationary elements are interchanged. With the field and quadrature axis rotating counterclockwise and the armature stationary, there is no electromagnetic difference between the two figures. Voltages induced in coils *a*, *b*, and *c* are sinusoidal and identical in amplitude and frequency but separated by 120 electrical degrees. If the ends of the coils were connected to slip rings, three-phase power could be extracted through brushes [Fig. 5-1(c)].

Obviously, construction of a machine with the interior static and the exterior rotating is undesirable, but the electromagnetic relationship of Fig. 5-1(b) can be maintained in either of two alternates: (1) returning to Fig. 5-1(a) but connecting the ends of each coil not to commutator segments but to slip rings, or (2) placing the armature windings in the stator and the field poles on the rotor. The latter arrangement is preferable for a number of reasons:

- As power increases, it becomes increasingly difficult to transfer it through sliding contacts. Field power is but a few percent of the armature power. A modern alternator with a rating of several hundred MVA may require excitation currents of 3000 to 4000 A; introducing the field power through slip rings itself is troublesome, and extraction of megawatts of power from a rotating armature is impractical.
- The armatures of larger alternators are force cooled with circulating gas or liquids. Force cooling a large rotating armature would be a formidable problem.
- The bulk and weight of the armature windings are substantially greater than the windings of the field poles. The size of the overall machine and the mechanical forces acting on the rotor are less with a rotating field.
- Windings must be braced against electromagnetic and, if rotating, centrifugal forces. Bracing is easier when the armature windings are in the stator.

Typical construction is illustrated in Fig. 5-2(a), which shows one quadrant of a four-pole generator with salient poles. For a two- or four-pole machine, the cylindrical rotor of Fig. 5-2(b) may be employed, but many alternators have too many poles to permit the cylindrical form.

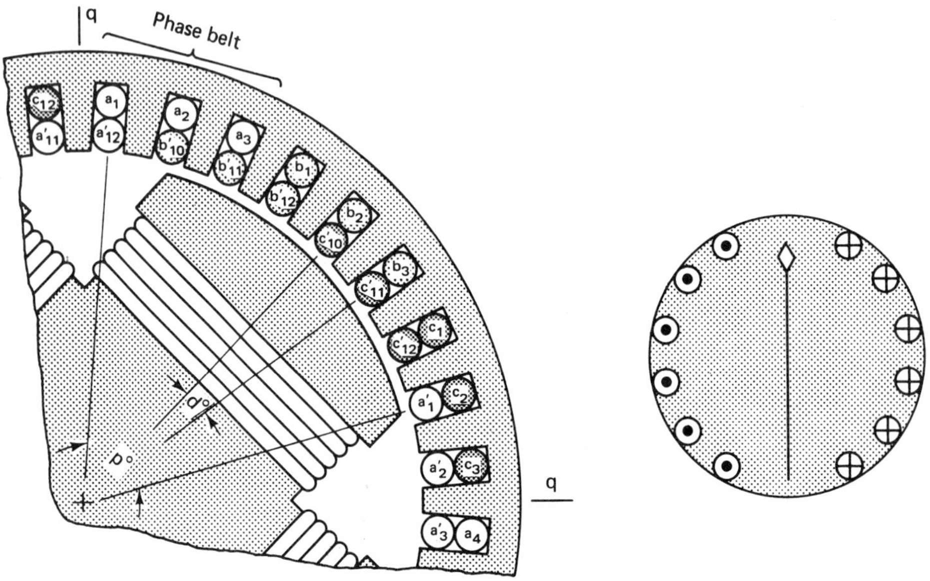

(a) Generator with salient-pole rotor (b) Cylindrical rotor

Figure 5-2 Synchronous Generator Construction

In Fig. 5-1 the concentration of the armature windings in six slots is a poor use of the available space; the normal distributed winding appears in Fig. 5-2. Before discussing this winding, a few definitions are in order.

The *pole pitch* is the distance between corresponding points (e.g., the centerline) of two adjacent poles. Often, the pole pitch is expressed in terms of the number of slots it spans. In electrical degrees, the pitch is 180°. The q axes on the illustration are separated by one pole pitch; from their location, it is evident that the pitch is nine slots.

The *coil pitch* is the distance between the two sides of a coil. Although often expressed as a per cent of the pole pitch, for our purposes it is more convenient to express it in electrical degrees. On the illustration, the angle $p°$ lying between the center of sides a_1 and a_1' is the coil pitch. Since this is a four-pole machine, each mechanical degree is 2 electrical degrees; the mechanical separation of a_1 and a_1' is 70°, but $p°$ is twice this, or 140 electrical degrees.

In this discussion, the *slot span* is the angle in electrical degrees between the centers of adjacent slots. On the illustration the angle is shown as $d°$; mechanically, the span is 10°, but $d°$ is 20 electrical degrees.

A *phase belt* is a group of adjacent coils that are connected in series to form one section of a phase of the winding. In the illustration, the machine has 36 slots with two coil sides per slot and, hence, 36 uniformly distributed coils. There are $\frac{36}{4}$, or 9, coils per pole. These coils are evenly divided among the number of phases for which the machine is wound. For this three-phase winding there are $\frac{9}{3}$, or 3, coils per phase belt. In the sector between the q axes, a_1–a_2–a_3 form one belt of phase A; an additional phase A belt lies in each of the second, third, and fourth quadrants. Phase A has four phase belts, connected in series to provide one of the three electrical phases of the machine.

One phase B belt and one phase C belt also lie in the representative quadrant.

In Fig. 5-1, the coil pitch is 180°, a *full-pitch* winding. A *fractional-pitch* winding has a pitch less than 180°; fractional pitch reduces the effects of distorting harmonics and thereby produces a truer sinusoidal waveform, shortens the ends of the winding, and simplifies the physical construction. Fractional pitch also reduces the magnitude of the voltage induced in a coil, but the disadvantage of a small voltage reduction is outweighed by the realized advantages. In determining the magnitude of the generated voltage, pitch must be considered by introduction of a *pitch factor*, k_p.

The voltages across each belt of a phase winding are identical in magnitude and phase angle and add arithmetically to provide the machine phase voltage. However, the angular separation of coils within the belt results in a similar angular separation between the coil voltages; the belt voltage is the *vector* sum of the coil voltages. This, too, must be recognized in an equation for E_p, the phase voltage, by application of a *distribution factor*, k_d.

In a given machine, k_p and k_d are constant multipliers to be applied to the phase voltage that is computed without regard to pitch and distribution. Combined, $k_p k_d = k_w$, the *winding factor*.

In Paragraph 2-1 it was shown that the flux through a loop rotating in a uniform static field is $BA \sin \omega t$. The same relationship exists for a static loop and a uniform rotating field. The product BA is Φ_m, the maximum flux through the loop; thus, the flux is $\Phi_m \sin \omega t$ and, from equation 1-2, the effective induced loop voltage is $4.44 f \Phi_m$. For a *concentrated, full-pitch* winding consisting of n_c series-connected coils per phase, each coil having n_t turns, the effective winding voltage is $4.44 N f \Phi$, in which $N = n_c n_t$. Φ, although not subscripted, is the maximum flux —that is, the flux in each pole.

Amended to account for coil pitch and distribution,

$$\boldsymbol{E_p = 4.44 N f \Phi k_w} \tag{5-1}$$

The frequency of the induced voltage in a machine with P poles was developed in Section 2-1:

$$\begin{aligned} f &= \frac{\omega P}{4\pi} \quad \text{[SI]} \\ &= \frac{PS}{120} \quad \text{[US]} \end{aligned} \tag{5-2}$$

A remaining requirement is the mathematical definition of k_w $(= k_p k_d)$.

$\boldsymbol{k_p}$. The pitch factor is the ratio of the resultant voltage induced in a fractional-pitch coil to that of a comparable full-pitch coil. In Fig. 5-3, e'_c is the effective voltage induced in each coil side. For a full-pitch coil, $e_c = 2e'_c$. A circle with radius e'_c is the locus of the tips of the coil-side voltage vectors for any pitch. The

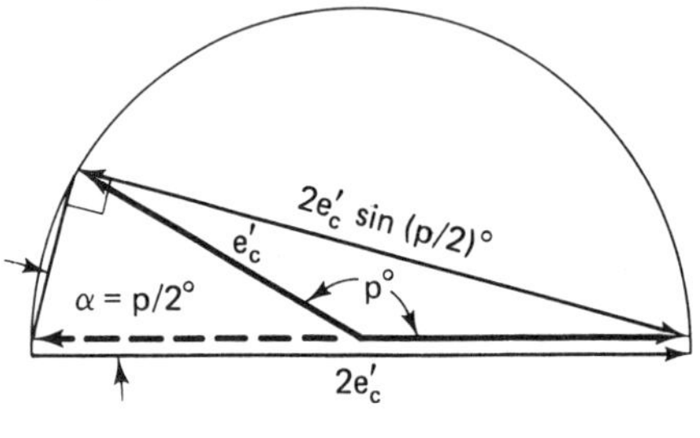

Figure 5-3 Pitch Factor

coil sides, separated by $p°$, intercept an arc of $p°$; angle α, formed by two chords, intercepts the same arc and is $(p/2)°$. From the right-triangle relationship, the resultant voltage is $2e'_c \sin(p/2)°$. Therefore,

$$\boldsymbol{k_p} = \frac{2e'_c \sin(p/2)°}{2e'_c} = \mathbf{sin}\,\frac{\boldsymbol{p}°}{\mathbf{2}} \tag{5-3}$$

$\boldsymbol{k_d}$. The distribution factor is the ratio of the ***vector*** sum of the voltages induced in the coils of a phase belt to the **arithmetic** sum of these coil voltages.

In Fig. 5-4, n vectors, each with the magnitude e_c but with the phase difference $d°$, add to produce the resultant shown as $\sum \bar{e}_c$. The distribution factor is $\sum \bar{e}_c/ne_c$.

A circle is drawn through the connecting points of the vectors. Radii to the head and tail of any vector form an isosceles triangle with base angle α. The

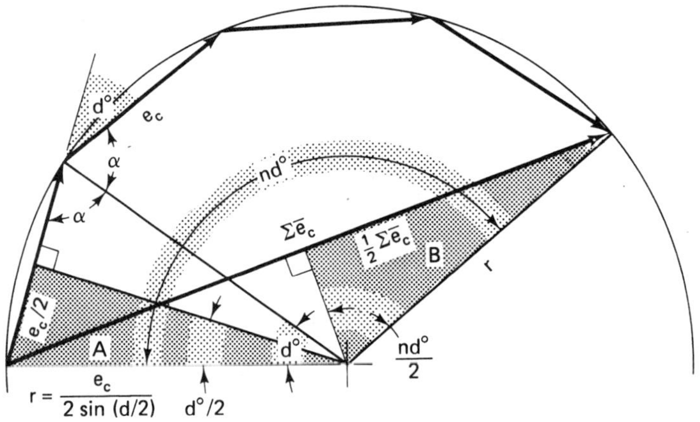

Figure 5-4 Distribution Factor

extension of the first vector forms the angle d° with the second vector; from the straight-line relationship, $2\alpha = 180 - d^\circ$. If a perpendicular now is drawn from the apex to the center of the base, triangle A is defined with the base $e_c/2$ and the angle $(d/2)^\circ$ as shown. From right-triangle relationships, the radius r is $(e_c/2)/\sin(d/2)^\circ$.

The arc subtended by the sum of the vectors is nd°; a perpendicular from the center of the circle to the resultant therefore forms the angle $(nd/2)^\circ$ shown in triangle B. From this triangle,

$$\tfrac{1}{2}\sum \bar{e}_c = r \sin(nd/2)^\circ = [(e_c/2)\sin(nd/2)^\circ]/\sin(d/2)^\circ$$

Therefore,

$$k_d = \frac{\sum \bar{e}_c}{ne_c} = (2)\frac{(e_c/2)\sin(nd^\circ)/2}{ne_c \sin(d^\circ/2)}$$

$$= \frac{\sin(nd^\circ/2)}{n \sin(d^\circ/2)} \tag{5-4}$$

It is helpful to note in both Figs. 5-3 and 5-4 that the magnitude of the resultant is less than the arithmetic sum of the vectors. Consequently, neither k_p, k_d, nor their product, k_w, can be greater than unity. Awareness of this limitation serves as one check on the correctness of a computed factor. It is unlikely that k_w will be lower than 0.85.

Example 5-1: A three-phase synchronous generator has 12 poles, 144 slots, and 144 eight-turn coils. It is driven at a speed of 600 r/min. The coil span is 10 slots. Rotor flux is 0.026 Wb/pole. Find: (a) frequency, (b) number of series turns per phase, (c) d°, (d) p°, (e) k_p, (f) number of coils per phase belt, (g) k_d, (h) k_w, (i) generated voltage per phase, and (j) V_o, the open-circuit terminal voltage, if the phases are wye-connected.

SOLUTION:

(a) $f = PS/120 = (12)(600)/120 = 60$ Hz.

(b) N, the series turns per phase, is the total number of turns in the armature divided by the number of phases (ϕ).

$$N = \frac{n_c n_t}{\phi} = \frac{(144)(8)}{3} = 384 \text{ turns per phase}$$

(c) In the armature there are $180P$ electrical degrees. Therefore,

$$d^\circ = \frac{180P}{n_s} = \frac{(180)(12)}{144} = 15^\circ$$

(d) Letting Y_s represent the coil span in slots,

$$p^\circ = Y_s d^\circ = (10)(15) = 150^\circ$$

(e) $K_p = \sin\left(\frac{p}{2}\right)^\circ = \sin\left(\frac{150}{2}\right) = 0.966.$

(f) $n = \frac{n_c}{P\phi} = \frac{144}{(12)(3)} = 4$ coils per phase belt.

(g) $k_d = \frac{\sin (nd/2)^\circ}{n \sin (d/2)^\circ} = \frac{\sin [(4)(15)/2]}{4 \sin (15/2)} = 0.958.$

(h) $k_w = k_p k_d = (0.966)(0.958) = 0.925.$

(i) $E_p = 4.44 N f \Phi k_w = (4.44)(384)(60)(0.026)(0.925) = 2460$ V.

(j) $V_o = \sqrt{3}\, E_p = \sqrt{3}\,(2460) = 4260$ V.

In a three-phase machine with uniform distribution of the slots, there are three phase belts per pole. Consequently, each belt spans 60° and this is the product nd°.

From the frequency equation, various numerical combinations of poles and driven speed will produce the standard 60 Hz. Rearranging the equation,

$$S = \frac{120f}{P} = \frac{(120)(60)}{P} = \frac{7200}{P}$$

P may be any even number; for standard alternators, the range is 4 to 52, with corresponding speeds ranging from 1800 to 138 r/min. An alternator powered by a steam turbine may have a speed of 1800 r/min, whereas the units at Hoover Dam are driven at 180 r/min.

5-2. ADVANTAGES OF POLYPHASE SYSTEMS

Suppose that the generator of Example 5-1 was connected for single-phase rather than three-phase power with no other change in the machine. How would the characteristics differ?

In the voltage equation, only N and k_w will change. For a single-phase machine, N increases by a factor of 3. In the winding factor, k_p is unchanged, but k_d becomes {sin [(12)(15)/2]/(12) sin (15/2)}. The winding factor is 0.617. Using ratios,

$$\frac{E_{1\phi}}{E_{3\phi}} = \frac{(Nk_w)_{1\phi}}{(Nk_w)_{3\phi}}$$

$$E_{1\phi} = \frac{3N}{N}\frac{0.617}{0.925}(2460) = 4920 \text{ V}$$

The single-phase line voltage is greater than the three-phase voltage, but this is of no consequence because of the ease with which an ac voltage can be transformed. However, any difference in *rated kVA* will be significant. Since it was stipulated that the only difference between the machines is the number of phases, the rated winding current is the same for both machines: the kVA ratings are

$$kVA_{1\phi} = V_R I_R \times 10^{-3} = (4920) I_R \times 10^{-3} = 4.92 I_R$$

$$kVA_{3\phi} = \sqrt{3}\, V_R I_R \times 10^{-3} = \sqrt{3}(4260) I_R \times 10^{-3} = 7.38 I_R = 1.5 kVA_{1\phi}$$

This result illustrates the first advantage of polyphase windings: *greater power may be extracted from a machine without changing the physical components.*

A second advantage of polyphase windings is *constancy of the developed power.* When power is pulsating, both the ability of the machine to withstand mechanical stresses and the system stability are endangered. Mechanical pulsations at a critical frequency can cause the prime mover to oscillate, leading to electrical oscillations. Such oscillations have occurred in the commercial power system as a result of coincidental load changes at the critical frequency.

At unity pf, currents are in phase with the phase voltages. The power curves for the single- and three-phase systems are shown in Fig. 5-5(a). In the single-phase system, $P = (V_m \sin \omega t)(I_m \sin \omega t) = (V_m I_m)(\sin^2 \omega t) = 0.5 V_m I_m (1 - \cos 2\omega t)$. This is a cosine curve with respect to the average value, $0.5 V_m I_m$, pulsating at a frequency of 120 Hz in the standard 60-cycle system.

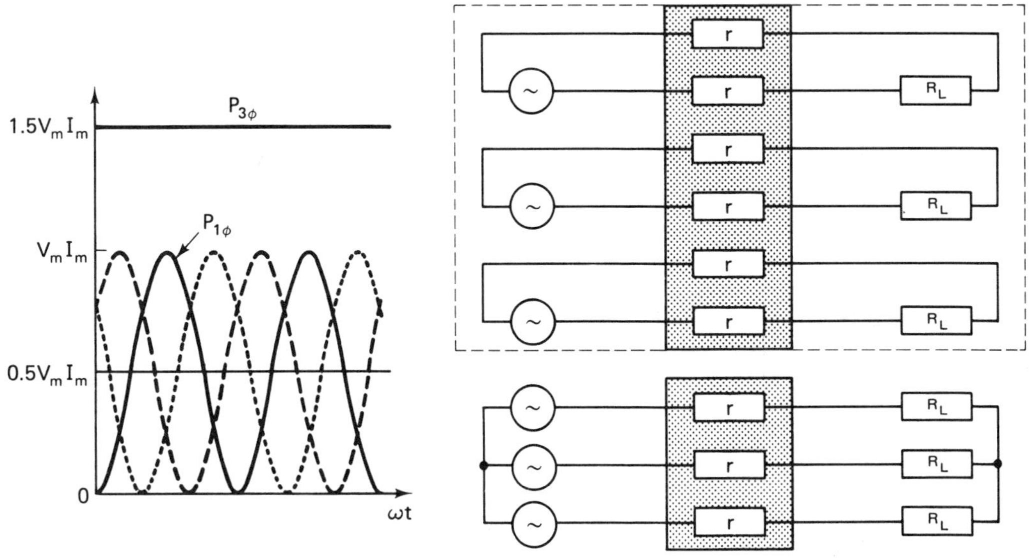

(a) Power (b) Transmission lines

Figure 5-5 Comparison of Single- and Three-Phase Systems

With three-phase power,

$$P_{3\phi} = \frac{V_m I_m}{2}\{(1 - \cos 2\omega t) + [1 - \cos(2\omega t + 240)] + [1 - \cos(2\omega t + 480)]\}$$

But

$$\cos(2\omega t + 480) = -0.5 \cos 2\omega t - 0.866 \sin 2\omega t$$

and

$$\cos(2\omega t + 240) = -0.5 \cos 2\omega t + 0.866 \sin 2\omega t$$

Substituting these latter expressions,

$$P_{3\phi} = 0.5V_mI_m[(1 - \cos 2\omega t) + (1 + 0.5 \cos 2\omega t + 0.866 \sin 2\omega t) + (1 + 0.5 \cos 2\omega t - 0.866 \sin 2\omega t)]$$

$$= 1.5V_mI_m$$

The three-phase power is *constant*; it does not pulsate.

A third advantage of a polyphase system is *reduction of the number of lines required to transmit a given amount of power.* In Fig. 5-5(b), the generated voltage in each phase of the polyphase generator is identical to that of each single-phase generator. The load resistances are identical in both systems and, therefore, will draw identical currents and power. But the number of transmission lines and the transmission line resistance, voltage drop, and power loss have been halved in the three-phase system.

A fourth, important advantage of polyphase power is its ability to create a *revolving flux.*

5-3. REVOLVING FLUX

The output of a synchronous generator is applied to an identical stator in Fig. 5-6(a). For the purposes of this discussion:

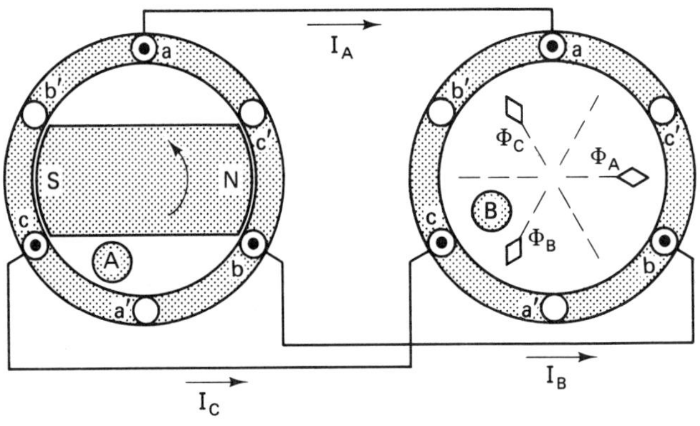

(a)

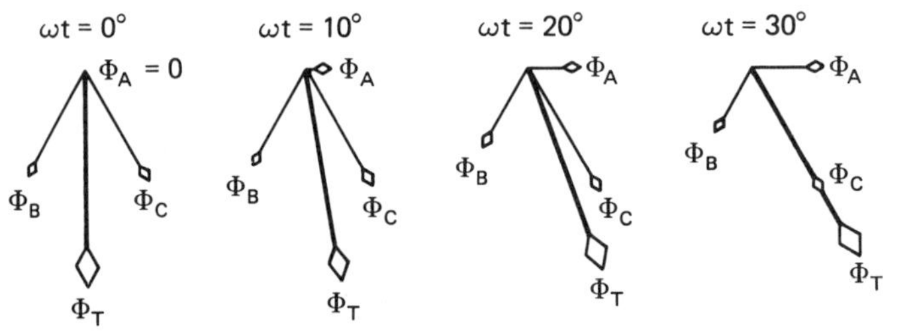

(b)

Figure 5-6 Revolving Flux

- The positive current directions are shown. These are not instantaneous directions; at any moment, either one or two currents would be in the opposite direction from those on the sketch.
- It is assumed that the currents are in phase with the generated voltages. Let I_A be $I_m \sin \omega t$; for the indicated direction of rotation, I_c lags I_A by 120° and, therefore, is $I_m \sin (\omega t - 120)$; similarly, I_B is $I_m \sin (\omega t - 240)$.
- In stator B the winding currents produce fluxes that always lie on the indicated axes but whose strength varies sinusoidally. Neglecting saturation, flux is proportional to and in phase with the current. Letting Φ represent the peak value of the flux,

$$\Phi_A = \Phi \sin \omega t \,\underline{/0^\circ}$$
$$\Phi_B = \Phi \sin (\omega t - 240) \,\underline{/240^\circ}$$
$$\Phi_C = \Phi \sin (\omega t - 120) \,\underline{/120^\circ}$$

The positive flux directions are shown in the figure.

The phase fluxes in machine B will combine vectorially to produce a resultant flux. To determine this resultant, each flux must be (1) expressed in terms of ωt by means of the relationship $\sin (\omega t - \alpha) = \sin \omega t \cos \alpha - \cos \omega t \sin \alpha$, and (2) converted from polar to rectangular form to permit summation.

$$\begin{aligned}
\Phi_A &= \Phi \sin \omega t\underline{/0} = \Phi(\sin \omega t)(1 + j0) \\
&= \Phi \sin \omega t + j0 \\
\Phi_B &= \Phi \sin (\omega t - 240)\underline{/240} = \Phi(-0.5 \sin \omega t + 0.866 \cos \omega t)(-0.5 - j0.866) \\
&= \Phi[(0.25 \sin \omega t - 0.433 \cos \omega t) + j(0.433 \sin \omega t - 0.75 \cos \omega t)] \\
\Phi_C &= \Phi \sin (\omega t - 120)\underline{/120} = \Phi(-0.5 \sin \omega t - 0.866 \cos \omega t)(-0.5 + j0.866) \\
&= \Phi[(0.25 \sin \omega t + 0.433 \cos \omega t) + j(-0.433 \sin \omega t - 0.75 \cos \omega t)]
\end{aligned}$$

Summing Φ_A, Φ_B, and Φ_C;

$$\begin{aligned}
\Phi_T &= \Phi[(1.5 \sin \omega t) + j(-1.5 \cos \omega t)] \\
&= 1.5\Phi\underline{/\omega t - 90}
\end{aligned}$$

Figure 5-6(b) shows the flux relationships for $\omega t = 0$, 10, 20, and 30°.

Interpreting this result (1) the strength of the flux is constant with a value of 1.5Φ, (2) the flux is revolving at an angular velocity ω, and (3) the direction of rotation is the same as that of the field of machine A, but the position of the flux lags the rotor flux of machine A by 90°. However, these results must be related to the assumption on which the calculations were based—specifically, that the currents are in phase with the generated phase voltages.

If the calculations are repeated with the current lagging the induced voltage by the angle α, the magnitude and velocity of the rotating flux are unchanged, but it will lag the rotor flux by $90° + \alpha$.

Revolving flux is an essential requirement for the polyphase synchronous and induction motors, discussed in Section 5-2 and Chapter Six.

5-4. EQUIVALENT CIRCUIT

Rotating flux also is significant in a synchronous generator. The windings of machines A and B of Fig. 5-6(a) are identically located and carry the same currents. Therefore, the windings of machine A also must develop a revolving flux, identical to that in the windings of machine B! This is armature reaction.

There are two ways of viewing the effect of armature reaction upon the voltages generated in the machine: (1) visualizing both the field flux and the armature reaction flux as generating independent voltage components which combine vectorially to provide the phase voltage V_p, or (2) vectorial combination of the two fluxes into a single resultant, the latter responsible for the generation of V_p. Either approach yields the same result. A fundamental relationship in either method is that each voltage always lags its flux by 90°. Armature reaction flux is in phase with the line current.

In Fig. 5-7, the pole flux (Φ_p) induces E_p; armature reaction flux (Φ_a) induces E_a; and the resultant (Φ_r) induces V_p. One caution is noted: I leads V_p in this sketch—that is, it is leading power factor loading. Also, the coincidence of I and E_p is unique to a particular power factor angle, θ.

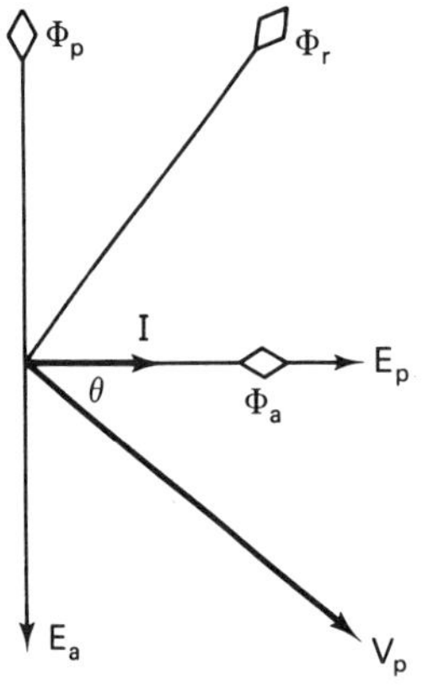

Figure 5-7 Flux and Voltage Vectors

In the circuit of Fig. 5-8, with the switch open the vector diagram is that shown in Fig. 5-8(b, 1). When the switch is closed the inductive load causes the current to lag the voltage. The armature reaction flux, in phase with the current, produces a voltage that vectorially subtracts from E_p and reduces the terminal voltage to V_p. When V_p is separated into the voltages across the load resistance and reactance, E_a clearly is in phase with V_x—that is, the armature reaction is inductive in character and can be represented by an inductor, X_a.

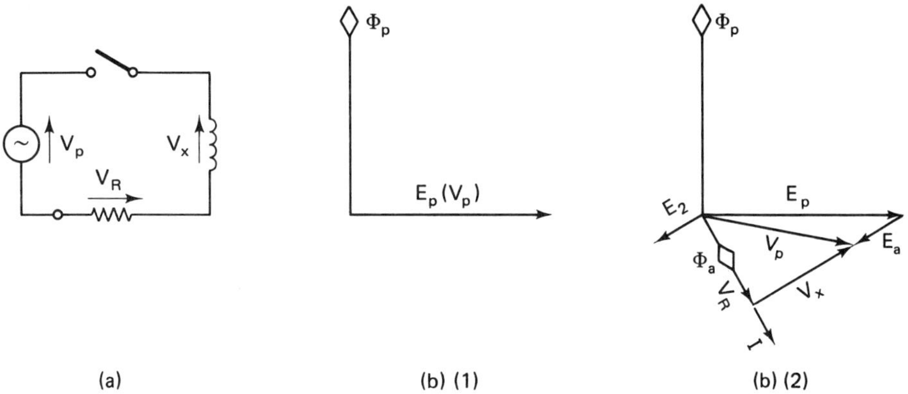

Figure 5-8 Armature Reaction

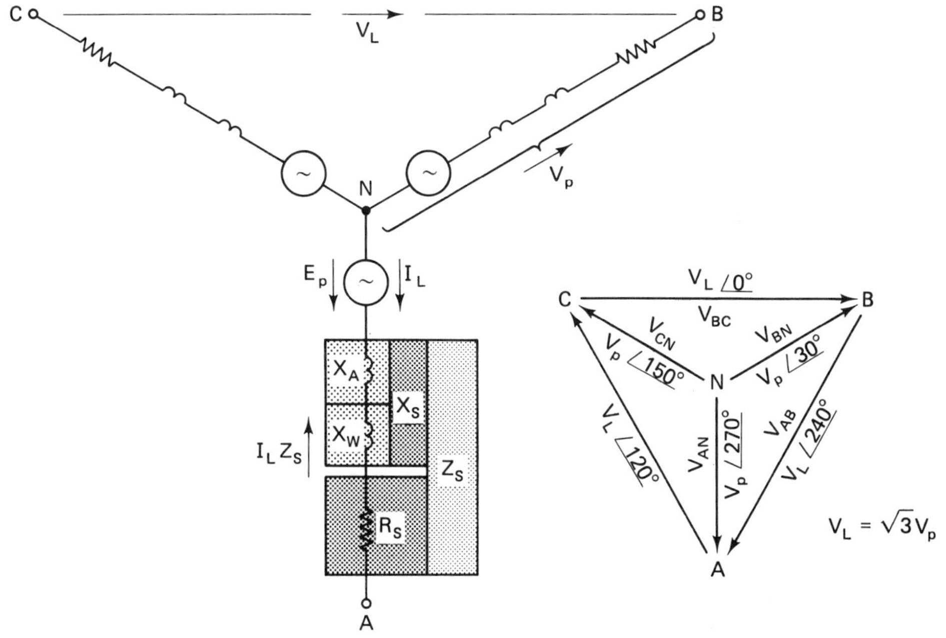

Figure 5-9 Alternator Circuit

The schematic of a three-phase synchronous generator is shown in Fig. 5-9.

The generator is shown wye-connected. Unless there is an interest in specific winding quantities, it is mathematically convenient to assume that the machine is wye-connected even though the connection may be delta. Normally, the machine is tested for winding impedance and the results used in determining voltage regulation and efficiency without concern for the actual connection; the computations, based on a wye, use the wye equivalents of an actual delta.

- The voltage generated in each winding by the field flux is E_p.
- Each winding has an inherent resistance, R_s, and reactance, X_w, due to leakage flux. The latter is added to X_a, the reactive equivalent of armature reaction, to produce the equivalent synchronous reactance, X_s. The synchronous impedance is $Z_s = R_s + jX_s$. In most machines, X_s is much larger than R_s. Also, X_w is small in contrast to X_s. Therefore, in analyzing the machine for characteristics other than efficiency it often is assumed that $Z_s = X_s$.
- When delivering a load, line current flows in the phase winding. An impedance drop, $I_L Z_s$, occurs in each winding and the terminal phase voltage is $V_p = E_p - I_L(R_s + jX_s)$.
- At the terminals, line and phase voltages are related by the voltage triangle.

5-5. EXCITATION SYSTEMS

A source of dc current is needed to excite the field of the generator. The amount of dc power required by the field varies with not only the kVA rating but also the driven speed; because the number of poles required for 60 Hz increases as the speed decreases, low speeds demand more excitation power. Typical requirements for a 125-kVA alternator are 2 kW at 1800 r/min and 9 kW at 180 r/min; a 7500-kVA alternator takes about 40 kW at 1800 r/min and 85 kW at 180 r/min. For standard machines, excitation voltages are 62.5, 125, 250, 375, and 500 V.

Concurrent with the need for dc excitation is a requirement for regulation of the voltage. In an industrial plant, machines are matched, subjected to a reasonably uniform loading patterm, and monitored by skilled technicians. But in the commercial power system, loading is determined to a considerable degree by customers rather than plant personnel, and although control centers monitor and alter power flow in response to changing demands, severe changes such as the taxing loads of a hot, summer afternoon and resumption of the industrial workday following an idle weekend impose variable requirements on the alternators. Additionally, numerous widely separated generators constantly are operating in parallel to supply a distribution grid covering a broad geographical area, and the alternator voltage must be consistent with the system voltage. The inherent voltage regulation of the synchronous generator is poor largely because of armature reaction.

For many machines, field power is supplied by a dc generator, the *exciter*, which itself may be separately excited by a smaller dc generator, the *pilot exciter*. The exciters may be driven by separate motors or their rotors may be on the same shaft as the field rotor of the alternator. In either case, the dc power must be extracted from the exciter armature through a commutator and brushes and then introduced to the alternator through brushes and slip rings. A system of this type is illustrated in Fig. 5-10; the pilot exciter is an amplidyne. Note that there are three rotating contacts in this system: amplidyne armature—exciter field; exciter

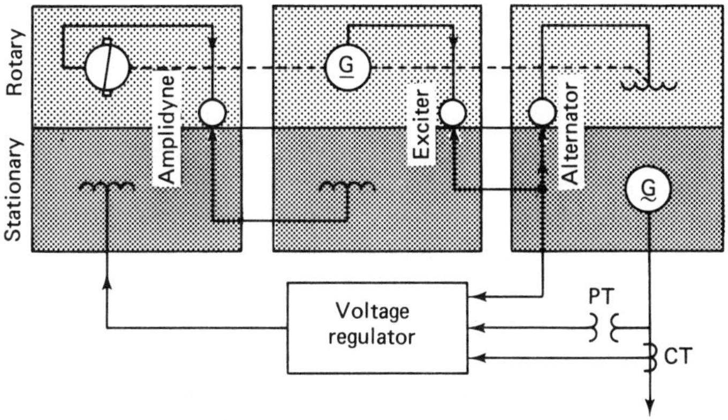

Figure 5-10 DC Generator Exciter

armature—external terminal; external terminal—alternator field. Both line voltage and line current are monitored, the former by a potential transformer (PT) and the latter by a current transformer (CT). The voltage regulator contains sensing devices as well as components such as magnetic amplifiers and transistors; it combines the functions of stabilizing the system when abrupt changes occur, compensating for the effect of line currents in the synchronous generator, and maintaining the line voltage at the desired level.

The brush contacts in this system are not ideal and are impractical for a synchronous generator with a rating above a few hundred MVA. The system can be somewhat improved by using an alternator–exciter rather than a dc generator–exciter. While direct exchange of an alternator for the dc generator in Fig. 5-10 does not reduce the number of sliding contacts, the output of the amplidyne is introduced to the exciter field through slip rings, and the commutator of the dc exciter is eliminated. The alternator–exciter armature is stationary, and its power must be rectified before transfer to the field of the main alternator.

When an adequate ac supply is already available at the plant site, it may be rectified to supply the main generator field and thereby eliminate the need for exciters.

One recently developed system is roughly comparable to the dc self-excited compound generator; in addition to the regular armature windings of the alternator, a separate winding of relatively few coils is added in some of the slots. The voltage generated in each phase of these windings is connected through a linear reactor to one of three windings on a phase excitation transformer; the linear reactor, in effect, converts the excitation windings of the alternator from a voltage to a current source.

A second winding on each excitation transformer is an output winding; the output windings of the three phase-transformers are interconnected along with power rectifiers to provide a dc output for the alternator field. Under open-circuit conditions, all the field power is supplied by the special windings in the armature.

The regular armature windings of the alternator are wye-connected; however, in forming the wye connection, the winding lead of each phase first passes through a window of its corresponding phase excitation transformer, thereby creating the third transformer winding, a single-turn winding sensitive to line current. Under load, this winding provides additional power to the output windings, power that is proportional to the line current. Since all the field power is developed in the stationary armature, slip rings are required to transfer it to the field.

Many large, modern synchronous generators have brushless excitation systems. One such system is illustrated in Fig. 5-11. The pilot exciter is a small, permanent-magnet alternator whose output, together with a feedback voltage from the main alternator, is the input to a magnetic amplifier. The output of the amplifier supplies the exciter, an alternator whose field is stationary with the armature rotating. The output of the exciting alternator enters an uncontrolled rectifier that is mounted on, and rotates with, the shaft common to the pilot, exciter, and main generator field. The rectifier supplies the main alternator field power. Regulating devices are external; magnetic coupling with the exciter–alternator provides the means for control of line voltage.

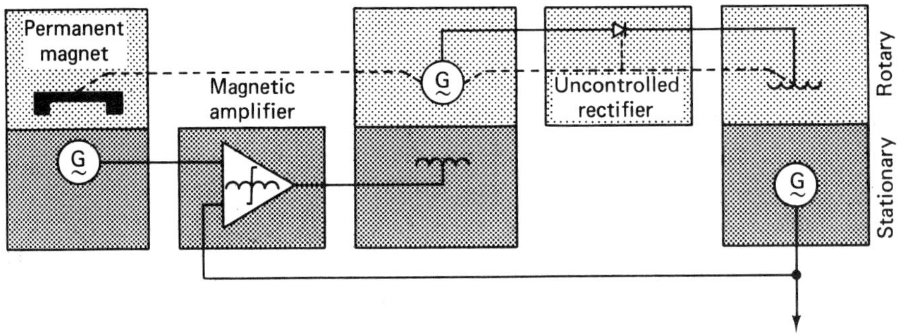

Figure 5-11 Brushless Excitation

5-6. VOLTAGE REGULATION

The best way to determine voltage regulation is to load a machine for rated current at rated voltage and the specified power factor. When the load is removed without changing the field current, the open-circuit voltage E_o can be measured, from which $\%\text{VR} = 100(E_o - V)/V$. This approach may be suitable for small generators but is impractical for alternators with large kVA ratings.

The approved methods of IEEE Standard No. 115 for determining voltage regulation require tests and computations that, because of equipment requirements and complexity, are beyond the scope of this text. The method described in the following paragraphs yields approximate results from tests than can be conducted in most learning laboratories. Tests are described in Section III; curves obtained from test data are used in the discussion.

The line voltage of most alternators is regulated by a feedback system that alters the field excitation in response to changed loading. Consequently, the relatively large voltage regulation based on constant field excitation is not indicative of usual voltage fluctuations in an operational system.

From the viewpoint of voltage regulation, the alternator differs from the transformer because of the large armature reaction voltage in quadrature with the line current. The vector diagrams of Fig. 5-12, although similar to those of a transformer in Fig. 4-5, differ in that E_p is *not* the open-circuit voltage of the alternator. However, E_p must be determined in order that the open-circuit voltage may be extracted from a curve.

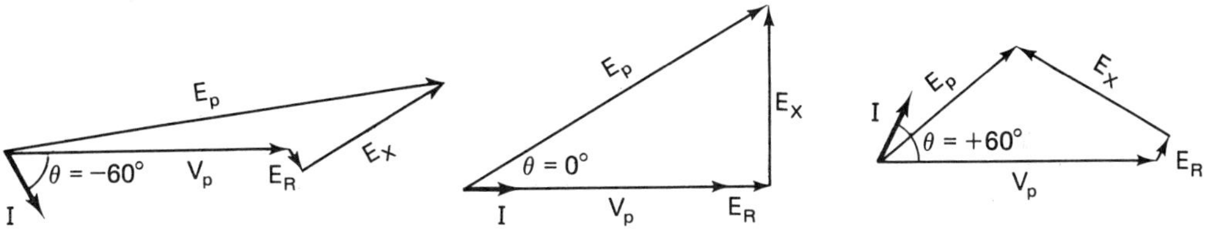

(a) Lagging pf (b) Unity pf (c) Leading pf

Figure 5-12 Voltage Regulation

In small generators of the type likely to be found in a learning laboratory, the magnitude of R_s relative to X_s may be significant. But in large alternators X_s is very much larger than R_s and the latter normally is ignored in computing voltage regulation (but not efficiency). In this development, only X_s is considered; the single effect of disregarding R_s is elimination of the small E_R vector in the diagrams of Fig. 5-12. (In the figures, E_R is exaggerated.)

The phase voltages of a loaded machine are shown in Fig. 5-13. V_p is the rated phase voltage, equal to $V_R/\sqrt{3}$. E_x, the phase armature reaction, is $I_L X_s$. E_p, the vectorial sum of V_p and E_x, is the voltage attributable to the field flux, Φ_p.

Turning now to flux components, Φ_p leads E_p by 90°. Φ_a is the component

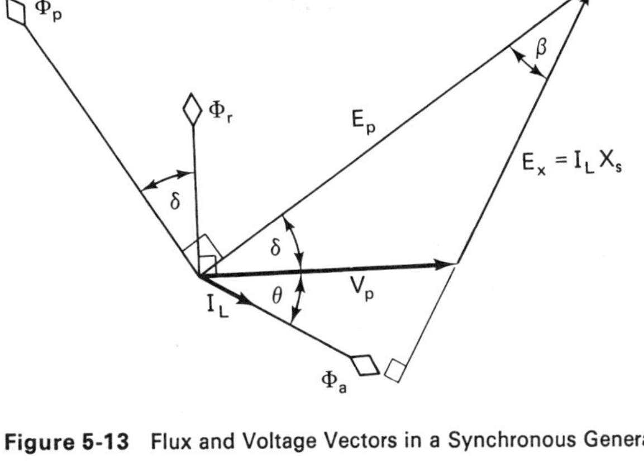

Figure 5-13 Flux and Voltage Vectors in a Synchronous Generator

that induces E_x—it leads E_x by 90°, but on the voltage diagram the form of the vector addition is $V_p + (-E_x) = E_p$ in contrast to the alternative, $V_p = E_p + E_x$, and E_x appears to lead Φ_a. The vector sum of Φ_a and Φ_p is Φ_R, the net flux responsible for V_p. The vector diagram does not consider saturation—all fluxes are related to their induced voltages by a machine constant.

If the load is now removed but the field current is unchanged, Φ_a and Φ_R disappear; Φ_p remains and generates an open-circuit phase voltage, E_o. With the desaturating effect of armature reaction eliminated, E_o will not rise to the magnitude E_p. The value must be extracted from a saturation curve.

The first step in computing voltage regulation is to determine the synchronous reactance. With R_s neglected, $X_s = Z_s$. The saturation curves of Fig. 5-14 include

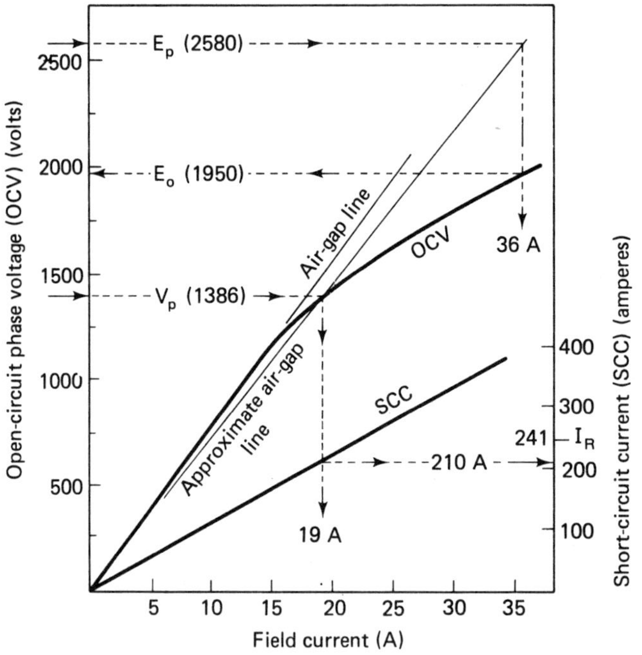

Figure 5-14 Synchronous Machine Saturation Curves

a plot of (a) open-circuit phase voltage as a function of field current (OCV), and (b) short-circuit line current, also as a function of field current (SCC). The *true air-gap line* is the linear portion of the open-circuit curve and its extension; this line is used in rigorous analysis, but an *approximate air-gap line* through the origin and rated phase voltage on the open-circuit curve allows reasonable accuracy in computations at rated conditions and simplifies the analysis. The approximate air-gap line is used in this discussion.

At any field current, the open-circuit voltage and the corresponding short-circuit current are related by Ohm's law to Z_s. To compute Z_s, the graph is entered at rated V_p to find I_f from the OCV curve; at the intersection of this I_f with the SCC curve, the armature current is extracted.

Example 5-2: The OCV and SCC curves for a 1000-kVA 2400-V four-pole three-phase synchronous generator are shown in Fig. 5-14. Determine X_s.

SOLUTION:

$$V_p = \frac{V_R}{\sqrt{3}} = \frac{2400}{\sqrt{3}} = 1386 \text{ V}$$

1386 V intersects the OCV curve at a field current of 19 A. This field current produces a short-circuit line current of 210 A. Therefore,

$V_p = I_L Z_s$

$$Z_s = \frac{1386}{210} = 6.6\ \Omega$$

$$X_s = Z_s = 6.6\ \Omega \qquad (R_s \text{ negligible})$$

The approximate air-gap line intersects the OCV curve at the point where Z_s was computed; Z_s is both a saturated value at V_p and a constant in linear computations related to this air-gap line. At rated current the voltage across Z_s is $E_x = I_R X_s$.

Example 5-3: (Continuation of Example 5-2)

Find the magnitude and phase angle of E_x at rated current and 0.866 lagging pf. Find the magnitude of E_p.

SOLUTION:

$$I_R = \frac{\text{kVA} \times 10^3}{\sqrt{3}\, V_R} = \frac{1000 \times 10^3}{\sqrt{3}\,(2400)} = 241 \text{ A}$$

$$\theta = -\cos^{-1} 0.866 = -30^\circ \quad \text{(negative for lagging pf)}$$

$$E_x = I_R X_s \angle 90 + \theta = (241)(6.6)\angle 90 + (-30)$$

$$= 1591 \text{ V} \angle +60^\circ$$

$$E_p = V_p \angle 0^\circ + E_x \angle 60^\circ$$

$$= (V_p + E_x \cos 60) + jE_x \sin 60$$

$$= [1386 + (1591)(0.5)] + j(1591)(0.866)$$

$$|E_p| = 2580 \text{ V}$$

The computed E_p is valid for the approximate air-gap line; the corresponding I_f is the field current for full load at the stated pf. When the load is removed without changing the field current, the open-circuit voltage is the voltage on the OCV curve for this field current; this accounts for saturation.

EXAMPLE 5-4: For the computed E_p of 2580 V at 0.866 lagging pf, what is the %VR?

SOLUTION: At 2580 V on the approximate air-gap line, the field current is 36 A. From the OCV curve at 36 A, E_o is 1950 V.

$V_p = E_p + E_x$

$$\%\text{VR} = \frac{100(E_o - V_p)}{V_p} = \frac{(100)(1950 - 1386)}{1386} = 40.7\%$$

5-7. POWER AND TORQUE

The voltage equation for a dc generator usually is expressed as $E_g = k\Phi\omega$; k is the numerical value of those elements in a particular machine that cannot be varied by the user ($k = PZ/2\pi a$). In the voltage equation for a specific synchronous generator, $E_p = 4.44Nf\Phi k_w$, N and k_w are constants. The frequency, $f = \omega P/4\,\pi$, contains constants except for ω, and the voltage equation then has only two user-variable terms, ω and Φ. E_p may be expressed as $k'\Phi\omega$, with k' representing the consolidation of coefficients and physical terms unique to the machine. But the generator normally is operated at rated speed to produce rated frequencies; ω is constant, and the generated voltage equation reduces to $k\Phi$.

On the vector diagram of Fig. 5-13, each voltage is k times the corresponding flux. When I_f is constant, Φ_p and E_p are constant, but the magnitudes of all other vectors change with load variations.

Noting that Fig. 5-13 is a representation of one phase of the machine, the power output per phase is $V_pI_L \cos\theta$. In terms of E_p rather than V_p, the power is $E_pI_L \cos(\theta + \delta)$; however, $\beta = 90 - (\theta + \delta)$ and $\cos(\theta + \delta) = \sin\beta$. Hence, the power is $E_pI_L \sin\beta$. From the law of sines, $(\sin\beta)/V_p = (\sin\delta)/I_LX_s$, from which $\sin\beta = (V_p \sin\delta)/I_LX_s$. Substituting this in the power expression,

$$P_\phi = E_pI_L \sin\beta = E_pI_L \frac{V_p \sin\delta}{I_LX_s}$$

$$= \frac{E_p}{X_s} V_p \sin\delta \tag{5-5}$$

Note that E_p and V_p are *scalar*—not vector—quantities.

Several significant points are apparent in this equation. First, for constant E_p the power is proportional to V_p and the sine of the angle δ, the angular separation of E_p and V_p. Second, inasmuch as Φ_p is perpendicular to E_p and Φ_r is perpendicular to V_p, the angle δ also occurs between the pole flux and the resultant flux. From $E_p = k\Phi_p$ and $V_p = k\Phi_r$, it follows that the power also is proportional to $\Phi_p\Phi_r \sin\delta$.

In mechanical terms, $P = \omega T$. Equating this to the electrical power of equation 5-5

$$\omega T = \frac{E_pV_p}{X_s} \sin\delta$$

and

$$T_\phi = \frac{E_p V_p}{\omega X_s} \sin \delta \quad [\text{SI}]$$
$$= \frac{7.04 E_p V_p}{S X_s} \sin \delta \quad [\text{US}] \tag{5-6}$$

With ωX_s and E_p constant, a change in electrical power must be accompanied by changes in either one or both of the remaining elements, V_p and δ. Electrical power always must be matched by mechanical power; T is responsible for the equality. The angle δ is the *torque angle*; it changes in response to electrical power changes.

Electrical power may be varied by changes in either the magnitude of the line current or the power factor angle. The character of the resulting vector changes is illustrated in Fig. 5-15 for (a) a change in I_L but not pf and (b) a change in pf but not I_L; in both instances, the change is with respect to the loading in Fig. 5-13, shown with dashed lines in Fig. 5-15.

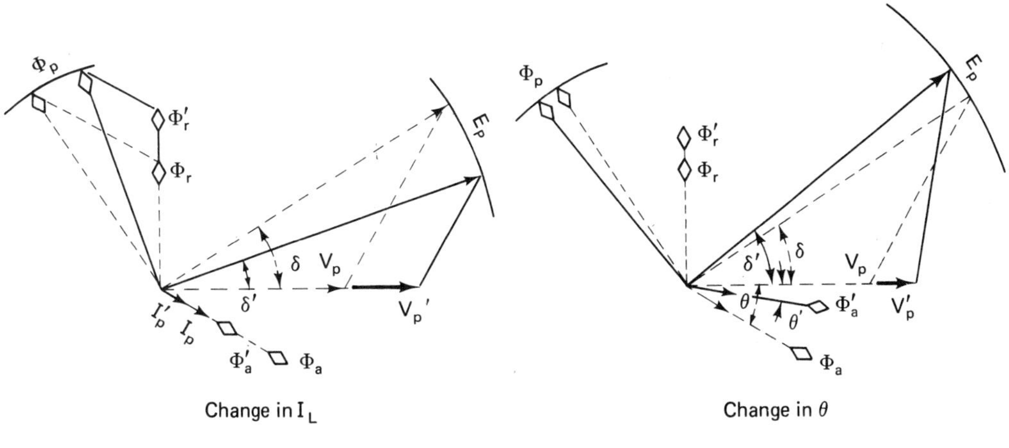

Figure 5-15 Effect of Varying I_L and θ

With no change in the pf, one expects a reduction in I_L to lead to lesser power but a probably increased V_p due to a reduction in IZ_s; both suggest a decrease in δ. This change is approximated in Fig. 5-15 (approximate primarily because magnetic saturation is ignored).

Maintaining the same line current but increasing the power factor (i.e., reducing θ) must result in increased power. The magnitude of the armature flux is unchanged, but at the lesser power factor angle, the vector difference, $\Phi_p - \Phi_a$, has increased, causing a rise in V_p. The torque angle also has increased.

In both Figs. 5-13 and 5-15, R_s and X_w have been neglected. If we now consider the voltage drops across these elements, the resulting vector diagram is that shown

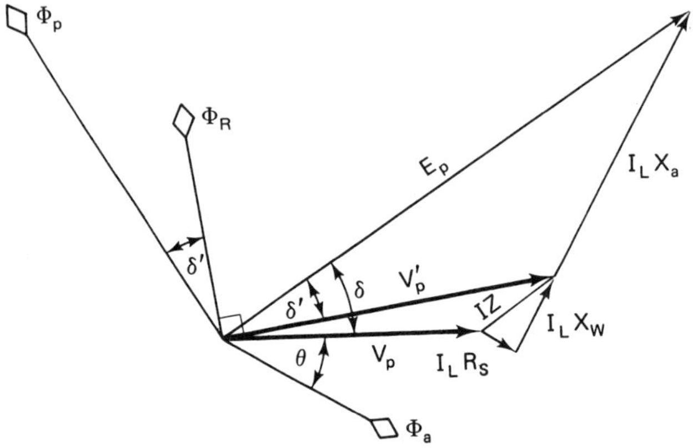

Figure 5-16 Diagram when R_s and X_w are included

in Fig. 5-16. Φ_a is proportional to $I_L X_a$; X_w accounts for localized leakage flux, whereas X_a is responsible for the pervasive armature flux. The resultant flux is perpendicular and proportional to V'_p, the generated voltage comparable to E_g in the dc generator. The IZ drop subtracts from V'_p and V_p appears at the terminals.

The torque angle also is known as the *power angle*. On this diagram, the angles differ; from the power viewpoint, δ applies, but the torque angle is δ'. The difference can be attributed to power losses in the winding. These losses, on the order of 1% in a large alternator, result in a small and generally negligible difference between δ and δ'.

5-8. EFFICIENCY To determine the efficiency of the synchronous generator at a specified power output and power factor, machine losses are added to the power output in order to determine the input power. The losses included in the computation are discussed next.

Armature Winding I^2R Loss—P_a. This loss is based upon the line current and phase resistance. Effective ac resistance is greater than dc resistance because of skin effects and eddy currents in the conductors; this difference is accounted for in stray power losses. R_s, the dc phase resistance, must be corrected to a temperature of (1) 25°C plus the observed temperature rise or, if the rise is not measured, (2) 75°C for class A insulation, 95°C for class B, 115°C for class F, and 130°C for class H.

The loss in each phase is $I_L^2 R_s$; in the three-phase machine, the net loss is $3I_L^2 R_s$.

Field I^2R Loss—P_f. This is based on the field current and measured field winding dc resistance, corrected for temperature like armature resistance. Losses in field rheostats used to adjust the generated voltage are not charged to the synchronous machine.

Core Loss—P_c. In the dc machine, with the assumption that flux remains essentially constant throughout the loading range, core losses are proportional to the speed. However, in the synchronous machine, the speed is constant, but there is considerable variation in the flux, owing to armature reaction. The voltage to which the net flux is proportional is $V_p + I_L R_s$; standard tests provide data for a plot of core losses as a function of voltage from which the proper loss can be extracted.

Friction and Windage Loss—P_{fw}. Since this type of loss is proportional to speed and the speed of a synchronous machine is constant, friction and windage losses are constant throughout the loading range. The magnitude of this loss is determined from tests.

Stray Load Losses—P_{stray}. The effective ac resistance of the armature winding is greater than the dc resistance because of skin effects. One component of the stray load loss accounts for this difference. A second component results from distortion of the flux by the load currents that increases the core loss. Neither of these losses is computed. Testing provides data for a plot of stray load losses as a function of line current.

Exciter Loss. If the exciter is an integral component of the alternator—that is, shares a common shaft or is permanently coupled—the losses in the exciter (I^2R, P_{fwc}) are included in the losses that determine the alternator efficiency, but non-integral exciter losses are not charged to the alternator.

From test data, loss curves are plotted on a single graph as illustrated in Fig. 5-17. These are machine—*not* per phase—losses. Core losses are plotted as a function of voltage. The short-circuit power curve, P_{sc}, is a function of I_L and consists of stray load losses plus the I^2R loss in the armature winding at the test temperature; the test temperature is unlikely to be the temperature upon which efficiency is based, and I^2R losses are not extracted from this curve. But if the I^2R loss at test temperature is computed and subtracted from the test data, the difference is the stray load loss, valid for any temperature. This difference appears on the loss chart as P_{stray}.

EXAMPLE 5-5: The 1000-kVA synchronous generator of the preceding examples is delivering rated power at 0.866 lagging pf. The field current is 36 A and the field resistance, corrected to 75°C, is 7.35 Ω. At 25°C, R_s is 0.0537 Ω. The insulation is class A. From Example 5-2, V_p is 1386 V; from Example 5-3, I_R is 241 A. The loss curves of Fig. 5-17 apply. Find the efficiency.

SOLUTION: For class A insulation, field and armature resistances must be corrected to 75°C. The field resistance is given for 75°, but the armature resistance is for 25°C.

$$R_{s(75)} = \frac{75 + 234.5}{25 + 234.5}(0.0537) = 0.064\ \Omega$$

P_a: $\quad P_a = 3I_L^2 R_s = (3)(241)^2(0.064) = 11.15\ \text{kW}$ 11.15 kW

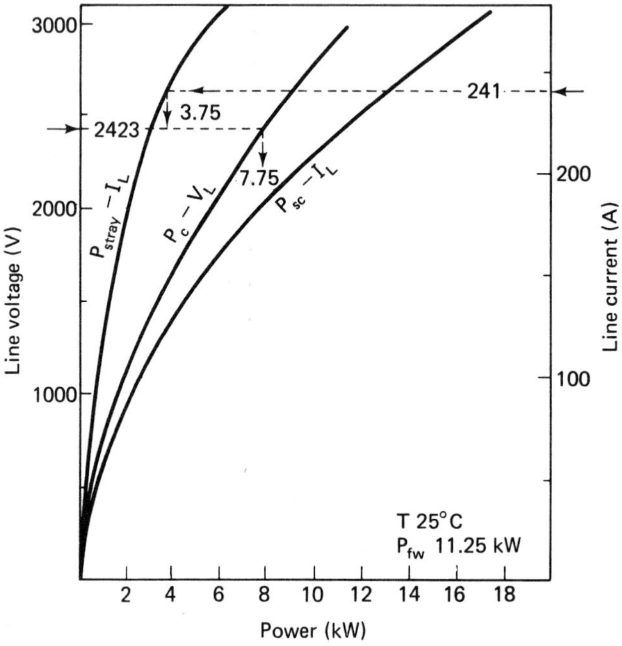

Figure 5-17 Power Losses

P_f: $P_f = (I_f)^2 R_f = (36)^2(7.35) = 9.53$ kW — 9.53 kW

P_c: this is extracted from the curve using an entry voltage of

$$V_c = \sqrt{3}(V_p\underline{/0} + I_L R_s\underline{/\theta}):$$

$$\theta = -\cos^{-1} \text{pf} = -\cos^{-1}(0.866) = -30° \qquad \text{(negative for lagging pf)}$$

$$V_c = \sqrt{3}[1386 + (241)(0.064)\underline{/-30}]$$

$$= 2423 \text{ V}$$

the corresponding P_c is 7.75 kW — 7.75 kW

P_{fw}: this appears on the loss chart — 11.25 kW

P_{stray}: entering the loss chart at rated line current of 241 A, the loss is extracted from the stray-load loss curve — 3.75 kW

$\sum$ losses: 43.4 kW

P_o: $P_o = \text{kVA} \times \text{pf} = (1000)(0.866) = 866$ kW — 866.0 kW

P_{in}: $P_{\text{in}} = P_o + \sum P_{\text{loss}}$ — P_{in} 909.4 kW

$$\%\eta = \frac{100P_o}{P_{\text{in}}} = \frac{(100)(866)}{909.4} = 95.2\%$$

SECTION 5-2. SYNCHRONOUS MOTOR

5-9. OPERATING PRINCIPLES In Fig. 5-18, two magnets, one bar and the other U-shaped, are mounted on and free to rotate about a common axle. With no external forces present, the tractive force between unlike poles will cause the magnets to align as illustrated. Represented as a vector through the pivot of the bar, the flux of the U-shape magnet is external—that is, the direction is from the north to the south pole; the bar flux is internal with a direction from south to north. When the bar and U are aligned, their flux vectors are coincident.

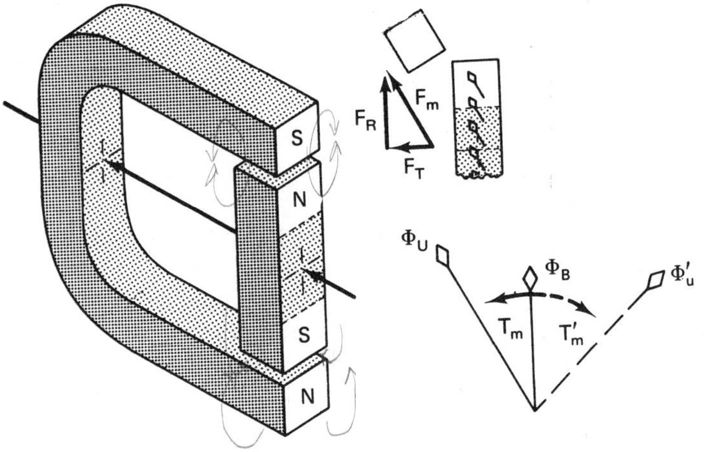

Figure 5-18 Magnetic Torque

When aligned, the magnetic tractive force on the bar is radial with respect to the pivot. If the bar now is restrained while the U is displaced, the consequence can be represented in several ways. From domain theory, the fluxes in the bar tend to realign with the flux of the U; the tractive force, F_m in the force diagram, has a radial component, F_r, and a torque component, F_t. Upon removal of the restraint on the bar, F_t will cause it to rotate to realignment with the U.

A second method of viewing the result is to regard the flux of each magnet as a vector that is fixed with respect to the body of the magnet. As illustrated, with an angular separation between the flux vectors, there is a magnetic torque, T_m, in a direction to restore alignment.

Assume that Φ_U is instantaneously accelerated to angular velocity ω, but acceleration of the bar is opposed by T_J, the inertial torque, and T_B, frictional torque (Fig. 5-19). The bar cannot immediately attain velocity ω. At t_1, Φ_U has completed a full revolution and arrived at the illustrated position while Φ_B has completed less than half of a revolution. Both T_m and T_B act to decelerate Φ_B. Shortly thereafter, Φ_U will pass and again accelerate Φ_B in the ω direction. Φ_B will rock back and forth but will not go into sustained rotation.

Now assume that both Φ_U and Φ_B are rotating at velocity ω [Fig. 5-19(b)].

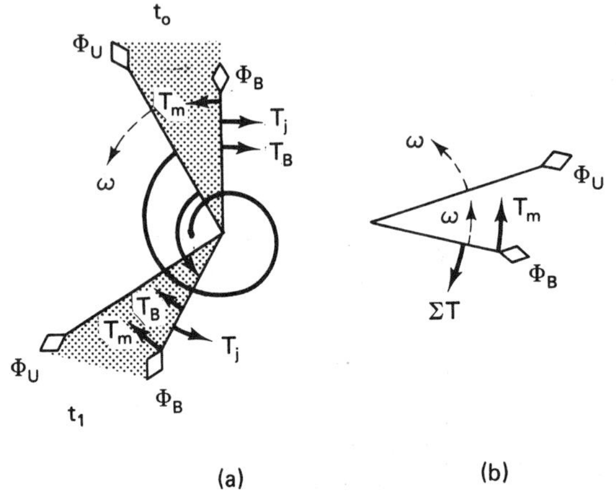

Figure 5-19 Torque with a Rotating Field

If T_m is equal to $\sum T$, rotation is sustained, but if $\sum T$ is greater than T_m, Φ_B will decelerate and resume the condition of the preceding paragraph.

The equivalents of Φ_B and Φ_U exist in a synchronous motor. In construction, the motor is identical to the synchronous generator. As illustrated in Fig. 5-6, an applied three-phase voltage creates a constant-strength flux rotating at a synchronous speed of $120f/P$, the analog of Φ_U. The rotor, excited by dc, is the counterpart of Φ_B.

Synchronous flux is produced electrically, has no inertia to overcome, and revolves at synchronous speed immediately upon energization of the armature windings, but the rotor cannot immediately accelerate to this speed. Therefore, the basic characteristics of a synchronous motor are that:

- It has no inherent starting torque.
- It can run only at synchronous speed.
- It will stall if, while running, the countertorque is increased beyond the maximum torque that the machine can develop.

5-10. MOTOR CIRCUIT ANALYSIS

The schematic of a synchronous motor is identical to that of a synchronous generator; however, in recognition of the different operating mode, the positive armature current is from the terminals into the winding.

The schematic and vector diagram for a synchronous generator are shown in Fig. 5-20(a). The diagram reflects the equations $E_p = V_p + I_L Z_s$ and $\Phi_R = \Phi_p + \Phi_a$. Note that I_L is shown flowing *out* of the machine inasmuch as power flow is *out* of the terminals. For the synchronous motor, power flow is *into* the terminals; a consistent interpretation is the reversal of current in the schematic, and the vector

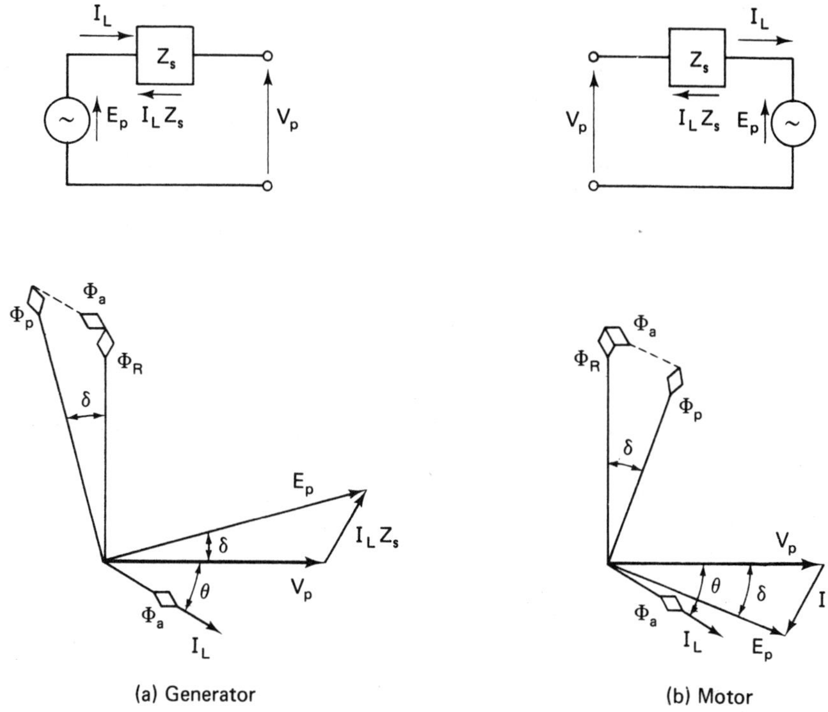

Figure 5-20 Synchronous Machine Diagrams

diagram reflects the equations $E_p = V_p - I_L Z_s$ and $\Phi_R = \Phi_p - \Phi_a$. Note that V_p and I_L remain separated by θ.

Inasmuch as there are no basic differences between the two machines, relationships developed for the generator will hold for the motor. Equations 5-5 and 5-6 are valid:

$$P_\phi = \frac{E_p V_p}{X_s} \sin \delta \tag{5-5}$$

$$T_\phi = \frac{E_p V_p}{\omega X_s} \sin \delta \tag{5-6}$$

The power and torque obtained from these equations are based on terminal power, as was the case for the generator. The output power of the motor will be less than this input by the I^2R and friction, windage, and core losses. For the purpose of developing features of the synchronous motor, these losses will be neglected.

The maximum torque will occur when the torque angle, δ, is 90°. V_p and X_s are constant, but the maximum torque will vary if the field current (and thereby E_p) is changed. Although losses are disregarded, torque computed from equation 5-6 is fairly representative. There are actually two torque components in a salient-pole rotor; to this point, only electromagnetic torque has been considered. In Fig. 5-18, if the bar magnet were replaced with an unmagnetized iron bar, the U-magnet flux would deviate from a direct pole-to-pole path to take a lower reluctance path through the iron; the contracting tendencies of the flux would then transmit torque to the bar. This is *reluctance* torque. Its magnitude will vary with the rotor position; for example, there can be no reluctance torque when the bar is aligned with the flux, nor will there be a torque when the bar is perpendicular to the flux, for this position does not present a defined, low-reluctance path through the bar axis. Also, note that the reluctance torque alternates in the same fashion as electromagnetic torque if the U is rotating at a speed greater than the speed of the bar.

When the reluctance torque is added to the electromagnetic torque, the maximum torque occurs at a torque angle of approximately 70°.

EXAMPLE 5-6: A 4000-V 5000-hp 60-Hz 12-pole synchronous motor is excited to produce unity pf at rated load. $X_s = 4.1\ \Omega$.

(a) What is the rated and the maximum torque in N·m and in lb-ft?
(b) For maximum torque, what is the armature current?

SOLUTION: (Fig. 5-21)

$$\text{(a)}\ P_\phi = \frac{\text{hp} \times 746}{3} = \frac{(5000)(746)}{3} = 1243\ \text{kW}$$

$$V_p = \frac{4000}{\sqrt{3}} = 2310\underline{/0^\circ}$$

$$I_R = \frac{P_\phi}{V_p} = \frac{1243 \times 10^3}{2310} = 538\text{A}$$

$$E_p = V_p - jI_R X_s = 2310\underline{/0} - (538)(4.1)\underline{/90} = 3194\underline{/-43.7^\circ}$$

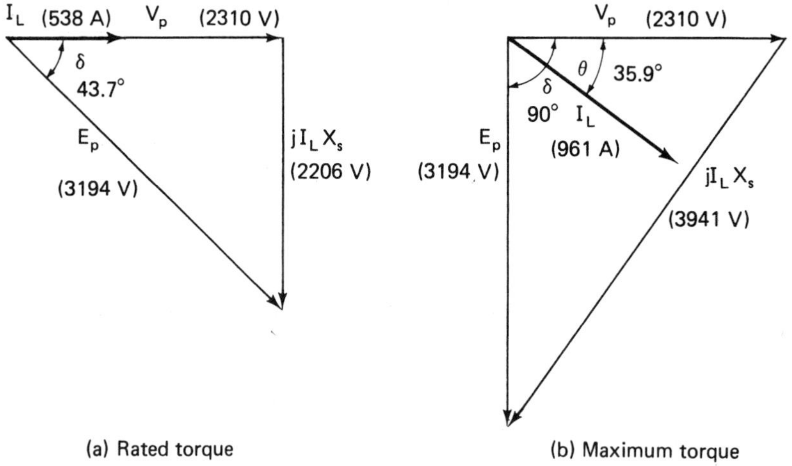

Figure 5-21 Example 5-6

$$\omega = \frac{4\pi f}{P} = \frac{4\pi(60)}{12} = 20\pi$$

$$T_\phi = \frac{E_p V_p \sin\delta}{\omega X_s} = \frac{(3194)(2310)(\sin 43.7)}{(20\pi)(4.1)} = 19.79 \times 10^3 \text{ N}\cdot\text{m}$$

The machine torque is $3T_\phi$ or 59.4×10^3 N·m.

$$T_{\text{lb-ft}} = \frac{T_{\text{N}\cdot\text{m}}}{1.355818} = 43.8 \times 10^3 \text{ lb-ft}$$

(For comparative purposes, you are encouraged to compute the torque from hp $= 2\pi ST/33000$.)

(b) For maximum torque, $\delta = 90°$:

$$T_{\max} = 3T_\phi = \frac{3(3194)(2310)(\sin 90°)}{(20\pi)(4.1)}$$

$$= 85.9 \times 10^3 \text{ N}\cdot\text{m}$$

$$= \frac{85.9 \times 10^3}{1.355818} = 63.4 \times 10^3 \text{ lb-ft}$$

$$I_L(X_s\underline{/90}) = V_p\underline{/0} - E_p\underline{/90} = 2310\underline{/0} - 3194\underline{/90}$$

$$= 3941 \text{ V } \underline{/54.1°}$$

$$I_L = 3941\underline{/54.1}/4.1\underline{/90} = 961 \text{ A } \underline{/-35.9°}$$

5-11. STARTING A SYNCHRONOUS MOTOR

Since the synchronous motor will not develop a starting torque, an auxiliary starting means must be provided. The machine may be started by driving it with another motor—sometimes the dc exciter in a reversed role—but more commonly by starter windings built into the machine. In either case, if the dc field were excited during the starting period, large pulsating countertorques would occur as the revolving flux lapped the rotor. Conceptually, starting requires acceleration of the unexcited rotor to a speed close enough to synchronous that, when dc is applied to the field, the resulting torque will raise the speed to synchronous before the rotating field can lap the rotor. The timing of excitation is dependent upon two conditions: (1) attainment of a required speed, and (2) a position of the rotor relative to the rotating flux that will result in a torque in the direction of rotation.

The prevalent method of starting synchronous motors employs an **amortisseur**, a permanently short-circuited winding embedded in the pole shoes and interconnected at opposite ends of the pole (Fig. 5-22). In Chapter Six, the induction motor with a squirrel-cage rotor is discussed; the amortisseur, in effect, is a partial squirrel-cage winding. Discussion of the functional principles is deferred to Chapter Six; at this point the significant characteristics are that (1) as long as the rotor is revolving at a speed less than synchronous, currents are induced in the amortisseur winding that produce torque in the direction of the rotating flux, but (2) no current

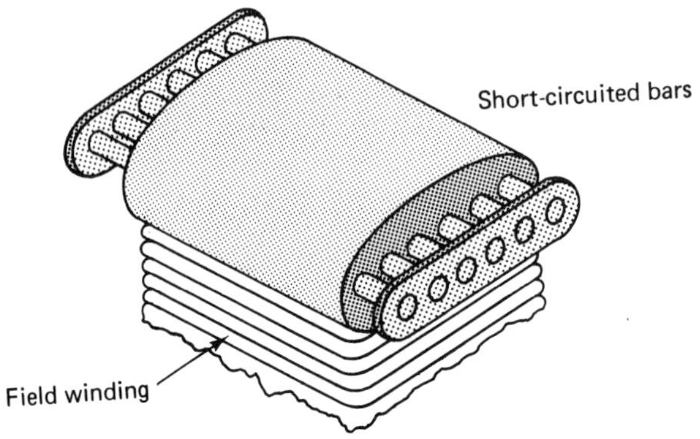

Figure 5-22 Amortisseur

or torque is developed at synchronous speed. The amortisseur accelerates the rotor to a speed on the order of 92 to 97% synchronous. The speed must be such that the torque developed at the moment dc excitation is applied exceeds the opposing torques of the motor and attached load by an accelerating torque sufficient to snap the rotor and load into synchronization.

To provide for application of dc excitation at the proper moment, the typical starter incorporates a *polarized frequency relay* (PFR). This relay is illustrated in Fig. 5-23. A static core supports two windings, one dc and the other ac. The direction of the flux produced by the dc coil is constant, but that of the ac coil alternates. In Fig. 5-23(a, 1) the flux pattern during the positive portion of the ac cycle is shown; for the negative portion, it is that of Fig. 5-23(a, 2). If a decaying voltage is applied to the ac coil, the air gap flux as a time function is typified by Fig. 5-23(b), with dc flux adding to and subtracting from the ac flux as illustrated. At some point, the flux in the armature no longer is adequate to hold it closed. The threshold at which this occurs is on the negative portion of the ac cycle; thus, both magnitude and position govern closure of the relay contacts.

If the armature windings are excited and the rotor is at rest, voltages are induced in the rotor coil. Also, if the rotor is revolving at a speed less than synchronous, voltages are induced in the field winding but at a lesser frequency. Representing the synchronous speed with S_S and the rotor speed with S_R, the speed of the flux relative to the rotor is $S_S - S_R$, and the frequency of the induced rotor voltage f_r is $P(S_S - S_R)/120$. At rest, $f_r = f_a$, the frequency of the applied voltage, but f_r decreases as S_R increases. Similarly, the magnitude of the induced rotor voltage decreases with increased speed. The induced rotor voltage is applied to the ac coil of the PFR, making the relay sensitive to both S_R and the relative positions of the rotor and the synchronous flux. The net relay flux thus is patterned as in Fig. 5-23(b).

A typical starter is illustrated in Fig. 5-24. The shading emphasizes the energized circuits that determine the moment of application of dc to the rotor. The ac

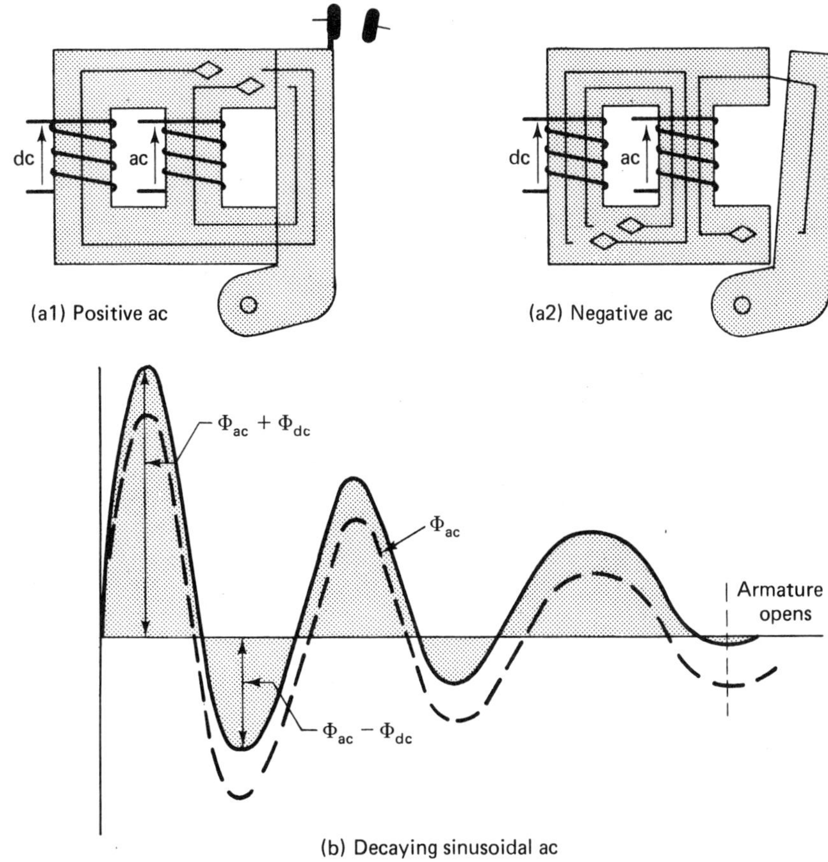

Figure 5-23 Polarized Frequency Relay

coil of the PFR is paralleled by a reactor; at high f_r (low rotor speed) the impedance of the reactor is large and most of the current flows through the PFR coil, but as speed rises and f_r drops, the impedance drops and the reactor takes proportionately more of the circuit current. The complete field circuit includes a current-limiting resistor and an out-of-step relay coil (OSR). The purpose of the OSR is to shut down the motor if synchronization does not occur within a preset time, perhaps 5 to 7 seconds, because the amortisseur winding is incapable of prolonged operation and must be protected.

In the dc circuit the normally closed PFR contacts in series with the FC coil open before the M contacts close; consequently, the field contactor relay, FC, does not energize until the PFR drops out near synchronous speed. When drop-out does occur, the unshaded, normally open FC contacts close to apply dc to the field and the motor pulls into step with the revolving flux.

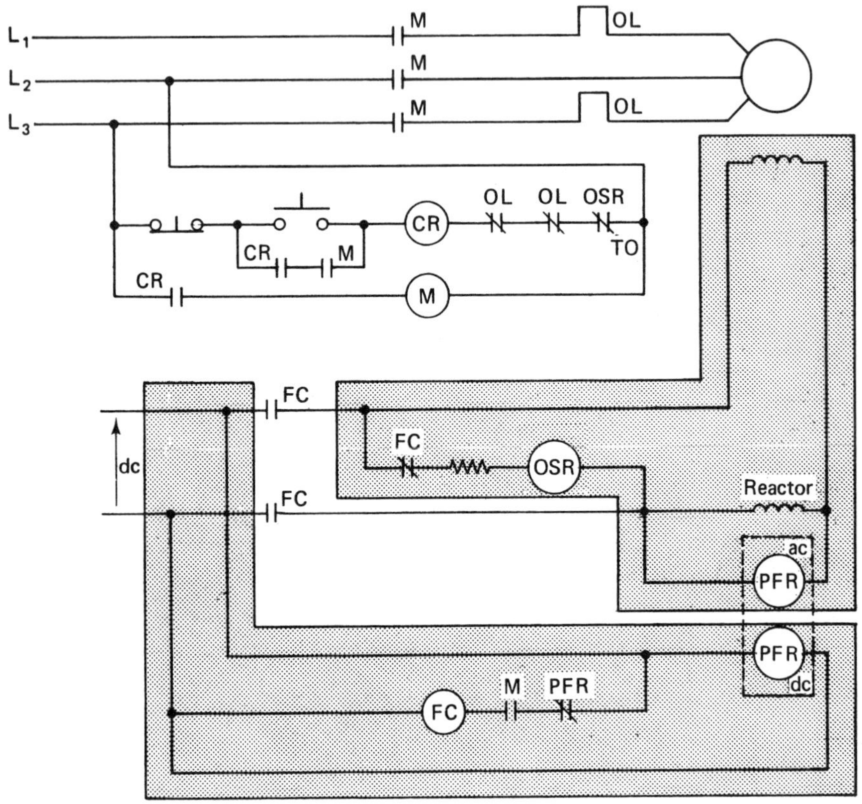

Figure 5-24 Synchronous Motor Starter

5-12. TORQUE In selecting a motor for a particular application, four torque characteristics must be considered: locked-rotor torque, pull-in torque, pull-out torque, and rated torque.

Locked-rotor torque is the minimum torque at any angular rotor position that is developed with the rotor locked (i.e., stationary) and rated voltage at rated frequency applied to the terminals. This torque is provided by the starter windings discussed in Paragraph 5-11. For rated speeds above 500 r/min, the minimum locked-rotor torques, expressed as a percentage of rated torque, range from 40% (1250 hp and larger) to 100% (200 hp and below). For speeds of 450 r/min and less, locked-rotor torque is at least 40%.

Pull-in torque is the maximum constant torque at rated voltage and frequency under which a motor will pull a connected load into synchronism when the dc rotor excitation is applied. This torque may range from 100% for high-speed to 30% for low-speed motors.

Pull-out torque is the maximum sustained torque which the machine will develop at synchronous speed with rated voltage applied at rated frequency and with normal excitation. This torque ranges from 150 to 200% of rated and is particularly important in applications where load torques in excess of rated may occur.

A motor cannot be chosen for a particular application solely on the basis of the rated horsepower requirement. It is important that the extreme loading conditions be identified to ensure that the motor can start and will continue to run under all anticipated circumstances (running load torque less than pull-out torque.)

5-13. MOTOR POWER FACTOR

One attraction of the synchronous motor is that it can be made to operate at a leading power factor. The benefits of this capability are discussed in Paragraph 5-20. This section is concerned with the means for altering the pf.

Two variables determine the power factor at which the motor will operate: E_p (which is controlled by I_f) and the load power. This investigation considers the separate circumstances of constant power and constant E_p.

Constant Power. The effect of a change in E_p while power remains constant is illustrated in Fig. 5-25(a). V_p is constant. If power is constant, $I \cos \theta$ must remain constant and the locus of the tip of the current vector will be the vertical line through the tip of the unity pf current vector. The vector IZ always is at the angle $-90 + \theta$; the vertical component is $IZ \sin(-90 + \theta) = -IZ \cos \theta$. Since $I \cos \theta$ is constant, the locus of the tip of E_p must be the horizontal line shown in the illustration. The vectors I_0, E_0, and $I_0 Z_s$ apply at unity pf. As illustrated, raising E to E_1 causes a leading current while lowering to E_2 results in a lag.

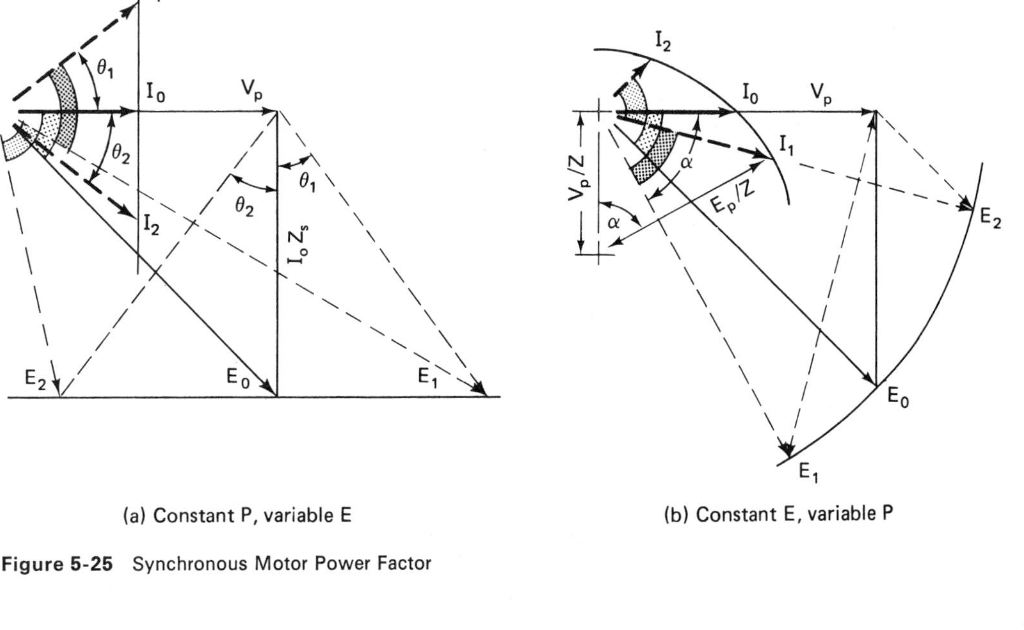

Figure 5-25 Synchronous Motor Power Factor

Constant E_p. With E_p at the angle α, current can be expressed as

$$I\underline{/\theta} = \frac{V_p\underline{/0} - E_p\underline{/\alpha}}{Z\underline{/90}}$$

$$= \frac{V_p}{Z}\underline{/-90} + \frac{E_p}{Z}\underline{/90 + \alpha}$$

A circle with the radius E_p/Z, centered at $(V_p/Z)\underline{/-90}$, will be the locus of I, with α the angle between the two current components [Fig. 5-25(b)].

From the diagram it is evident that reduction of the line current causes an increased leading pf, whereas increasing line current results in a more lagging pf.

The ratings of synchronous motors specify the pf to which the ratings apply; usually, the specification is 1.0 or 0.8 leading. For either, the maximum rated excitation is that which produces rated power at the rated pf.

EXAMPLE 5-7: The motor of Example 5-6 is rated at 0.8 pf leading. What is the rated E_p?

SOLUTION:

$$\theta = \cos^{-1} 0.8 = 36.9°$$

$$E_p = V_p\underline{/0} - I_R X_s\underline{/90 + \theta}$$

$$= 2310\underline{/0} - (538)(4.1)\underline{/90 + 36.9}$$

$$= 4039 \text{ V}\underline{/-25.9°}$$

5-14. HUNTING When a motor is operating at steady state and there is a sudden and substantial increase or decrease in the load torque, the rotor must adjust to a new torque angle. In doing so, the motor is momentarily accelerated or decelerated; because of inertia, the rotor overrides the correct new angle and must now react in the opposite direction. Consequently, there is a transient adjustment period characterized by damped oscillation. This is termed "hunting"—it occurs not only in the synchronous motor but also in the generator upon abrupt change in the loading.

During hunting, the amortisseur winding will have induced currents in a direction to create torques opposing the motion of the rotor relative to the synchronous flux—that is, torque in the rotational direction if the rotor is slowing down but in the opposite direction when the rotor speeds up. Thus, it serves to dampen the oscillation, and so the amortisseur is also known as the "damper winding." For damping purposes, the winding is placed in synchronous generators.

SECTION 5-3. TESTING

As with the dc machine, the complete range of tests for a synchronous machine are extensive, with alternative procedural methods for measuring specific types of data. The tests described in this section are limited to those requiring minimal and common test equipment in order to determine machine quantities from which motor and generator efficiency and generator voltage regulation may be computed. The tests are identical for both motor and generator. A drive motor is required for the open- and short-circuit tests; a dc shunt motor is particularly desirable for this purpose, since all testing should be done at the rated speed of the machine undergoing test. It may be possible to use a machine's own exciter as the prime mover during the test. A synchronous or induction motor is an acceptable drive.

5-15. RESISTANCE MEASUREMENT

Both the armature and field resistance must be measured. The test should be performed with the machine at ambient temperature and, of course, static. Either a bridge or voltmeter–ammeter method may be used, but for the latter the value of the test current should be low to minimize the impact of temperature rises associated with large currents. Measurements are made at the terminals of the armature and, if possible, the slip rings of the field in order to eliminate the brushes.

For the armature, measurements are taken between each pair of terminals. The average of these three readings is R_t. Assuming a wye connection, R_t is the resistance of two phases in series; therefore, $R_s = R_t/2$.

5-16. OPEN-CIRCUIT TEST

The open-circuit test identifies (1) friction and windage losses, (2) core losses, and (3) points on the open-circuit saturation curve.

The test requires determination of the mechanical power input to the tested machine while driven at synchronous speed. This is done by measuring the power input to the drive motor and then subtracting the motor losses. When the term "input" is used in the remainder of this discussion, it refers to the power input to the tested machine, not the drive motor.

For any value of field current, the open-circuit power input consists of P_{fw} and P_c. P_{fw} is constant since it is proportional to speed and the speed is constant. However, the core loss is proportional to flux; the open-circuit voltage also is proportional to the flux, and therefore the core loss is proportional to the voltage. When $I_f = 0$, $P_c = 0$, and the input is P_{fw}. For any other value of I_f, P_c for the corresponding voltage is $P_{in} - P_{fw}$.

The test circuit is shown in Fig. 5-26(a). Initially, the machine is driven at

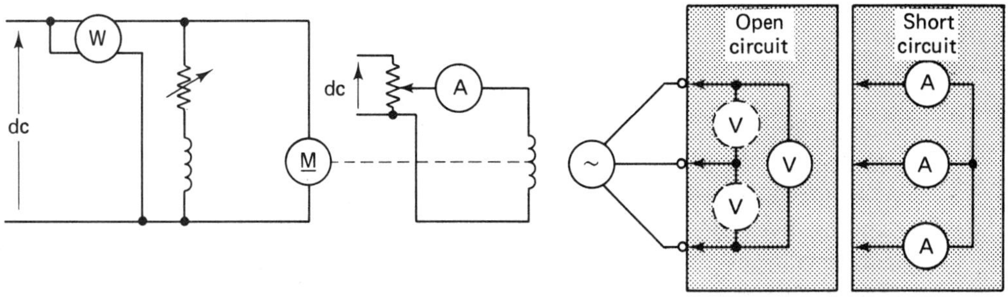

Figure 5-26 Open- and Short-Circuit Test Schematic

rated speed without excitation and the terminals are checked for residual voltage. If a voltage is present, the rotor must be demagnetized before proceeding. This can be done by repetitive excitation of alternating polarities, successively decreasing the magnitude of the field current on each change. After demagnetization, P_{fw} is measured with no current in the field.

A series of concurrent readings of P_{in}, V_L, and I_f are taken between $I_f = 0$ and a maximum I_f that is the lesser of (1) rated I_f or (2) the I_f producing V_{max}, a V_L equal to approximately 130% V_R. For each reading, P_c is computed.

Results of the open-circuit test are plotted on the two separate graphs, saturation and loss. On the saturation graph, either V_L or V_p is plotted vs. I_f (Fig. 5-14). On the loss chart, P_c is plotted as a function of V_L (Fig. 5-17). The usual procedure of showing the independent variable on the X-axis and the dependent variable on the Y-axis has been reversed. To avoid hysteresis effects during the test run, field current never is raised beyond, then returned to, the desired value; if a change for a new reading results in a larger increase than is desired, the field current should be reduced to zero, then gradually increased to the desired level. The machine should run long enough at a new setting to ensure stabilization of the speed and voltage before recording the readings.

At least 12 sets of data with the following distribution are required:

1 set: $V_L = 0$

3 sets: $V_L < 0.6V_R$

2 sets: $0.6V_R < V_L < 0.9V_R$

4 sets: $0.9V_R < V_L < 1.1V_R$, including one at approximately V_R

1 set: $1.1V_R < V_L < V_{max}$

1 set: V_{max}

Following the test run, the excitation is removed and P_{fw} rechecked; if the difference from the original reading is not more than 5%, the average of the two readings is the recorded P_{fw}. A difference between 5 and 10% should be prorated in determining P_c:

$$\Delta P_{fw} = P_{fw2} - P_{fw1}$$

$$c = \frac{\Delta P_{fw}}{V_m}$$

For each set of data, $P_{fw} = P_{fw1} + cV_L$. If ΔP_{fw} is greater than 10%, the data are discarded and the test is repeated.

V_L may be measured between one set of terminals only, except for one reading at approximately V_R, where all phases are checked to verify balance. If preferred, the three voltages can be measured and averaged at each setting. Either line voltage or phase voltage may be used in plotting data. Since voltage regulation is computed by phase, phase voltage was used on the OCV curve of Fig. 5-14. V_L was used in the P_c loss curve of Fig. 5-17 since efficiency is computed on a machine basis.

5-17. SHORT-CIRCUIT TEST

The circuit for this test is Fig. 5-26(a), with the shaded portion replaced by Fig. 5-26(b). Three ammeters, wye-connected to read line currents, provide a short circuit across the line terminals of the tested machine. The short-circuit impedances are extremely low; to maintain balance, meters should have identical impedances and the connecting leads should be of uniform length and gauge.

The impedance of the short-circuiting ammeters and leads is much less than the synchronous impedance of the machine undergoing test. The windings are highly inductive and cause the current to lag the generated voltage by an angle approaching 90°. The resulting fluxes and generated voltage are related as illustrated in Fig. 5-27. A value of I_f that produces a relatively high open-circuit voltage will develop only a small short-circuit voltage. Since the net flux is small, the core loss is negligible; power input consists of P_{fw}, I^2R, and stray load losses.

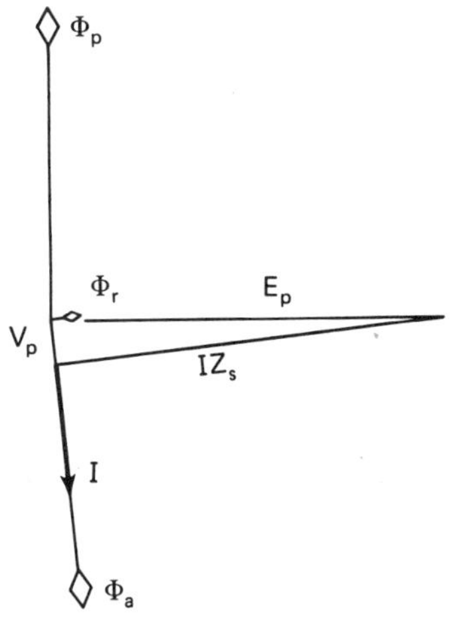

Figure 5-27 Vector Diagram of Short-Circuited Armature

Measurements of the power, field current, and the three line currents are made at approximately 125, 100, 75, 50, and 25% rated line current. To minimize the effect of temperature changes, readings begin with the high current and are taken promptly. After beginning the run, field currents should be decreased only; if an adjustment results in a too-low line current, excitation should be returned to the starting point and then lowered to the desired setting.

At the beginning and conclusion of the run, the power input to the tested machine is measured with $I_f = 0$ to check P_{fw}. If the readings differ, corrective action is the same as that for variances in the open-circuit test.

When starting the short-circuit test, the shorting connection must be in place before the field is excited. At no time should the short-circuit connection be initiated while the machine is generating an open-circuit voltage.

The short-circuit saturation curve is a plot of I_L as a function of I_f. For each set of data, I_L is the average of the three ammeter readings. The data are plotted on the same graph as the open-circuit saturation curve. Because of the large and lagging armature reaction, the iron does not saturate during the short-circuit test, and the plotted data should be a straight line through the origin.

For each set of observed data, P_{fw} is deducted from the power input; the remainder is the short-circuit loss consisting of copper and stray-load losses. The copper losses, $3I^2R_s$, are deducted from the short-circuit loss to determine the stray-load loss. Stray-load and short-circuit losses are plotted as a function of I_L on the same graph as the core loss.

The use of test data in determining the voltage regulation and efficiency of a synchronous generator was illustrated in Paragraphs 5-6 and 5-8. The losses in a synchronous motor are identical to those of the generator; loss curves are used in the same manner in determining the motor efficiency.

SECTION 5-4. RATINGS AND APPLICATION

5-18. SYNCHRONOUS MACHINE RATINGS

Generator. Standard ratings range from 1.25 to 75,000 kVA and 208Y/120 to 14,400 V. The power ratings of customized units in the utility system are in hundreds of MVA and some have voltage ratings greater than 14.4 kV.

Standard generators are rated at a power factor of 0.8 lagging; thus, the power range of standard generators is 1 to 60,000 kW. The power factor defines limitations in the rotor excitation. To preclude excessive temperature rise, not only the armature current but also the field current must be limited to rated values. Rated field current is that current which produces rated terminal voltage when the generator is delivering rated kVA at the rated pf. Terminal voltage may be maintained at rated value for power factors greater than 0.8 by a voltage regulator, but at lesser power factors, it cannot be maintained without exceeding the rated field current. The reasons can be understood by a study of Fig. 5-28.

For 0.8 power factor the relationship is shown by solid voltage and flux vectors. If rated line current is maintained but the power factor is increased, the magnitude of I_RX_s remains constant though its angle changes; to maintain a constant V_p, E_p (and Φ_p) must be reduced by a regulator. This range of operation is indicated by lightly shaded areas.

For power factors less than 0.8, E_p and Φ_p can be no greater than their magnitude at 0.8 pf. The relationship at 0.5 pf is shown with short dashed lines; with I_RX_s unchanged, both Φ_R and V_p have declined. The dark-shaded areas show the range of conditions between 0.0 and 0.8 pf.

The operating range at leading power factors is not shown; leading pf loading of an alternator is uncommon, but from the diagram it is apparent that it would require further reduction of Φ_p and E_p in order to maintain a constant V_p.

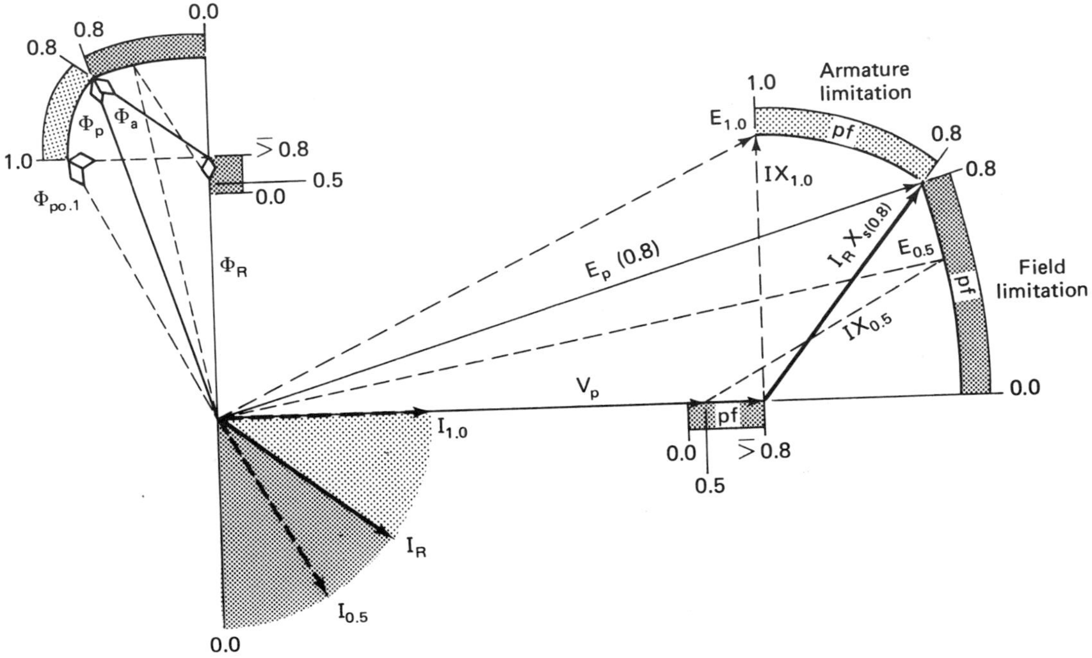

Figure 5-28 Operating Limitations

By NEMA standards, a synchronous generator will operate successfully at rated kVA, frequency, and power factor with terminal voltages not more than 5% above or below rated voltage. With the field adjusted for rated conditions, it can provide 150%I_R for a period of 1 minute, and it can withstand a short circuit for 30 seconds. In emergencies, it will sustain a 25% overspeed without mechanical injury.

Motor. Standard power and speed ratings range to 100,000 hp and 80 to 1800 r/min. Voltage ratings are 200, 230, 460, 575, 2300, 4000, 4600, and 13,200 V, applicable only at 60 Hz. Frequency ratings are either 50 or 60 Hz, and the rated pf is either 1.0 or 0.8 leading.

The ability of a motor to start and synchronize a load varies with both the horsepower and speed rating of the machine. For standard motors, the load inertias upon which rated pull-in torque is based are specified in NEMA MG-1.

Motors will perform successfully, although not according to machine ratings, at deviations from rated voltage and frequency providing that (1) V does not vary more than ±10% from rated, (2) f does not vary more than ±5%, and (3) the sum of the absolute values of the V and f variations does not exceed 10%. In emergencies, motors will withstand 25% overspeed without mechanical injury.

A motor rated at unity power factor may be operated at 0.8 pf and vice versa, but when this is done, the performance ratings must be amended. The crucial factors are armature current and field excitation. To operate a 0.8-pf motor at unity pf, the field excitation must be lowered; the absolute value of the armature current is unchanged and the power rating thereby will increase by the ratio $V_pI_p/0.8V_pI_p = 1.25$. This also means the full-load torque has been increased by a factor of 1.25. However, with the decrease in E_p the pull-out torque will decrease to approximately 75% of the corresponding torque at 0.8 pf.

A motor rated at unity pf cannot be made to operate at 0.8 pf by raising E_p—this would require increasing the field current beyond the rated value. The 0.8 pf can be obtained only by reducing the armature currents. The rated power and torque at 0.8 pf, influenced not only by the power factor but also by the reduced magnitude of the line current, is about 35% of the original rating. The absolute value of the pull-out torque is the same for both power factors since E_p is unchanged.

The frequency of starting a synchronous motor is limited by the heat developed during the starting period. With the motor initially at ambient temperature, two successive starts, coasting to rest between starts, are possible. With the motor at rated-load operating temperature, one start is permissible.

5-19. PARALLEL OPERATION OF GENERATORS

The commercial distribution grid is supplied by many generators operating in parallel. As load demands fluctuate and in order to permit repair and maintenance, units are added and removed from the line. Although the rating of an individual generator might be 500 to 600 MVA, it nevertheless is puny with respect to the grid system supplied by probably more than a hundred alternators. Under these circumstances, a change of conditions in one alternator will have negligible effect upon the line characteristics, and analysis of the generator is based upon the assumption that it is connected to an *infinite bus*, that is, a terminal connection at which the voltage and frequency is absolutely constant.

Contrasting with the infinite bus, if only two generators of comparable ratings are paralleled, changes in one generator will affect the terminal characteristics of both and the internal conditions of the second.

The operation of a generator when its electromagnetic characteristics are varied cannot be isolated from the operation of its prime mover. All practical drives have drooping power–speed characteristics; any change in the power output of the generator requires a matching change in the power output of the prime mover. Thus, to maintain a constant frequency a change in generator power requires an adjustment of the controls of the prime mover.

In addition to investigating the operation of a generator while paralleled, this section considers the problems and procedures in synchronizing an on-coming generator to the line before closing the switch that completes the paralleling connection.

Operation on an Infinite Bus. In Paragraph 5-13 and Fig. 5-25, a synchronous motor was analyzed under conditions where the line voltage was constant but (1) excitation was varied while maintaining constant power, and (2) excitation was constant but the power was varied. On an infinite bus, a generator responds to variations in these elements in a fashion comparable to the motor but with the significant difference that leading and lagging power factor relationships are reversed. In Fig. 5-29(a), power is constant as the excitation is varied; increasing the excitation causes the current to displace in a lagging direction. With constant

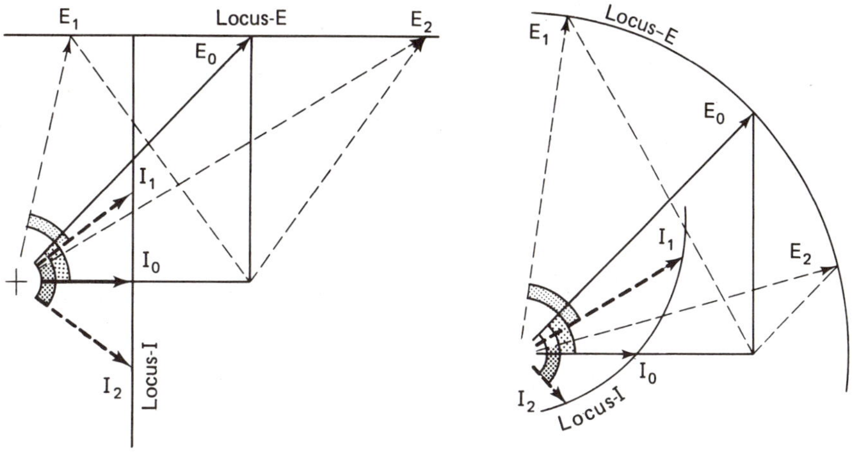

(a) Change in excitation (b) Change in power

Figure 5-29 Synchronous Generator on an Infinite Bus

excitation [Fig. 5-29(b)], increasing the power output causes the current to be more leading.

Two Paralleled Generators. When two generators operate in parallel, the load division is determined by two factors: the level of excitation of each generator and the power–speed characteristic of each prime mover.

Consider two identical generators operating in parallel with the stipulation that line voltage and current remain constant and the power output of each generator also must remain constant. Writing a current equation,

$$\frac{E_1 - V_p}{Z_s} + \frac{E_2 - V_p}{Z_s} = I_L$$

from which

$$E_1 + E_2 = 2V_p + I_L Z_s$$

To meet the stipulations, the right side of this equation is constant. This vector appears in Fig. 5-30.

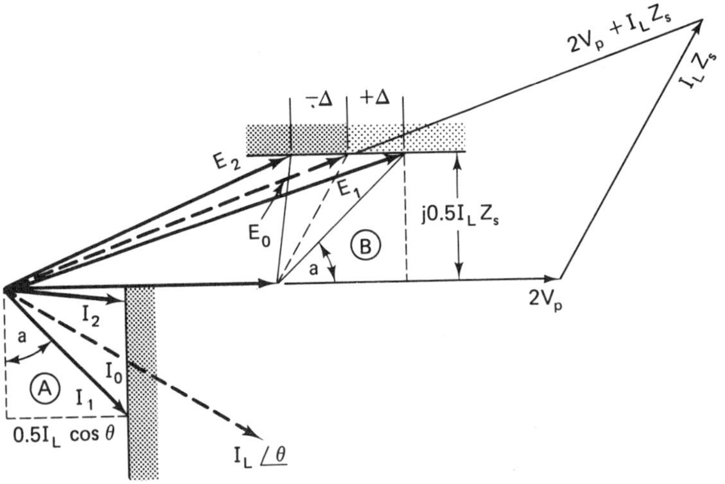

Figure 5-30 Two Paralleled Generators—Constant Power, Varied Excitation

When operating at identical levels, the generated voltage and current in each machine are E_o and I_o on the diagram. If the excitation is adjusted, the requirements are that:

- The sum of the horizontal components of E_1 and E_2 remain constant. Thus, if E_1 gains a horizontal increment Δ, E_2 must lose the same increment.
- In either generator, the horizontal component of the current must remain constant since its power may not change.
- The tip of the vectors E_1 and E_2 always lie on the horizontal line $j0.5I_L X_s$.

The resulting vectors are shown for an increase in E_1 and the concurrent decrease in E_2, with currents reflecting these changes. Note that the effect is a redistribution of reactive power only.

To redistribute active power between generators, the controls of the prime movers must be adjusted.

In Fig. 5-31, S_o is the initial speed–power curve for both G1 and G2, each of which is supplying P_{10}. If the prime mover of G1 is adjusted to the curve S_1, it delivers more power with an accompanying decrease in the power demanded from G2. The speed of the latter will rise, resulting in the power levels P_1'–P_2'. To restore the original frequency, G2 must be lowered to the S_2 curve, resulting in the division P_{11}–P_{21} in which the power gained by G1, P_{11}–P_{10}, equals the power lost by G2, P_{10}–P_{21}. Note that the change was affected by adjustments in the prime movers only; no changes were made in the excitation of either generator.

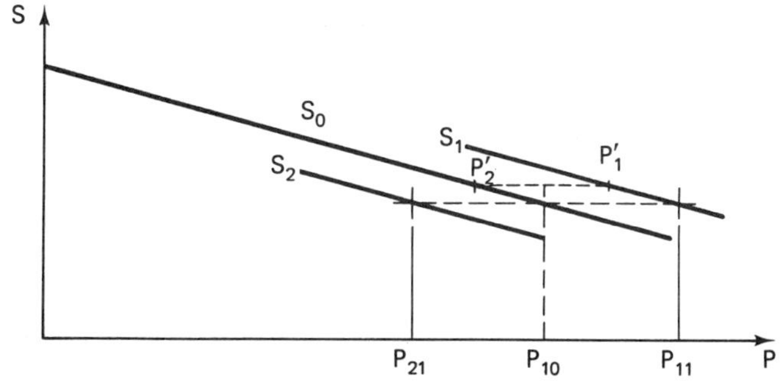

Figure 5-31 Two Paralleled Generators—Constant Excitation, Varied Power

Synchronization. At the terminals of a generator, three voltage characteristics can vary: magnitude, frequency, and phase sequence. When the machine is to be paralleled with a second alternator or system, a fourth variable is introduced—the phase angle of the voltage referred to a time base common to both machines.

The consequence of improper synchronization is illustrated by considering the paralleling of two synchronous generators. Assuming that (1) both are wye-connected with a grounded neutral, and (2) the loading is balanced, identical relationships occur in all phases, differing only by the 120° angular separation. Therefore, the system can be analyzed on a phase basis from the circuit schematic of Fig. 5-32. The generated phase voltage of each machine is represented by a single-phase alternator. G1 is supplying the line and G2 is to be added to the system.

The focus of this analysis is on *trends* when paralleling is completed without proper synchronization. Any dramatic change in the operating condition of a generator initiates a transient period that is complicated by electrical and mechanical time constants in the generator and prime mover and by nonlinear magnetic saturation. The electrical analysis in this discussion ignores these complications

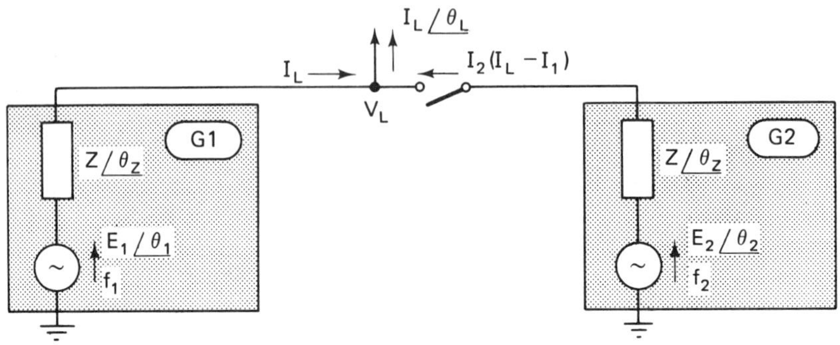

Figure 5-32 Synchronization

but does identify the problems. The currents developed in the following discussion do not represent actual values that would occur, but they do affect the envelope of transient currents when all facets are considered.

Excepting frequency, all the characteristics shown on the illustration are vector quantities. For comparable machines, the synchronous impedances are unlikely to differ significantly; it can be assumed that $Z\underline{/\theta_z}$ is identical in both with θ_z approximately 85°. Angles are not shown on I_1 and I_2 but are developed below.

At the line node with the switch closed, V_L may be expressed in the quantities of either machine. Equating the two,

$$E_1\underline{/\theta_1} - I_1(Z\underline{/\theta_z}) = E_2\underline{/\theta_2} - (I_L - I_1)(Z\underline{/\theta_z})$$

Solving for currents,

$$I_1 = \frac{(Z\underline{/\theta_z})(I_L)}{Z\underline{/\theta_z} + Z\underline{/\theta_z}} + \frac{E_1\underline{/\theta_1} - E_2\underline{/\theta_2}}{Z\underline{/\theta_z} + Z\underline{/\theta_z}}$$

$$= \frac{I_L}{2} + \frac{E_1\underline{/\theta_1} - E_2\underline{/\theta_2}}{2Z\underline{/\theta_z}}$$

$$I_2 = \frac{I_L}{2} - \frac{E_1\underline{/\theta_1} - E_2\underline{/\theta_2}}{2Z\underline{/\theta_z}}$$

In these expressions, each current has two components. The first, $I_L/2$, hereafter referred to as the *normal* component, is the current division that would occur across identical parallel impedances or, in this case, when both the magnitudes and phase angles of E_1 and E_2 are equal. With θ_z identical for both machines, the normal components of I_1 and I_2 are in phase with I_L.

The second component is a *circulating* component superposed upon the normal. Note that the magnitude of the circulating component is identical in both currents but additive in I_1 and subtractive in I_2.

Two inequalities may occur in the current equations: $|E_1| \neq |E_2|$ and/or $\theta_1 \neq \theta_2$. For the discussion, each will be considered separately.

CASE 1: $E_1 \neq E_2$, $\theta_1 = \theta_2$

A trace around the closed loop of Fig. 5-32 shows the voltage and current vectors of G2 in opposition to those of G1. On the vector diagram of Fig. 5-33 the voltage vectors and the normal current components are shown in opposition. If $E_1 \neq E_2$, a difference voltage exists; as a vector, it always is coincident with the E_1 and E_2 axis and has the magnitude $\Delta E = E_1 - E_2$. In the diagram, only instances of E_2 with a magnitude less than E_1 are illustrated.

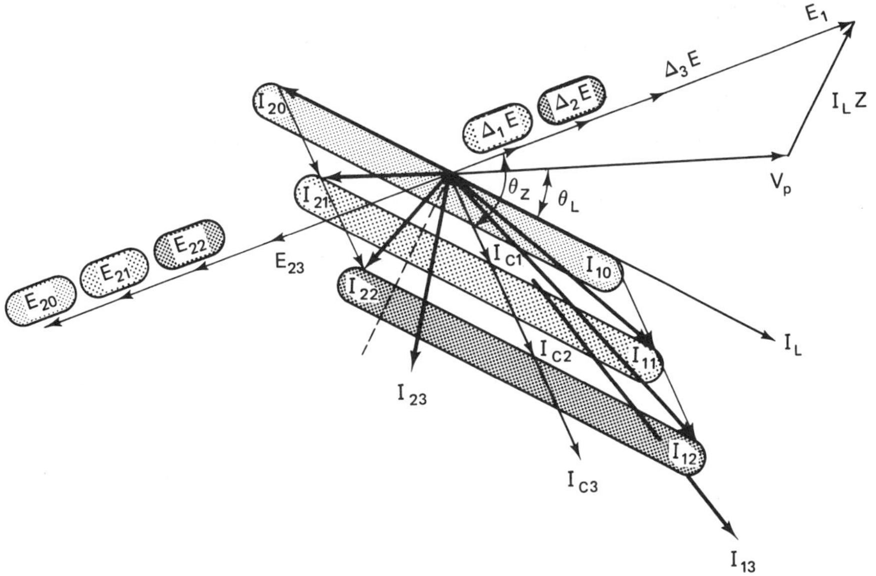

Figure 5-33 Synchronizing—$E_1 \neq E_2$

Initially, G1 is supplying the line current I_L. If the switch is closed when $E_2 = E_{20} = E_1$, I_L divides into the normal components I_{10} and I_{20}. The magnitude and angle of these normal components remain constant; any circulating component is added to each to produce the generator currents.

The circulating current will lag the difference voltage by the impedance angle, 85° in the illustration. Although the magnitude of the circulating current I_c will vary as the difference voltage changes, the angle of the current remains constant. Thus, when $E_2 = E_{21}$, the difference voltage $\Delta_1 E$ results in the circulating current I_{c1}, which combines with the normal components to identify I_{11} and I_{21}; similarly, E_{22} results in I_{12} and I_{22}.

On the diagram, a dashed line is shown perpendicular to I_L. I_2 can be separated into two vectors, one along the perpendicular and the other on the I_L axis. If I_2 lies clockwise from the perpendicular, the projection on I_{20} is in the G2 positive direction—that is, G2 is supplying a portion of the line current. But when $E_2 = E_{23}$, the current I_{23} lies counterclockwise from the perpendicular, the projection on I_{20} is negative, and G2 is not supplying line current but drawing current from G1—that is, G2 is now a motor being driven by G1.

Procedurally, it is neither practical nor desirable that $E_2 = E_1$ at the moment the switch is closed. Instead, E_2 is matched in magnitude and phase angle to V_p; after closing the switch, adjustments are made to cause the desired load division.

CASE 2: $E_1 = E_2$, $\theta_1 \neq \theta_2$ (Fig. 5-34)

If $\theta_2 = \theta_{20} = \theta_1$, the normal current division, I_{10} and I_{20}, again occurs. However, θ_2 may vary through 360°; consequently, the difference voltage vector is at the base of E_1, but the head can lie anywhere on a circle centered on the head of E_1 with a radius E_2 (= E_1). The normal components, I_{10} and I_{20}, remain on the axis of I_L but the circulating components no longer are confined to a single axis; two disparate resolutions are illustrated for θ_{21} and θ_{22}.

The maximum circulating current occurs at 180° separation, represented on the diagram by $\Delta_m E$ and the currents I_{cm}, I_{1m}, and I_{2m}.

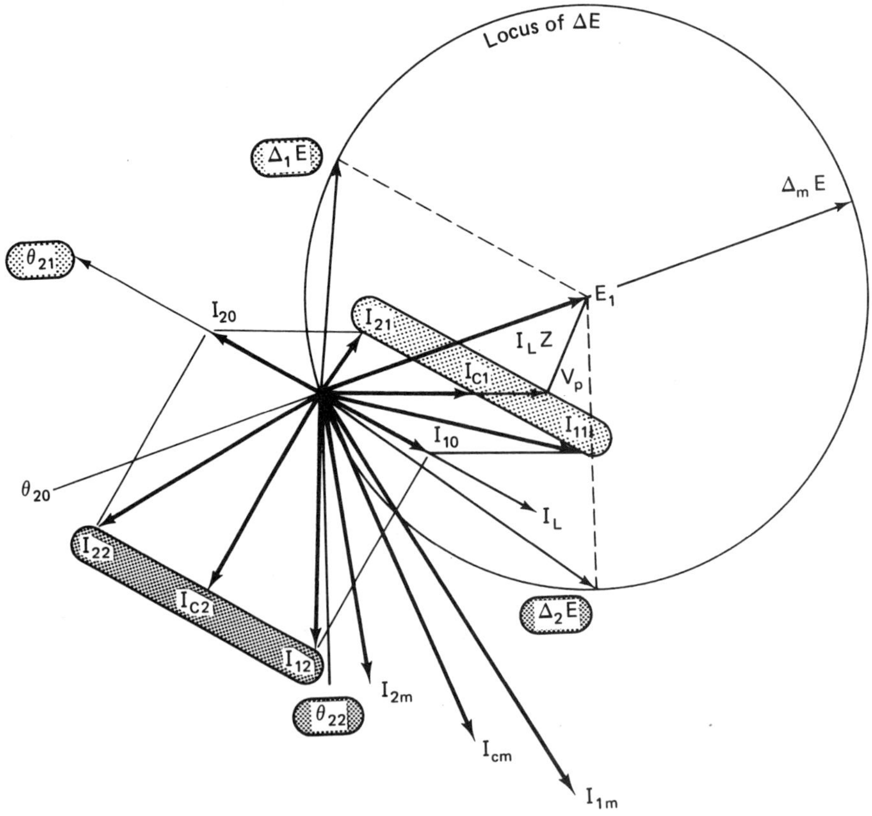

Figure 5-34 Synchronizing—$\theta_1 \neq \theta_2$

There are three circumstances that will cause differences in θ_1 and θ_2:

1. $f_1 \neq f_2$: If $f_2 = f_1 + \Delta f$, the angular velocity of the phasor E_2 is $\omega_2 = \omega_1 + \Delta\omega$. T is the period of E_1—that is, the increment of time during which E_1 rotates through 360°—in the period T, E_2 will rotate through $(\omega_1 + \Delta\omega)T = 360° + \Delta\omega T$. In Fig. 5-35(a), E_1 and E_2 both are located at zero degrees at t_0. At $t_1 = 1$ T, E_2 is at location E_{21}; at $t_2 = 2$ T it is at E_{22}. For each cycle of E_1, E_2 gains an additional angular increment and, at $t_n = nT$, is located at E_{2n}. The difference voltage traces the circular locus of Fig. 5-35(a) at the frequency $(f_2 - f_1)$. Once paralleled, the frequencies f_1 and f_2 cannot differ; any tendency toward a difference results in load shifting to maintain equality.

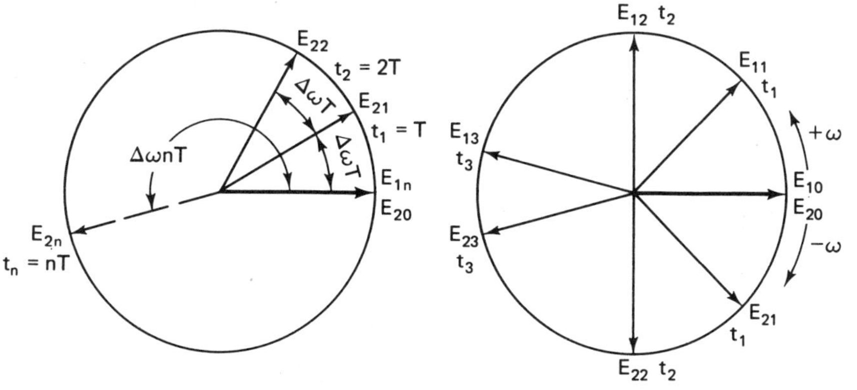

(a) $f_1 \neq f_2$ (b) Differing phase sequences

Figure 5-35 Causes of Phase-Angle Differences

2. *Differing phase sequences*: If like-lettered terminals of the polyphase generator are interconnected—that is, A_1–A_2, B_1–B_2, and C_1–C_2—any angular difference will be constant providing the phase sequence is the same for both generators. But if one generator has the sequence *ABC* and the other is *CBA*, the corresponding condition in the single-phase schematic is that of the phasors E_1 and E_2 rotating in opposite directions, illustrated in Fig. 5-35(b). In one full cycle, the maximum difference voltage occurs twice—at the positions E_{12} and E_{22} and, 180° later, when the positions of the vectors are interchanged. If sequences differ, interchange of any two of the three connections will result in identical sequences. The interchange must be made at the terminals of the oncoming generator—changing the line sequence would create havoc in many types of loads.

3. *Instantaneous angle:* With the frequencies and sequences matched, the instantaneous rotor positions in the two machines still may differ to the extent that there is a constant angular separation between the terminal voltages. Again,

once paralleling is completed this separation cannot occur, but at the moment when the switch is closed, the phase angles should be the same.

Synchronizing Procedure. A method for synchronizing an on-coming generator, often used in the laboratory and comparable in principle to the more sophisticated methods used in the power industry, is illustrated in Fig. 5-36.

A lamp bank is connected between each pair of corresponding terminals of G1 and G2. When the terminal voltages of the generator are identical, no voltage appears across the lamps and they are extinguished, but any disparity will cause the lamps to glow. If $E_1 = E_2$, the peak voltage that can occur across each bank is $2E_p$, approximately $2V_L/\sqrt{3}$.

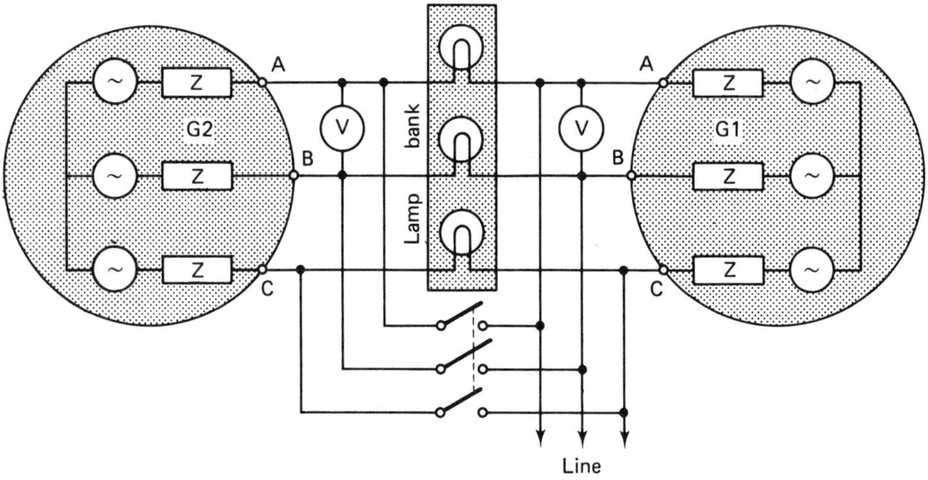

	Indication	Cause	Correction
	Rise and fall— Not in unison	Different phase sequences	Interchange two lines of oncoming generator
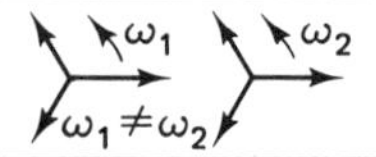	Rise and fall— In unison	Differing frequencies	Raise or lower speed of oncoming generator
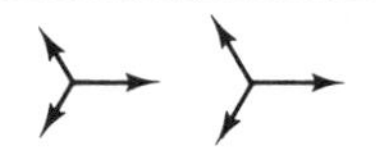	Steady glow	Unequal voltages	Adjust excitation of oncoming generator
		Phase separation	Slight change in speed of oncoming generator
All dark—correct condition for paralleling			

Figure 5-36 Synchronizing Procedure

When the phase sequences differ, the voltage across each bank will rise and fall, but peak values do not occur simultaneously in all phases, and the lamps will not vary in unison. If the sequences are correct but frequencies differ, the rise and fall is simultaneous in all phases. Finally, if the lamps glow steadily there are two possible causes: unequal voltages, observable on the voltmeters, or a phase-angle difference.

It would be a difficult if not impossible task to get complete correspondence between G1 and G2 before closing the switch. With the sequences matched and the voltages equal, the speed of G2 can be adjusted until the frequencies very nearly match and the rise and fall in the lamps occurs over a relatively long time span. The switch then may be closed at the moment when all lamps are extinguished.

5-20. POWER FACTOR CORRECTION

Most electrical loads are inductive. Unimproved industrial power factors range from 0.35 to 0.85 lagging, and raising the pf is an objective of both the plant and the supplying utility.

With voltage as the reference vector, a current lagging the voltage by the power factor angle θ may be separated into two components, one in phase and the other in quadrature with the voltage. The in-phase component, $I \cos \theta$, is active current; $I \sin \theta$ is reactive current largely attributable to magnetizing current in electromagnetic equipment. When the current and its components are multiplied by the circuit voltage, power relationships may be represented by a power triangle (Fig. 5-37).

Apparent power, S, is the product of the circuit voltage and current as registered on a voltmeter and ammeter. For three-phase power, $S = \sqrt{3}\ VI$. The unit is volt-amperes (VA or kVA).

True, or *active power*, P, is the power registered on a wattmeter. The unit is watts (W or kW). $P = S \cos \theta = \sqrt{3}\ VI \cos \theta$ (3-phase).

Reactive power, Q, is the product $S \sin \theta = \sqrt{3}\ VI \sin \theta$ (3-phase). The unit is reactive volt-amperes (VAr or kVAr); it may be measured with a varmeter.

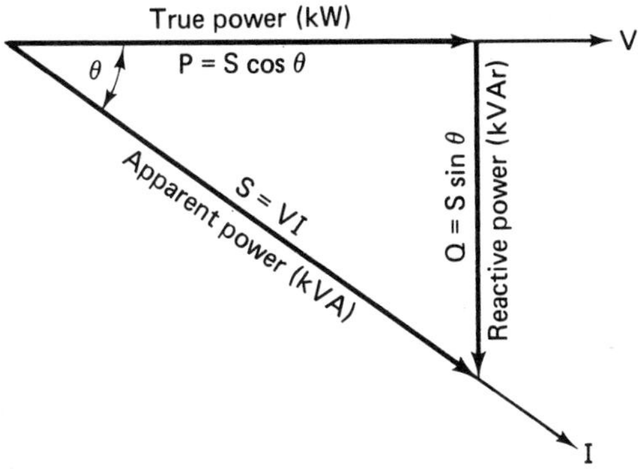

Figure 5-37 Power Triangle

True power is the useful converted power plus the unavoidable losses; at unity power factor ($\theta = 0°$) all of the current is active. Reactive current, although vital in electromagnetic conversion processes, does not itself convert to useful power and is detrimental in the distribution system. Equipment ratings are based on thermal properties that limit their permissible currents; it is for this reason that ac equipment is rated in kVA rather than kW. Equipment is underutilized when, for example, supplying 1000 kW at 0.8 pf, for at unity pf it could deliver 1250 kW. Switchgear, cables, connectors, alternators, transformers—all equipment in the generation and distribution system—require increased capacity as the power factor decreases, with voltage drops and I^2R losses similarly increased. Power factor correction is directed at elimination of reactive power not in the functional equipment, where it is essential, but in the distribution system; ideally, the correcting device is located at the terminals of the equipment responsible for the reactive power.

Correction requires introduction of leading kVAr to eliminate lagging kVAr. Leading kVAr may be supplied by capacitors; standard three-phase capacitors are available in ratings up to 13.2 kV 600 kVAr and may be banked to provide greater capacity.

An alternative method of improving the pf is addition of synchronous motors, overexcited to operate at a leading pf. Occasionally, a synchronous motor is installed exclusively for power factor correction; it drives no load, develops little torque, and consequently has a very light frame. This form of the motor is termed synchronous capacitor (condenser).

Selection between capacitors and synchronous motors for power factor correction is a complex problem. Economics must be balanced against the differing system characteristics for capacitors and motors. Usually, correction to unity pf requires unjustified expenditures. On the other hand, normal utility rates may be based upon a minimum pf as high as 0.9 or 0.95, with severe penalties at lesser pf.

Example 5-8: The machining operations in an industrial plant require 500 kW at lagging 0.5 pf. (a) What is the required kVA rating of a synchronous capacitor to improve the pf to 0.9? (b) If a 500-hp 89% efficient synchronous motor operating at 0.8 leading pf is added in lieu of a capacitor, what is the resulting pf and kVA?

SOLUTION: (Fig. 5-38) (subscripts are *p*, plant; *c*, capacitor; *m*, motor)

(a) $S_p = \dfrac{P}{\text{pf}} = \dfrac{500}{0.5} = 1000 \text{ kVA}$

$\theta_p = \cos^{-1} 0.5 = 60°$

$Q_p = S \sin\theta = 1000 \sin 60 = 866 \text{ kVAr}$

The power consumed by a synchronous capacitor is small and generally neglected. Thus, the power in the improved system remains 500 kW. For 0.9 pf,

$$S'_p = \frac{500}{0.9} = 556 \text{ kVA}$$

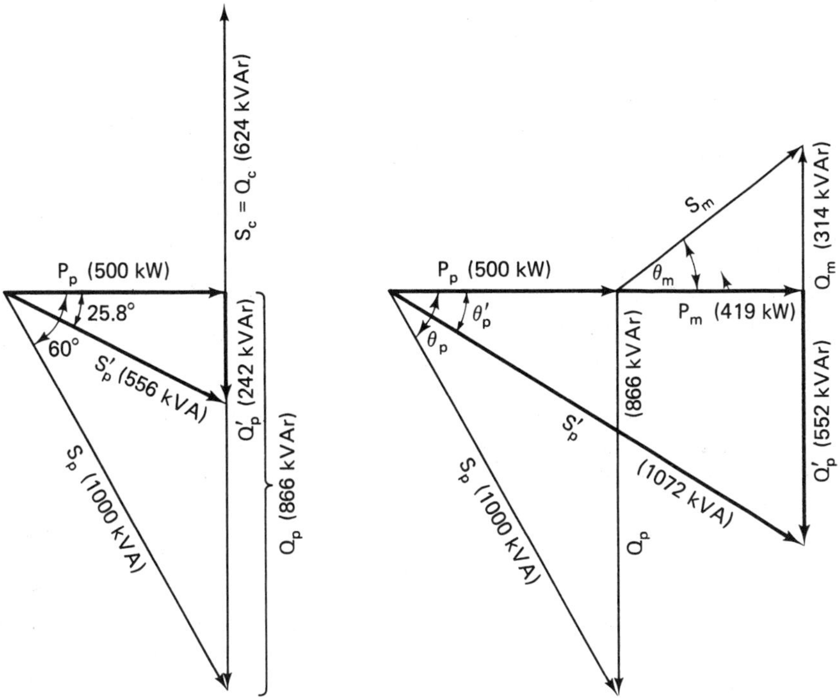

Figure 5-38 Example 5-8

$$\theta'_p = \cos^{-1} 0.9 = 25.8°$$
$$Q'_p = 556 \sin 25.8 = 242 \text{ kVAr}$$

The required synchronous capacitor kVA is

$$S_c = Q_c = Q_p - Q'_p = 866 - 242 = 624 \text{ kVA}$$

(b) $P_m = \dfrac{P_o}{\eta} = \dfrac{(500)(0.746)}{0.89} = 419 \text{ kW}$

$$S_m = \frac{P_m}{\text{pf}} = \frac{419}{0.8} = 524 \text{ kVA}$$
$$\theta_m = \cos^{-1} \text{pf} = \cos^{-1} 0.8 = 36.9°$$
$$Q_m = S_m \sin \theta_m = 524(\sin 36.9) = 314 \text{ kVAr}$$

The plant power triangle becomes

$$P'_p = P_p + P_m = 550 + 419 = 919 \text{ kW}$$
$$Q'_p = Q_p - Q_m = 866 - 314 = 552 \text{ kVAr}$$
$$S'_p = \sqrt{P^2 + Q^2} = 100\sqrt{9.19^2 + 5.52^2} = 1072 \text{ kVA}$$
$$\text{pf} = \frac{P}{S} = \frac{919}{1072} = 0.857$$

PROBLEMS AND QUESTIONS

The machine to which Problems 5-1 to 5-11 apply has the following characteristics: three-phase Y-connected 60 Hz, 72 slots, 72 coils, 60 turns per coil, eight poles, and 6.62 × 10^{-3} Wb per pole.

5-1. What is the rotor speed?

5-2. Find the number of belts per phase and the number of coils in each belt.

5-3. How many turns are there in each phase?

5-4. What is the slot span in electrical degrees ($d°$)?

5-5. What are the possible coil spans (in slots) if the pitch in electrical degrees must be greater than 120 but less than 180? For each, what is the pitch factor?

5-6. Find the distribution factor.

5-7. The coil span is eight slots. Find the winding factor.

5-8. Find the generated voltage per phase.

5-9. What is the terminal voltage?

5-10. Rated current is 260 A per coil. What is the rated kVA?

5-11. If the connection is changed from Y to Δ, what is the rated line voltage, line current, and power to a 0.75-pf load?

5-12. A 10-MVA three-phase 60-Hz alternator is delta-connected. The machine has 90 slots, 90 coils, 60 turns per coil, six poles, 1.58 × 10^{-2} Wb per pole, and a coil span of 12 slots. Find the generated voltage per phase, line voltage, and coil current.

5-13. An alternator is rated 100 MVA, 1800 r/min, 13.8 kV, 60 Hz. If neither the rated armature or field current is changed, what are the ratings with the speed reduced to 1500 r/min?

5-14. The terminal board of a six-phase laboratory generator has numbered terminals for the ends of each stator winding; no windings are interconnected internally. The rating of each winding is 64 V 11.8 A. A vector diagram [Fig. 5-39(a)] identifies phase angles and terminal markings—for example, one winding has the terminals 5 and 6 with V_{65}

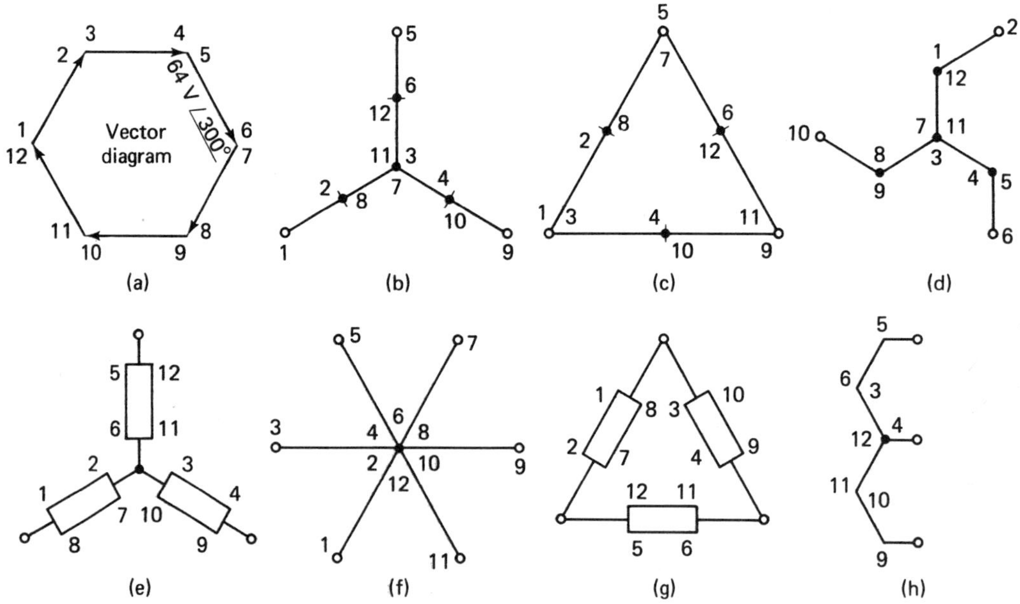

Figure 5-39 Problem 5-14

$= 64\text{ V }\underline{/300°}$. The remaining subfigures of Fig. 5-39 show a number of optional connection diagrams. For each, determine the line voltage, line current, kVA rating, and number of phases.

5-15. A two-phase system with line voltages $V\underline{/0°}$ and $V\underline{/90°}$ is supplying a balanced resistive load. (a) In terms of V_m and I_m, what is the load power? (b) What is the power in terms of rms V and I? (c) Is the power pulsating or constant?

5-16. A stator has two independent, mutually perpendicula windings. Φ_m is the maximum flux created in each winding. (a) With two-phase voltages $V\underline{/0°}$ and $V\underline{/90°}$ applied to the vertical and horizontal windings, respectively, what is the net stator flux? Is this flux rotating or confined to one axis? (b) If both windings are excited by $V\underline{/0°}$, what is the net stator flux? Is it rotating or confined to one axis?

5-17. The power supplied to a balanced three-phase unity pf load is $\sqrt{3}\ V_L I_L$. In Paragraph 5-2 the power is shown as $1.5 V_m I_m$. Are these expressions consistent? Explain.

5-18. (a) If at rated load the field current of the alternator of Fig. 5-14 is 30 A and $V_p = 1390$ V, with the load removed and no change in the field current, what is the open-circuit phase voltage? (b) For this loading, what is the %VR? (c) To what value must the field current be reduced at no load in order to achieve 0% VR?

5-19. A synchronous generator is rated 3750 kVA 0.8 pf 3000 kW 6900 V. The open-circuit voltage and short-circuit current characteristic curves appear in Fig. 5-40.
(a) Find the approximate %VR and field current for rated power at 0.8 pf lagging.
(b) Repeat part (a) for unity pf.
(c) Repeat part (a) for 0.75 pf lagging.
(d) The field current may not exceed the value for rated power at rated pf. Can the generator deliver rated kVA at both unity and 0.75 lagging pf without exceeding the field current rating?

5-20. In the expression for instantaneous voltage,

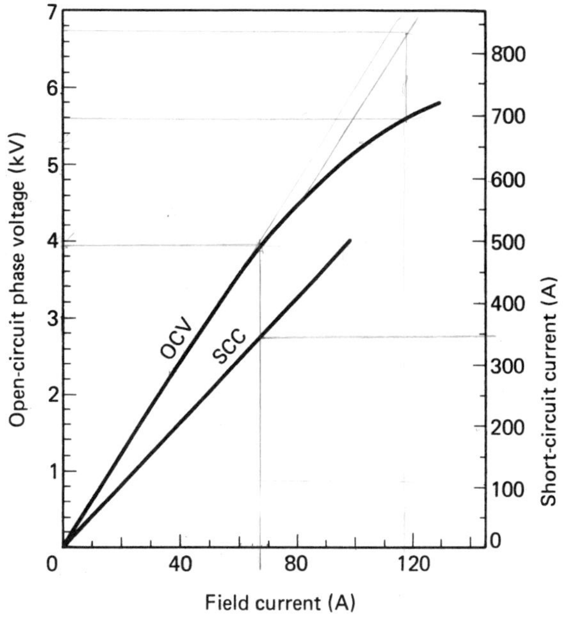

Figure 5-40 Problem 5-19

$V_m \sin \omega t$, ω is defined as the angular frequency. In a rotating machine, ω is the angular velocity of the rotor. Both terms have the unit rad/s. Denoting the former by ω_f and the latter by ω_r, how are they related in a synchronous machine?

5-21. The power developed by each of two alternators is identical. One alternator has four poles and the other has 40. (a) If the frequency of both alternators is 60 Hz, what is the ratio of the developed machine torques? (b) If the frequency of the 4-pole machine is 60 Hz but that of the 40-pole is 50 Hz, what is the torque ratio?

5-22. The loss curves for the generator of Problem 5-19 are given in Fig. 5-41. The field voltage is 250 V; when delivering rated power at 0.8 pf, the field current is 116 A. By test, the friction and windage loss is found to be 37 kW. At 20°C the dc phase resistance is 0.127 Ω; the operating temperature is 75°C. What is the full-load efficiency for a 0.8-lagging-pf load?

5-23. The motor of Example 5-6 is developing 2500 hp. Field excitation is unchanged—E_p remains at 3194 V. Losses are disregarded. Find the (a) torque, (b) torque angle, (c) power factor angle, and (d) line current.

5-24. A synchronous motor operates on a 60-Hz supply to drive an alternator whose output frequency is 25 Hz. (a) What is the minimum possible number of poles in each machine? (b) With the minimum number of poles, what is the rotational speed? (c) Is the speed of part (b) the minimum or the maximum speed at which this frequency conversion can be affected with no restrictions on the number of poles?

5 25. A claw-type rotor is used in some alternators (e.g., automobile, synchronous generator excitation systems). The rotor, illustrated in Fig. 5-42, is two identical ferromagnetic shells with projecting fingers that are fitted to cup a doughnut-shaped coil, the latter excited by dc. In the illustration, each claw has seven fingers. The rotor is used in a con-

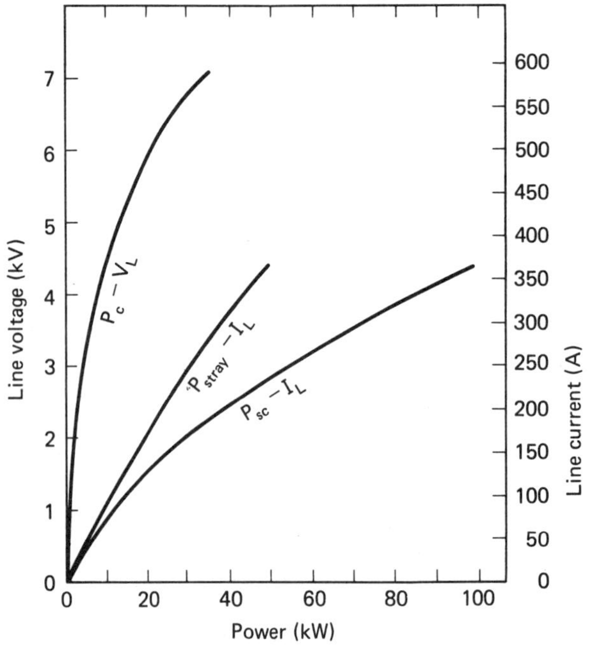

Figure 5-41 Problems 5-22 and 5-26

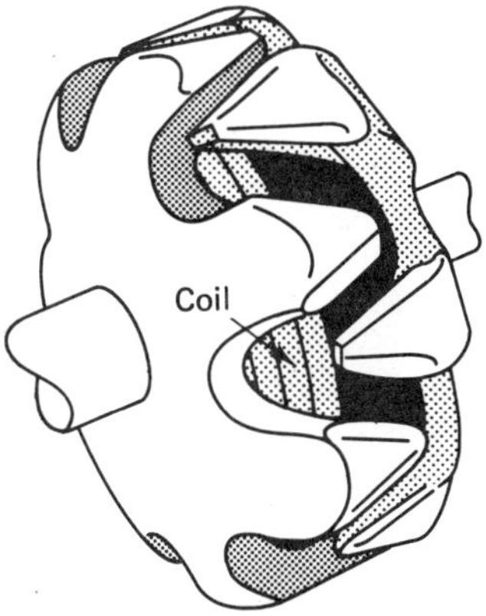

Figure 5-42 Problem 5-25

ventional three-phase stator. (a) Analyze the magnetic circuit and explain the fundamental principles. (b) How many poles are on this rotor? (c) Must the coil rotate or may it be static? (d) Could this rotor be used in a synchronous motor?

5-26. The curves of Figs. 5-40 and 5-41 apply to a 4000-hp 6.9-kV 60-Hz 900-r/min 0.8-pf (lead) synchronous motor. At operating temperature, $R_s = 0.155\ \Omega$ and $X_s = 11.0\ \Omega$. At rated load, the line input is 3200 kW at 0.8 pf and the field input is 33 kW. $P_{fw} = 37$ kW. (a) Find the power output (hp), efficiency, and load torque. (b) How many leading kVArs are supplied by the motor?

5-27. A 1250-kVA 4160-V synchronous generator is supplying a line current of 150 A to a wye-connected unity pf load. $E_p = 3000$ V, $V_p = 2400$ V, and $Z_s = 6.4\underline{/86^\circ}\ \Omega$. For the purpose of this problem, assume that E_p (3000 V) and the load current (150 A) remain constant. (*Note:* The line current of this generator *may* vary with problem conditions.)

(a) A second, identical generator is paralleled with the on-line generator. For each of the following conditions in the on-coming generator at the moment of completion of the paralleling connection, find the theoretical circulating current and the line current of each generator.

(1) $V_L = 4160$V in phase with V_L of the on-line generator.

(2) $E_p = 2400$ V lagging E_p of the on-line generator by 60°.

(3) $E_p = 3000$ V lagging E_p of the on-line generator by 30°.

(b) Discuss the validity of the assumption.

5-28. The uncorrected power factor of a plant operation is 0.65. The power requirement for the operation is 600 kW. Find the capacitive kVArs required to raise the power factor to (a) 0.9, and (b) 1.0.

5-29. An industrial plant requires 1500 kW at 0.7 lagging pf. The plant is expanded with the addition of a synchronous motor whose input is 1500 kVA at 0.8 leading pf. (a) What is the power factor of the plant after the addition? (b) What is the net increase in plant kVA? In plant power?

5-30. The NEC specifies the following: "The total kVAr rating of capacitors that are connected on the load side of a motor controller shall not exceed the value required to raise the no-load power factor of the motor to unity." The no-load and full-load inputs to a 50-hp motor are 2 kW at 0.1 lagging pf and 44 kW at 0.85 lagging pf, respectively. (a) If the pf is improved by capacitors to unity at no-load, what is the corrected full-load pf? (b) If the full-load pf were corrected to unity, what would be the no-load pf?

6

Polyphase Induction Motor

When electricity first was put to use on a relatviely broad scale, it was in the form of direct current. By the mid-1880s, 110- and 220-V dc systems were fairly common but had the accompanying disadvantages of low-voltage distribution—poor regulation and efficiency. Compounding the problem, dc voltage could not (and cannot) be transformed to higher voltages. Practical alternators and transformers were known and a desire for conversion to ac was common, but the missing link was an effective alternating current motor. This need was met in 1888 when Nicola Tesla invented the polyphase induction motor. From that point on, ac predominated; today, the induction motor, with its simplicity of construction, relatively constant speed, ruggedness, reliability, and comparatively low cost, is the principal industrial motor.

6-1. OPERATIONAL PRINCIPLES

Three basic principles from the study of synchronous machines are reiterated:

1. *A flux rotating at angular velocity ω will induce voltages in a polyphase winding*. The function of the rotor in a synchronous generator is merely to provide the rotating flux; an identical flux produced by other means will generate identical voltages.

2. *If currents are permitted to flow as a result of the generated voltages, they produce a flux that also rotates at velocity* ω *but lags the inducing flux*. The angular lag is $90° + \alpha$, where α is the angle by which the current lags the voltage.

3. *Application of three-phase voltages to a three-phase winding results in a flux rotating at the angular velocity* $4\pi f/P$ [SI] *or speed* $120f/P$ [US].

In Fig. 6-1 a three-phase voltage is applied to a stator identical to that of a synchronous machine, creating a rotating flux, Φ_s. A rotor with windings identical to those of the stator is inserted in the latter but blocked so that it cannot rotate. The stator flux will induce polyphase voltages in the rotor. If the rotor winding is short-circuited, the resulting currents produce a rotor flux that follows the stator flux but lags it by $90° + \alpha$.

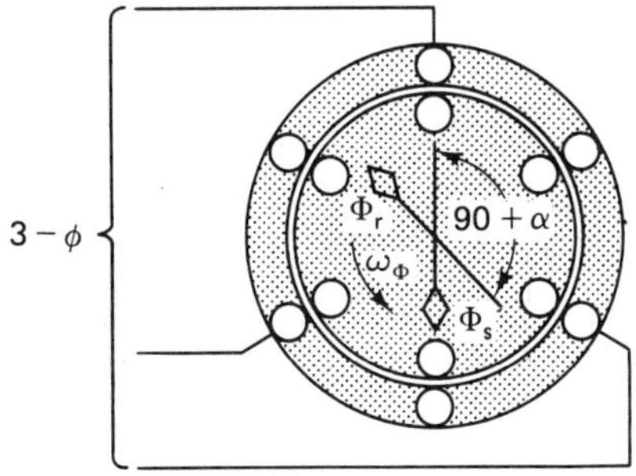

Figure 6-1 Induction-Motor Principle

In the salient pole of a synchronous motor, the flux created by dc excitation obviously is fixed within the iron of the rotor. Less obvious but true, the rotor flux in the illustration, although constantly changing its position within the rotor iron, nevertheless is fixed in it *at any instant*. Consequently, the rotor flux reacts with the stator flux to impart torque to the rotor. The torque angle is $90° + \alpha$. If the rotor now is freed, it will rotate in the direction of the revolving stator flux.

To induce a voltage, there must be relative movement between the stator flux and the rotor. If the rotor revolves at the speed of the stator flux, no voltage is induced, no rotor flux is created, and no torque is developed. However, at any rotor speed less than that of the flux, voltages are induced, currents flow, a rotor flux is created, and torque is produced. Thus, the freed rotor will accelerate to a speed where the resulting electromagnetic torque is in balance with countertorques acting on the rotor, and it will continue to rotate at that speed as long as there is no further change in the system. This is an induction motor—a motor in which power is introduced to the rotor not by conduction but by induction which, in many respects, is comparable to the transfer of power from the primary to the secondary of a transformer. In fact, the similarity permits borrowing transformer theory for application to the motor.

6-2. EQUIVALENT CIRCUIT

The synchronous speed of an induction motor, S_S, is the speed of the revolving flux, $120f/P$, where f is the frequency of the applied voltage. The rotor speed S_R typically is on the order of $0.95S_S$ at rated load.

The frequency of the voltage induced in the rotor is determined not by S_S or S_R but rather by their difference, $S_S - S_R$, for this is the speed of the flux with respect to the rotor windings. Therefore, the rotor frequency, f_r, is $P(S_S - S_R)/120$. Applying ratios,

$$\frac{f_r}{f} = \frac{P(S_S - S_R)/120}{PS_S/120} = \frac{S_S - S_R}{S_S} = s$$

$S_S - S_R$ is termed *slip speed*, and s is *slip*. The motor could be analyzed on the basis of slip speed, but it is more convenient to work with slip. Thus, at any speed,

$$\mathbf{f_r = sf} \tag{6-1}$$

$$\mathbf{S_R = (1 - s)S_S} \tag{6-2}$$

When the rotor is blocked, $S_R = 0$, $s = 1$, and $f_r = f$.

Suppose that a rotor, constructed from ferromagnetic material but having no winding, is inserted into a stator and blocked to prevent rotation. In Fig. 6-2(a), the stator and phase currents are identical to those of Fig. 5-6. From Fig. 5-6(b), at $t = 0$, Φ_s will be at the location shown on the sketch. This discussion focuses on phase A and assumes that the machine is ideal; there is no power conversion and the stator current consists of only a magnetizing component, lagging the applied voltage by 90°. In the accompanying vector diagram, I_s' is the magnetizing current, oriented at 0°, with the phase voltage, V_p, at +90°.

It must be remembered that Φ_s is rotating at synchronous speed—the location on the sketch is an instantaneous position. The rotating Φ_s will induce a counter emf, E_p, in the stator windings. From Lenz's law, the counter emf must be out of phase with V_p by 180°, placing it at −90° on the vector diagram. This location can be verified from the sketch by considering currents induced in the phase A winding; as Φ_s continues to rotate, flux through coil a_s–a_s' increases to the right and requires that induced currents flow *into* the plane of conductor a_s. However, the basis for this figure and Fig. 5-6 is a positive current out of the plane; the voltage is therefore negative and, from the location of the flux with respect to the coil, at its maximum value—that is, E_p is at −90°.

The rotor of Fig. 6-2(a) now is replaced with a rotor having windings identical to those of the stator [Fig. 6-2(b)]. The rotor remains blocked. The voltage induced in rotor coil a_r–a_r' must be equal to E_p, since the coils are identically situated and the same flux is responsible for both voltages. This circumstance is duplicated in phases B and C; therefore, if the rotor conductors are short-circuited, again assuming only magnetizing reactance is present, a rotor flux is created which directly opposes the stator flux. In the ideal machine, E_p essentially equals V_p; to retain a constant E_p, the stator must supply an additional flux, equal but in opposi-

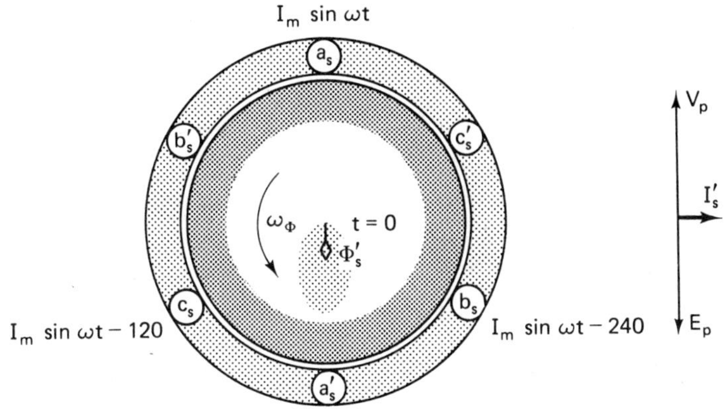

(a) Stator only

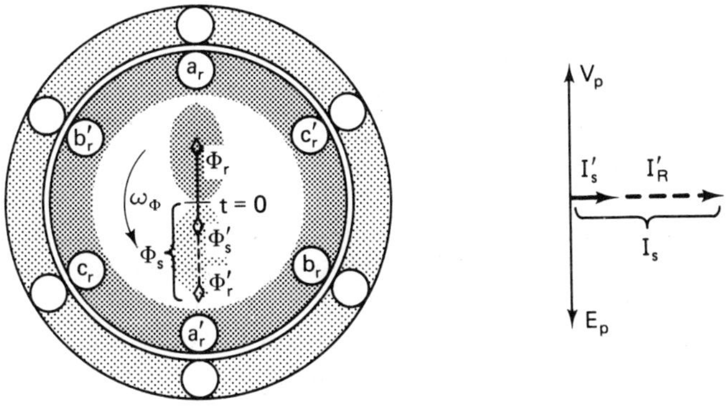

(b) Stator and rotor

Figure 6-2 Fluxes and Stator Voltages

tion to Φ_r, and the total stator flux is $\Phi'_s + \Phi'_r$ with a corresponding current $I_s = I'_s + I'_r$.

When Fig. 6-2 is compared with Fig. 4-2, it is seen that the stator–rotor relationship duplicates the relationship between the primary and secondary of an ideal transformer.

Will this duplication continue if the rotor now is unblocked and rotating at speed S_r? The answer to this question is complicated by two factors; (1) because of rotation, a_r–a'_r no longer remains opposite a_s–a'_s, and (2) the speed of the flux relative to the rotor coils is not the speed of the rotating flux but rather sS_S.

Let the synchronous flux rotate for time t_1. The angle through which it rotates is $\beta = 360 S_S t_1$ degrees (Fig. 6-3). From equation 6-2, with the rotor freed

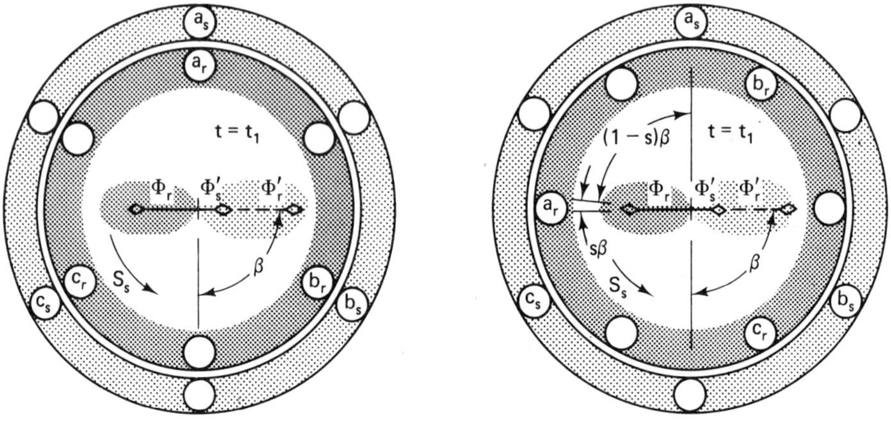

(a) Rotor blocked (b) Rotor freed

Figure 6-3 Stator Flux

$S_r = (1 - s)S$; in the period t_1 the rotor will revolve through $360(1 - s)S_S t_1 = (1 - s)\beta$. Within the rotor, the speed of the rotor flux relative to the rotor is sS_S. The angular rotation of the rotor flux *with respect to the rotor* in the interval is $360(sS_S)t_1 = s\beta$. When related to the stator, the displacement of the rotor flux is the sum of the rotor displacement and the angular movement of the flux within the rotor—that is, $(1 - s)\beta + s\beta = \beta$ degrees. Hence, despite rotation, Φ_r remains in direct opposition to Φ_s. The parallelism of stator–rotor and transformer primary–secondary relationships is reaffirmed.

If the above analytical approach is repeated for a practical machine with power components introduced in the form of resistance in the stator and rotor windings, the results parallel those of the realistic transformer. It therefore is possible to adapt the equivalent circuit of the transformer to the induction motor.

The utility of a transformer requires access to the secondary voltage. However, in most cases there is neither a need nor, in many motors, a capability to access the rotor voltage. The aim is to provide transformed electrical power to the rotor for conversion to mechanical power that becomes available at the rotor shaft. Consequently, actual circuit quantities in the rotor usually are unimportant; performance is readily evaluated from an equivalent circuit with rotor parameters referred to the stator.

The rotor flux in the induction motor is attributable to the mmf in the rotor windings. To this point, the discussion assumed the rotor windings were identical to the stator windings but short-circuited. In a short-circuited phase belt consisting of n single-turn coils separated by $d°$, the mmf is ni_c, where i_c is the current in the coils. The mmf is therefore

$$\mathfrak{F} = n\left[\frac{\underline{e/0} + \underline{e/d} + \underline{e/2d} + \cdots + \underline{e_n/(n-1)d}}{nZ}\right]$$

$$= \frac{\sum \bar{e}}{Z}$$

in which e is the effective coil voltage, $\sum \bar{e}$ the vector sum of the coil voltages, and Z the impedance of each coil.

Suppose that we now isolate and short-circuit individual coils rather than the series-connected belt. The mmf now is

$$\mathfrak{F} = \frac{e/0}{Z} + \frac{e/d}{Z} + \frac{e/2d}{Z} + \cdots + \frac{e/(n-1)d}{Z}$$
$$= \frac{\sum \bar{e}}{Z}$$

The mmf is the same in both cases. This fact allows departure from the conventional polyphase winding in the rotor structure.

A rugged, easily constructed, economical, and highly effective rotor is illustrated in Fig. 6-4. This is a *squirrel-cage rotor*. Laminations, punched from magnetic steel and including slot cutouts, are assembled on the shaft. Copper or aluminum is cast in the slots or bars are inserted to provide conductors, and cast end rings form a low-resistance short circuit. A realistic squirrel-cage will have more slots than appear on the illustration, and the shape of the slots differs; in fact, operational characteristics may be changed by altering slot geometry.

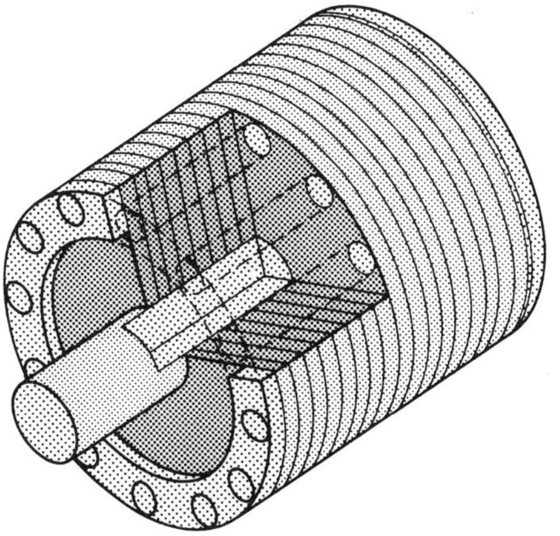

Figure 6-4 Squirrel-Cage Rotor

A second type of rotor is the *wound rotor*, in which insulated conductors are inserted to form a polyphase winding, with the ends of each phase connected to slip rings. The rotor must be wound for the same number of poles as the stator, but although the rotor normally is wound for the same number of phases as the stator, this is not essential. The number of rotor turns seldom match the number of stator turns. (Regarding phases, the squirrel-cage rotor may be viewed as a rotor with a large number of phases, each having phase belts of but one turn.) Later, it will be seen that access to the wound rotor by means of the slip rings permits control of the running characteristics. From the viewpoint of building the rotor, the wound rotor is more costly than the squirrel cage.

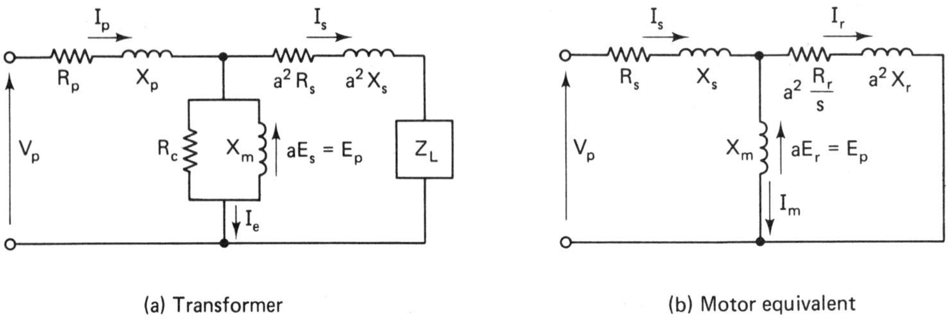

Figure 6-5 Transformer-Motor Analogy

The schematic of a realistic transformer is shown in Fig. 6-5(a). Although it appears as a single-phase transformer, it also is representative of one transformer in a wye-connected three-phase bank. In the transformer, the primary and secondary are static. However, when translating the transformer schematic to a phase representation of the induction motor, the stator is static but the rotor is not. The diagram must be adapted to reflect speed or, more conveniently, slip.

For any line current, E_p is $V_p - I_s(R_s + jX_s)$. (*Note*: The subscript p is retained in the motor schematic to identify phase voltages in contrast to line voltages.) With the rotor locked, the *reflected* rotor voltage, aE_r, must equal E_p. E_p remains constant and the rotor voltage at any speed must be $s(aE_r)$, for voltage is proportional to frequency and $f_r = sf$. By the same token, the rotor reactance at any speed is related to the locked-rotor reactance by the frequency ($X = 2\pi fL$); thus, at any speed, the reflected rotor reactance is sa^2X_r, where X_r is the reactance with the rotor locked. Finally, in representing the rotor circuit, Ohm's law must be properly satisfied:

$$\frac{I_r}{a} = \frac{s(aE_r)}{a^2R_r + jsa^2X_r} = \frac{aE_r}{a^2(R_r/s) + ja^2X_r}$$

From this relationship, the rotor may be represented as shown in Fig. 6-5(b).

In the realistic transformer, the exciting current, $I_e = E_p(1/R_c + 1/jX_m)$, is small, permitting elimination of R_c and X_m from the schematic with the understanding that the core loss is not thereby excluded from efficiency calculations. The magnetizing component of I_e in the transformer is negligibly small because the magnetic circuit consists only of low-reluctance steel; however, in the induction motor the magnetic circuit includes high-reluctance air gaps that require a large I_m to achieve a satisfactory flux. I_m cannot be disregarded, but the core-loss component, I_c, remains small enough to be negligible. R_c may be eliminated, retaining the stipulation that core loss must be included in efficiency calculations. However, X_m must be retained. Thus, the motor circuit is that shown in Fig. 6-5(b).

This circuit has two deficiencies: (1) the turns ratio is included, yet it seldom is known and cannot be determined by normal testing, and (2) the power in the reflected resistance, a^2R_r/s, is the sum of the rotor I^2R loss and the developed power (power converted from electrical to mechanical form).

The first deficiency is eliminated by regarding the turns ratio as unity—that is, assuming that the rotor has the same number of turns as the stator. This, in essence, is the achievement when a turns ratio is applied—the actual rotor is replaced with an equivalent rotor for which the turns ratio is unity. E_r, R_r, and X_r are the assumed rotor quantities.

The second deficiency may be remedied merely by subtracting R_r from R_r/s:

$$\frac{R_r}{s} - R_r = \frac{R_r - sR_r}{s} = \frac{(1-s)R_r}{s}$$

R_r is the component in which I^2R losses occur; therefore, $(1 - s)R_r/s$ must be the resistive equivalent of developed power.

The final equivalent is shown in Fig. 6-6. The difference between I_R and I_s is substantial and prevents a combination comparable to R_{ep} and X_{ep} in the simplified transformer schematic. The core loss is not represented in the circuit; further, friction, windage, and stray-load losses are not separately identified but are a part of the developed power.

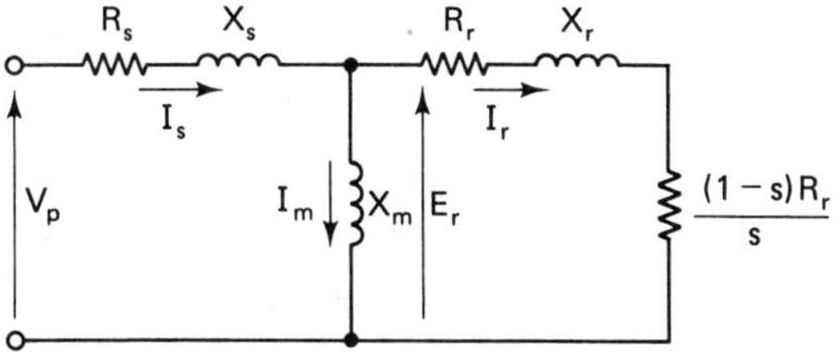

Figure 6-6 Motor Schematic

6-3. PERFORMANCE ANALYSIS

When the currents in Fig. 6-6 are known, the power in any portion of the circuit may be identified. For the three phases of the machine,

$$P_s = 3I_s^2R_s \qquad (I^2R \text{ loss in the stator})$$

$$P_r = 3I_r^2R_r \qquad (I^2R \text{ loss in the rotor})$$

$$P_d = \frac{3I_r^2(1-s)R_r}{s} \qquad (\text{developed power})$$

The rotor power input, RPI, is

$$\text{RPI} = \frac{3I_r^2R_r}{s} = 3I_r^2R_r + \frac{3(1-s)I_r^2R_r}{s}$$

From this relationship,

$$P_r = s\text{RPI}$$
$$P_d = (1 - s)\text{RPI}$$

For output power, the friction, windage and core losses must be deducted at some point. These losses literally occur at several places in the circuit—that is, stray-load losses in R_s and R_r, core loss at the point where R_c is removed in simplifying the circuit, and friction and windage losses as part of the developed power. But to compute these losses at the point where they occur would require extensive testing of the motor to identify individual machine quantities. Further, there would be little difference in the computations from those obtained if *all* these losses are assessed against *developed power*—in fact, the difference in power output computed by the two methods would in all probability be less than the tolerance in either computation when the precision of test measurements is considered. IEEE Std 112A assesses $P_{fwc+\text{stray}}$ against P_d. The resulting power balance is

$$P_{\text{in}} - P_s = P_r + P_d = P_r + (P_o + P_{fwc+\text{stray}}) = \text{RPI}$$

Two complementary analytical techniques permit evaluation of the phase circuit of the motor: (1) combining the parallel magnetizing and rotor branches to produce a series circuit that accents stator quantities and permits identification of line currents, power factors, power inputs, and efficiencies; and (2) conversion to a Thévenin equivalent with R_r/s as a variable load resistance to stress the rotor and, in particular, relate slip to motor performance. Either method permits determination of output power, torque, and slip.

The first approach is illustrated in Example 6-1.

EXAMPLE 6-1: A 550-V 200-hp three-phase 4-pole 60-Hz induction motor has the following characteristics:

$$X_s = 0.1\ \Omega \qquad X_r = 0.286\ \Omega \qquad X_m = 7\ \Omega$$
$$R_s = 0.052\ \Omega \qquad R_r = 0.062\ \Omega \qquad P_{fwc+\text{stray}} = 4.5\ \text{kW}$$

For a slip of 0.038, find the motor power factor, output horsepower, speed, load torque, and efficiency.

SOLUTION: The schematic with the given quantities appears on the left of Fig. 6-7. To find I_L the circuit is reduced to the equivalent series circuit on the right:

$$R_t + X_t = (R_s + jX_s) + \frac{jX_m(R_r/s + jX_r)}{R_r/s + j(X_m + X_r)}$$
$$= (0.052 + j0.1) + \frac{(j7)(0.062/0.038 + j0.286)}{0.062/0.038 + j(7 + 0.286)}$$
$$= 1.486 + j0.696 = 1.641\ \Omega\ \underline{/25.1}$$

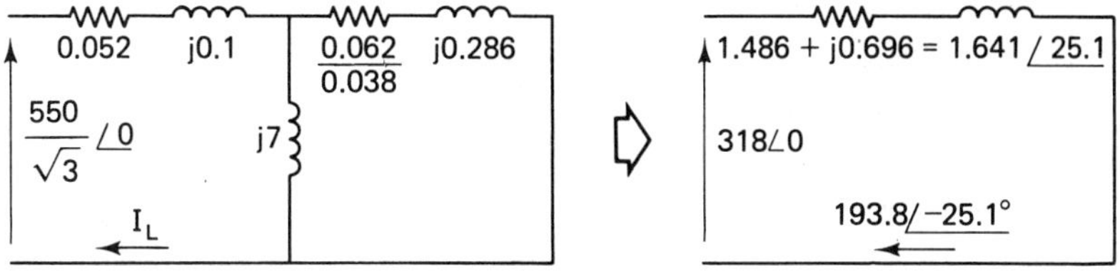

Figure 6-7 Example 6-1

$$V_p = \frac{V_L}{\sqrt{3}} = \frac{550}{\sqrt{3}} = 318 \text{ V}$$

$$I_L = \frac{V_p}{Z_t} = \frac{318\underline{/0^\circ}}{1.641\underline{/25.1^\circ}} = 193.8 \text{ A } \underline{/-25.1}$$

The motor pf is cos 25.1° = 0.906.

$$P_{\text{in}} = 3V_pI_L \cos\theta = (3)(318)(193.8)(0.906) = 167.5 \text{ kW}$$

$$P_o = (1 - s)(P_{\text{in}} - 3I_L^2R_s) - P_{fwc+\text{stray}}$$

$$= (1 - 0.038)[167.5 - (3)(193.8)^2(0.052)(10^{-3})] - 4.5$$

$$= 151.0 \text{ kW}$$

$$\text{hp}_o = \frac{P_o}{746} = \frac{(151)(10^3)}{746} = 202.4 \text{ hp}$$

The speed of the motor is S_R.

$$S_R = (1 - s)\left(\frac{120f}{P}\right) = (1 - 0.038)\frac{(120)(60)}{4} = 1732 \text{ r/min}$$

$$T_L = \frac{33{,}000 \text{ hp}}{2\pi S} = \frac{(33{,}000)(202.4)}{2\pi(1732)} = 614 \text{ lb-ft}$$

$$\%\eta = \frac{100P_o}{P_{\text{in}}} = \frac{(100)(151)}{167.5} = 90.1\%$$

Note that the usefulness of this schematic is in determining the line current, power factor, and input power. If input power and line current are measured, the only impedance required in the computations is R_s.

The torque developed in any motor is $7.04P_d/S$. In the induction motor, $P_d = (1 - s)\text{RPI}$ and the motor speed is $(1 - s)S_S$. Therefore, the developed torque is

$$T_d = \frac{7.04(1 - s)\text{RPI}}{(1 - s)S_S} = \frac{7.04 \text{ RPI}}{S_S} \tag{6-3}$$

When RPI is expressed in terms of slip, the torque is related to slip. One method for relating slip to machine quantities is by converting the phase schematic

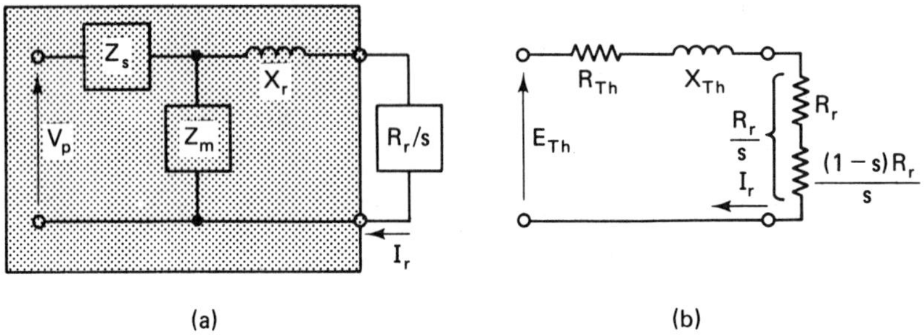

Figure 6-8 Thévenin Rotor Circuit

to its Thévenin equivalent. In Fig. 6-8(a) the circuit is represented with Z_s substituted for $R_s + jX_s$ and Z_m in lieu of X_m. The shaded portion of the circuit must be converted to a Thévenin source. The Thévenin voltage is the voltage at the terminals with R_r/s removed, and the Thévenin impedance is X_r, in series with the parallel combination of Z_s and Z_m:

$$E_{\mathrm{Th}} = \frac{V_p Z_m}{Z_s + Z_m}$$

$$\frac{Z_s Z_m}{Z_s + Z_m} = Z_p = R_p + jX_p \qquad \text{(parallel combination)}$$

$$Z_{\mathrm{Th}} = Z_p + X_r = R_p + j(X_p + X_r) = R_{\mathrm{Th}} + X_{\mathrm{Th}}$$

The Thévenin circuit appears in Fig. 6-8(b). From this circuit, the rotor current is

$$|I_r| = \frac{E_{\mathrm{Th}}}{\sqrt{(R_{\mathrm{Th}} + R_r/s)^2 + X_{\mathrm{Th}}^2}}$$

and the rotor power input is

$$\mathrm{RPI} = \frac{3I_r^2 R_r}{s} = \frac{3E_{\mathrm{Th}}^2 R_r/s}{(R_{\mathrm{Th}} + R_r/s)^2 + X_{\mathrm{Th}}^2} \tag{6-4}$$

Torque has a maximum value, T_m, when RPI is maximum. From the maximum power transfer theorem, power in a variable load resistance is maximum when the resistance equals the absolute value of the source impedance. This requires that

$$\frac{R_r}{s_{mT}} = \sqrt{R_{\mathrm{Th}}^2 + X_{\mathrm{Th}}^2} \qquad \text{where } s_{mT} \text{ is the slip for maximum torque}$$

$$s_{mT} = \frac{R_r}{\sqrt{R_{\mathrm{Th}}^2 + X_{\mathrm{Th}}^2}} \tag{6-5}$$

The maximum developed power does not occur at s_{mT}. Developed power is the power in the resistive equivalent of the load; it has a maximum value when the resistive equivalent equals the Thévenin impedance plus R_r. Thus,

$$\frac{(1 - s_{mP})R_r}{s_{mP}} = \sqrt{(R_{\text{Th}} + R_r)^2 + X_{\text{Th}}^2}$$

$$s_{mP} = \frac{R_r}{R_r + \sqrt{(R_{\text{Th}} + R_r)^2 + X_{\text{Th}}^2}} \tag{6-6}$$

EXAMPLE 6-2: Find the (a) maximum torque (T_m), and (b) maximum power (P_m) developed by the motor of Example 6-1.

SOLUTION: The Thévenin equivalent circuit must be determined:

$$E_{\text{Th}} = \frac{X_m V_p}{R_s + j(X_m + X_s)} = \frac{(7/90)(318\underline{/0°})}{0.052 + j(7 + 0.1)}$$
$$= 314 \text{ V } \underline{/0°}$$

$$Z_{\text{Th}} = \frac{(R_s + jX_s)(jX_m)}{R_s + j(X_s + X_m)} + X_r = \frac{(0.052 + j0.1)(j7)}{(0.052 + j(7 + 0.1))} + j.286$$
$$= (0.0505 + j0.385)\ \Omega$$

(a) $$s_{mT} = \frac{R_r}{\sqrt{R_{\text{Th}}^2 + X_{\text{Th}}^2}} = \frac{0.062}{\sqrt{0.0505^2 + 0.385^2}} = 0.16$$

$$\text{RPI}_{mT} = \frac{3I_R^2 R_r}{s_{mT}} = \frac{3E_{\text{Th}}^2 R_r/s_{mT}}{(R_{\text{Th}} + R_r/s_{mT})^2 + X_{\text{Th}}^2}$$
$$= \frac{3(314)^2(0.062)/(0.16)}{(0.0505 + 0.062/0.16)^2 + 0.385^2} = 337.0 \text{ kW}$$

$$T_m = \frac{7.04\text{RPI}_{mT}}{S_S} = \frac{(7.04)(337)(10^3)}{1800} = 1318 \text{ lb-ft}$$

(b) $$s_{mP} = \frac{R_r}{R_r + \sqrt{(R_{\text{Th}} + R_r)^2 + X_{\text{Th}}^2}}$$
$$= \frac{0.062}{0.062 + \sqrt{(0.0505 + 0.062)^2 + 0.385^2}}$$
$$= 0.134$$

$$\text{RPI}_{mP} = \frac{3E_{\text{Th}}^2 R_r/s_{mP}}{(R_{\text{Th}} + R_r/s_{mP})^2 + X_{\text{Th}}^2} = \frac{(3)(314)^2(0.062)/0.134}{(0.0505 + 0.062/0.134)^2 + 0.385^2}$$
$$= 332.5 \text{ kW}$$

$$P_m = (1 - s)(\text{RPI}_{mP}) = (1 - 0.134) = (332.5) = 287.9 \text{ kW}$$

6-4. EFFECT OF ROTOR RESISTANCE ON PERFORMANCE

From equation 6-5, torque is maximum when

$$\frac{R_r}{s} = \sqrt{R_{\text{Th}}^2 + X_{\text{Th}}^2} = |Z_{\text{Th}}|$$

Substituting this value in equation 6-4 and simplifying,

$$\text{RPI}_{mT} = \frac{3E_{\text{Th}}^2 |Z_{\text{Th}}|}{(R_{\text{Th}} + |Z_{\text{Th}}|)^2 + X_{\text{Th}}^2}$$

$$= \frac{3E_{\text{Th}}^2}{2(R_{\text{Th}} + |Z_{\text{Th}}|)}$$

In this equation, R_r does not appear. Consequently, with all parameters constant except R_r, varying the latter changes s_{mT} but *not* the magnitude of T_m.

In Fig. 6-9 the solid curve is a plot of the torque-slip characteristics of the motor in the example problems. With only the rotor resistance varied, the characteristic curves change to those illustrated for $R_r = 0.5Z_{\text{Th}}$, $1.0Z_{\text{Th}}$, and $1.2Z_{\text{Th}}$.

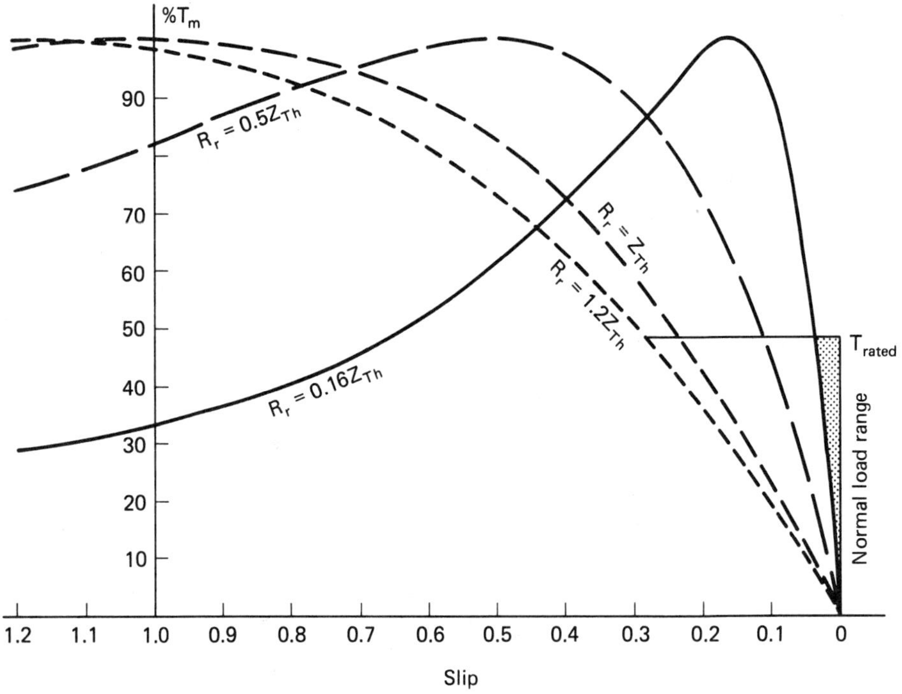

Figure 6-9 Torque-Speed Characteristics with R_r Varied

From a study of these curves, the following conclusions may be drawn:

1. *The starting torque is maximum when* $R_r = |Z_{\text{Th}}|$. As the ratio $R_r/|Z_{\text{Th}}|$ increases or decreases from 1.0, the starting torque is degraded.

2. *Speed regulation is best when the ratio* $R_r/|X_{\text{Th}}|$ *is small.* Assuming T_{rated} is unchanged as R_r is varied, the rated speed for each version is the point where the curve crosses the T_{rated} line. The no-load speed is the same in all cases; therefore, with increased R_r and the consequent reduction in full-load speed, %SR increases.

3. *Rated power decreases as* R_r *increases.* Again, this observation assumes T_{rated} is constant. Since power is proportional to the product ST, the lower speed at which T_{rated} occurs when R_r is increased results in decreased power.

4. *Efficiency decreases with increased R_r.* The power output is less than the developed power, $(1 - s)$RPI, and RPI is less than P_{in}. Consequently, the efficiency must be less than $(1 - s)$.

Ideally, a general-purpose motor would have a relatively large rotor resistance when starting but a low resistance while running. With a wound rotor, external resistance may be connected to the rotor circuit not only to vary the starting characteristics but also to permit adjustment of the running speed. More subtle methods of enhancing starting and running charactersitcs of the squirrel-cage rotor are based on the cage design.

A double-cage rotor slot is ilustrated in Fig. 6-10. Also shown is the general pattern of the leakage flux which is the cause of X_r. The flux density at the base of the slot is substantially greater than the density near the circumference; consequently, the reactance of the inner bar is greater than that of the outer bar. However, since resistance is inversely proportional to area, the resistance of the inner bar is less than the resistance of the outer bar.

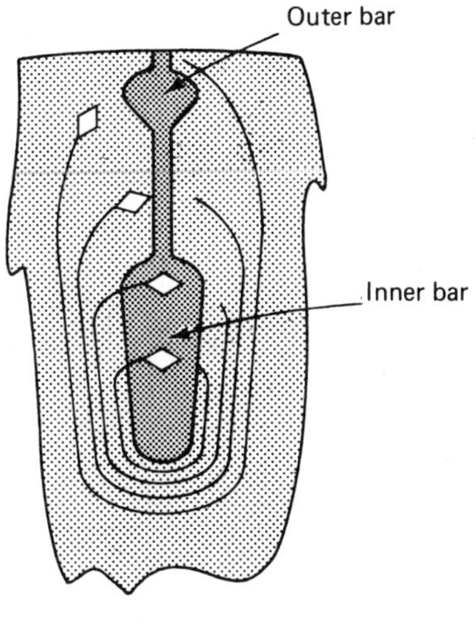

Figure 6-10 Double-Cage Rotor Slot

The inductive effect is proportional to frequency. At starting, the frequency is that of the line, but when running it is sf. The impedance of the inner bar is several times that of the outer bar at starting, most of the current will flow in the outer cage, and the rotor has a high effective resistance. But as speed builds, the reactance decreases and more of the current flows in the inner cage; at running speed, the effective resistance is low.

A similar effect can be achieved without shaping an inner and outer cage—that is, by extending the width of the narrow connecting channel between the inner and outer bar of Fig. 6-10 to form a deep-bar winding, with the slot roughly rectangular in cross section.

For standard induction motors there are four lettered design classes, A to D, distinguished by locked rotor current, torques, and slip at rated load. Four torques are definitive, all based upon application of rated voltage: (1) locked rotor torque (T_{LR}), (2) pull-up torque (T_{PU}), (3) breakdown torque (T_{BD}), and (4) rated torque (T_R). The first three torques are expressed as percentages of T_R. T_{LR} is the minimum starting torque, T_{PU} is the minimum torque developed during the acceleration period from rest to the speed at which T_{BD} occurs, and T_{BD} is the maximum developed torque.

Characteristics within a design class vary with power and speed ratings. Typical torque–speed curves based on NEMA MG-1 are shown in Fig. 6-11. For classes A, B, and C, slip at rated load is less than 5%, but it ranges upward to about 12% in class D. Locked rotor currents are on the order of $8I_R$ for class A and 5 to $6I_R$ for classes B, C, and D. Class B uses a deep-bar or double-cage rotor and class C is double cage. In double-cage rotors and the single-cage rotor of class D, brass may be used in the bars to achieve higher resistance.

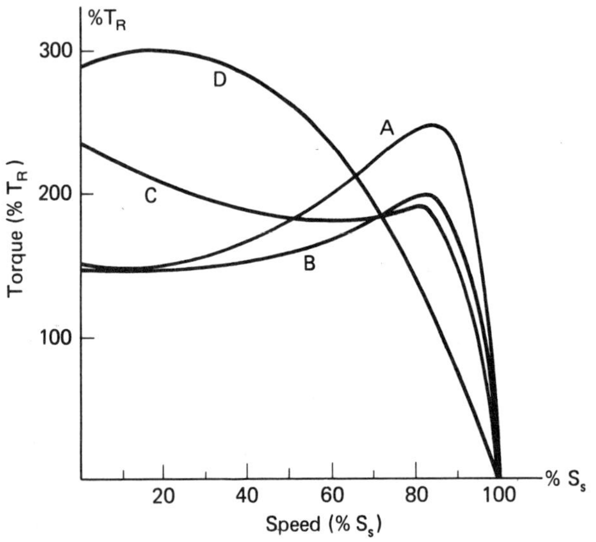

Figure 6-11 Nema Design Classes

6-5. STARTING Starting current may be found from the equivalent circuit of Fig. 6-6 with $s = 1$.

$$I_{ST} = \frac{V_p}{Z_s + \dfrac{Z_m Z_r}{Z_m + Z_r}}$$

Example 6-3: Find the starting current of the motor of Example 6-1.

SOLUTION:

$$Z_s + \frac{Z_m Z_r}{Z_m + Z_r} = (0.052 + j0.1) + \frac{(j7)(0.62 + j0.286)}{(0.062 + j(7 + 0.286)}$$

$$= 0.391\ \Omega\ \underline{/73.7}$$

$$I_{\text{ST}} = \frac{318\underline{/0}}{0.391\underline{/73.3}} = 813\ \text{A}\ \underline{/-73.7}$$

Starting torque may be determined from equations 6-3 and 6-4 with $s = 1$.

Example 6-4: Find the starting torque of the example motor.

SOLUTION:

$$T_{\text{ST}} = \frac{7.04\ \text{RPI}_{ST}}{S_S} = \left(\frac{7.04}{S_S}\right)\left[\frac{3E_{\text{Th}}^2 R_r}{(R_{\text{Th}} + R_r)^2 + X_{\text{Th}}^2}\right]$$

$$= \left(\frac{7.04}{1800}\right)\left[\frac{(3)(314)^2 0.062}{(0.0505 + 0.062)^2 + 0.385^2}\right] = 446\ \text{lb-ft}$$

All standard induction motors can withstand the currents of full voltage starting. However, the effect of these currents on the plant distribution system usually is unacceptable; the momentary voltage drop in the feeders interferes with other equipment and operations, and the currents may be excessive for components of the plant distribution system.

Large motors also have a significant effect upon the utility distribution system and must conform to rules and regulations of the power company; these restrictions may be in the form of limitations on starting current per motor or supply feeder and on the rate of change of current (e.g., 250 A per half-second).

To avoid the objectionable effects of across-the-line starting, reduced-voltage starting commonly is used for squirrel cage motors. From the equations above, starting currents are proportional to the applied voltage, but starting torque is proportional to the square of the voltage. Thus, a starting voltage of 50% V_p would result in $I_{ST} = 0.5I_{LR}$ and $T_{ST} = 0.25T_{LR}$, with V_p, I_{LR}, and T_{LR} the values at rated voltage.

Reduced voltage may be obtained by inserting resistors or reactors in series with the motor terminals during starting, but a more common and effective method uses autotransformers. Either three wye-connected or two open-delta connected transformers can be used.

Closure of the S or R contacts in Fig. 6-12(a) may be manually or automatically controlled. For starting, the S contacts are closed, the transformers are across the line, and the voltage at the motor terminals is the secondary voltage of the transformers. Starters usually are constructed with 50, 65, and 80% taps; the illustration uses 65% resulting in $I_{ST} = 0.65I_{LR}$ and $T_{ST} = 0.42T_{LR}$. From the transformer relationship $\text{kVA}_{\text{in}} = \text{kVA}_{\text{out}}$, $V_p I_L = (0.65V_p)(0.65I_{LR})$, from which

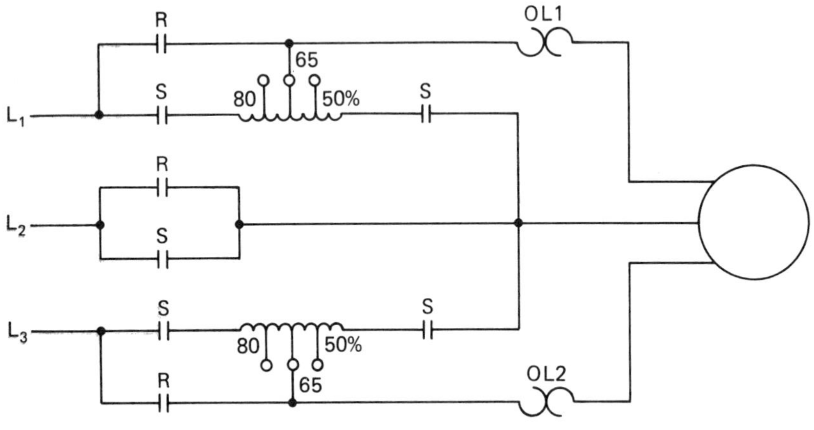

(a) Open transition

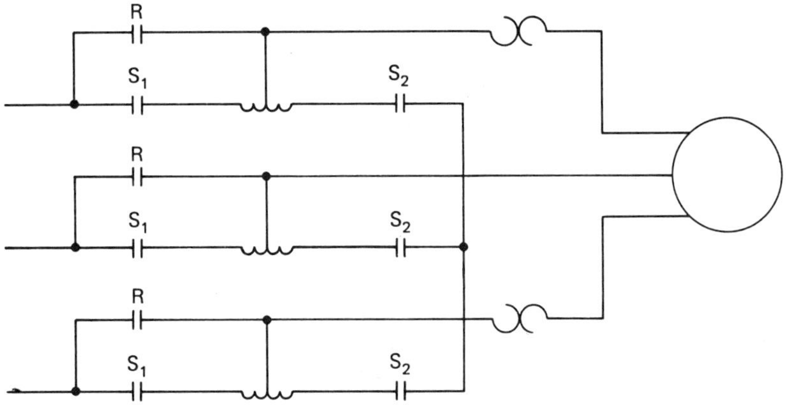

(b) Closed transition

Figure 6-12 Autotransformer Reduced-Voltage Starting

$I_L = 0.42I_{LR}$. Achievement of this same line current by use of series resistors or reactors would result in a motor current of $0.42I_{LR}$ and a torque of only $0.18T_{LR}$, a substantial reduction that argues in favor of the autotransformers.

After building to an appropriate speed, the S contacts are opened, following which the R contacts are closed and full voltage is applied. Since the circuit is momentarily interrupted, this is termed *open-circuit transition*; it has the objectionable feature of transient current spikes caused by momentary interruption of continuity.

Three autotransformers allow *closed-circuit transition* [Fig. 6-12(b)]. Initially, S1 and S2 contacts are closed and the autotransformers supply the motor. During the transition, S2 contacts open but S1 remain closed and a portion of the transformer windings become series reactors. The closure of R shorts these segments, following which S1 opens.

In some motors, reduced starting currents are possible without the use of auxiliary voltage-reduction equipment because of the construction of the stator windings. One method, *part-winding*, requires that the stator winding be constructed with two segments per phase; this construction is common in dual-voltage motors, where two identical sections may be paralleled for low voltage or series-connected for high voltage (e.g., 220/440 V). A second method, *wye–delta* starting, requires that the running connection of the phases is delta.

Part-winding starting is possible only if (1) the running mode has the segments connected in parallel, and (2) the distribution of the segmented winding in the stator slots is compatible with part-winding starting. A dual-voltage motor may or may not meet the distribution requirement—it cannot be assumed that a dual-voltage motor operated at the lower voltage may be started by this method. Where compatible, during the starting period only one segment of each phase winding is employed; after attaining an appropriate speed the other segment is added in parallel. Motors adapted for this method draw starting currents of approximately 60% I_{LR} and develop starting torques slightly less than 50% T_{LR}.

Wye–delta achieves reduced starting currents by an initial wye connection of the phases that subsequently is switched to delta for the normal running mode. The wye will draw only 33% of a delta current; in the wye, the phase voltage is reduced from the delta phase voltage by the factor $1/\sqrt{3}$, resulting in a torque of 33%T_{LR}. Note that this method requires that both ends of each phase winding be connected to separate terminals.

A wound-rotor motor can be started with reduced line currents by inserting balanced external impedance in series with the rotor phase windings. Although both reactance and resistance may be used, the former is less common, requires tailoring to a particular rotor, and is costly. Addition of resistance for starting purposes has a significant impact on starting torques—in fact, this latter consideration is more important than line-current reduction. In Fig. 6-13, the torque–speed

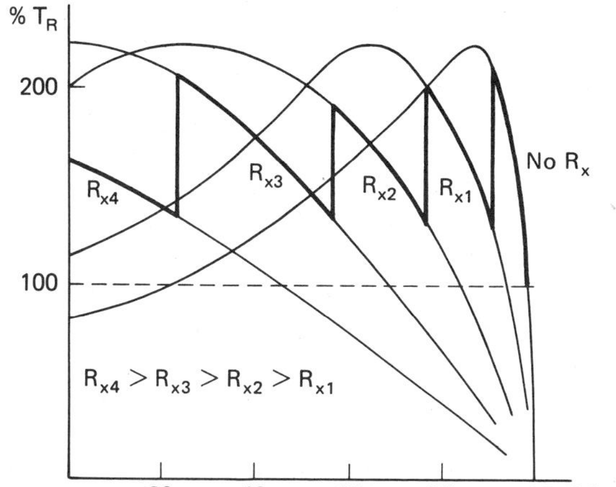

Figure 6-13 Starting—External Resistance in Rotor (Wound Rotor)

characteristics of a wound-rotor motor are shown with only inherent resistance and with four different external resistances inserted in the rotor circuit. During the starting period, resistance is eliminated by steps much in the fashion of the dc motor starter; the heavy curve illustrates the torque obtained as each increment is eliminated. The starting resistors may be designed to meet any required torque pattern; stepless resistors also may be employed.

The wound rotor with external resistance allows better starting torques at lower line currents than comparable squirrel-cage rotors. In addition, much of the heat due to I^2R losses occurs in the external resistance rather than within the machine. The wound rotor is favored in applications requiring acceleration of high-inertia loads such as elevators and hoists.

6-6. SPEED CONTROL There are several classifications based on variability of the speed:

1. *Constant-speed*: Speed is constant or practically constant. Most induction motors, synchronous motors, and dc shunt motors are in this category.

2. *Varying-speed*: Speed varies with loading, usually decreasing as the load increases. A series-wound dc motor fits this category.

3. *Adjustable-speed*: The speed may be adjusted for a constant load but, once adjusted, varies as the load changes. The dc compound motor with field control and the ac wound rotor with a rheostat in the rotor circuit are in this group.

4. *Multispeed*: One of two or more speeds may be selected, each being essentially constant-speed for load variations. Induction motors with provisions for changing the number of stator poles fit this category.

The speed control possible with a wound-rotor induction motor may be appreciated by referring to Fig. 6-13. The dashed horizontal line is the rated torque of the motor. If a three-phase rheostat is connected to the rotor circuit, the torque–speed curve can be positioned with T_r occurring at any desired speed (e.g., the four points where the R_{X1} to R_{X4} curves cross the dashed line). For all these curves, the no-load speed is the same—very nearly S_S—from which it is apparent that adding resistance increases the speed regulation of the motor; the speed regulation approximately equals the slip at which rated torque is obtained. Also, it earlier was noted that efficiency always is less than $(1 - s)$—a 50% speed reduction by rotor resistance is accompanied by less than 50% efficiency.

Numerous methods have been devised to apply the power normally lost in a wound rotor to the production of additional mechanical power. One such method is illustrated in Fig. 6-14. A wound-rotor induction motor is connected to the three-phase supply. A squirrel-cage motor is connected to the rotor of the wound-rotor motor. The rotors are mechanically coupled and, therefore, must run at identical speeds. In the following development, the subscripts w and q apply to the wound-rotor and squirrel-cage rotor, respectively.

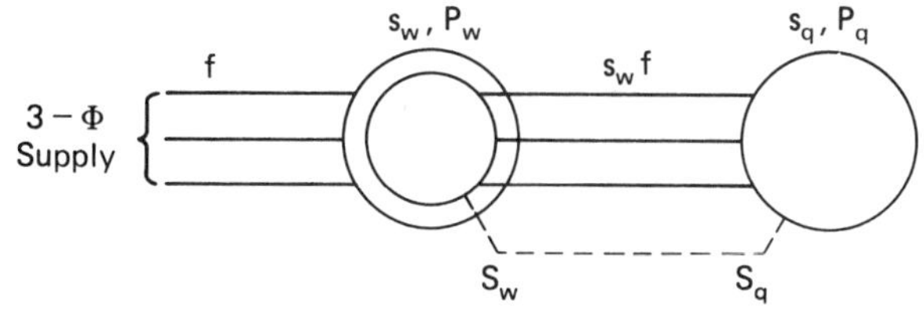

Figure 6-14 Linked Wound and Squirrel-Cage Rotors

$$S_w = \frac{(1 - s_w)(120f)}{P_w}$$

$$S_q = \frac{(1 - s_q)(120 s_w f)}{P_q}$$

Since $S_w = S_q$,

$$\frac{1 - s_w}{P_w} = \frac{s_w - s_q s_w}{P_q}$$

$$s_w = \frac{P_q + P_w s_q s_w}{P_q + P_w}$$

The product $s_q s_w$ is small; disregarding the second term in the numerator, the synchronous speed of the squirrel-cage motor is

$$S_q = \frac{120 s_w f}{P_q} = \frac{120[P_q/(P_q + P_w)]f}{P_q} = \frac{120f}{P_q + P_w}$$

This system allows the choice of three speeds using either motor, alone, or the two coupled according to the diagram—for example, at 60 Hz with $P_w = 4$ and $P_q = 6$, the choice of synchronous speeds is 1800, 1200, and 720 r/min.

The speed of a squirrel-cage motor may be varied in several ways. From the speed equation, $(1 - s)120f/P$, the slip, frequency, and poles all are quantities that can be changed in order to vary the speed.

Slip. RPI is proportional to the square of the applied voltage. If line voltage is varied, the torque–speed characteristics will vary. The curves for 0.5, 0.7, and $1.0V_L$ for a typical motor are shown in Fig. 6-15. Also shown is a common type of load whose torque is proportional to speed. The operating point is the intersection of the load curve with the curve for the applied voltage, with differing speeds at different voltages.

It must be understood that voltage control cannot be applied indiscriminately. A standard induction motor will perform satisfactorily at ±10% rated voltage. With this limitation, a 10% increase beyond rated voltage decreases the slip by about 17%—that is, if the slip at rated voltage is 5%, the slip with a 10% voltage increase is (0.05)(1 − 0.17) = 0.0415. For 10% decrease, slip increases about 21%—to 6.05% for a normal 5% motor. With the 10% variation from rated voltage at rated horsepower, heating may increase and cause accelerated deteriora-

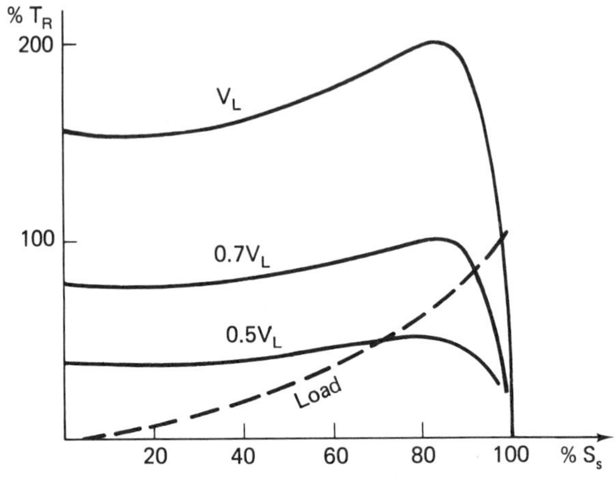

Figure 6-15 Speed Variation by Voltage Control

tion of the insulation if operation is prolonged. Therefore, a general-purpose motor seldom is adaptable to this type of control; this method primarily applies to small-horsepower motors driving fans.

Pole-changing Winding. The stator winding of a motor may be designed to permit alternate connections, with the effect of providing two differing number of poles, the ratio always being 2 : 1. Two coils of one phase of a winding are shown in Fig. 6-16. The schematic shows two methods of interconnecting the coils in order to produce two or four poles with no physical change in the windings. This principle is applied in three NEMA standardized windings, which offer different output characteristics to match differing loads.

The *constant torque* configuration of Fig. 6-17(a) results in a delta connection at low speed with line voltage applied to T1, T2, and T3, and a parallel-wye connection at high speed, with T1, T2, and T3 interconnected and voltage applied to T6, T4, and T5. Markings of the terminals are standard and the line must be

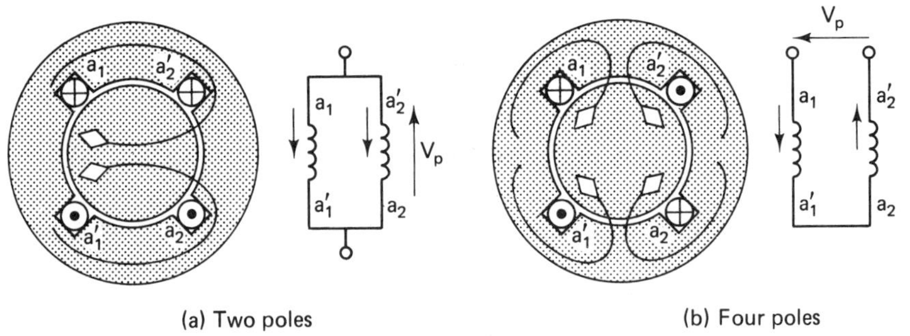

Figure 6-16 Pole Changing

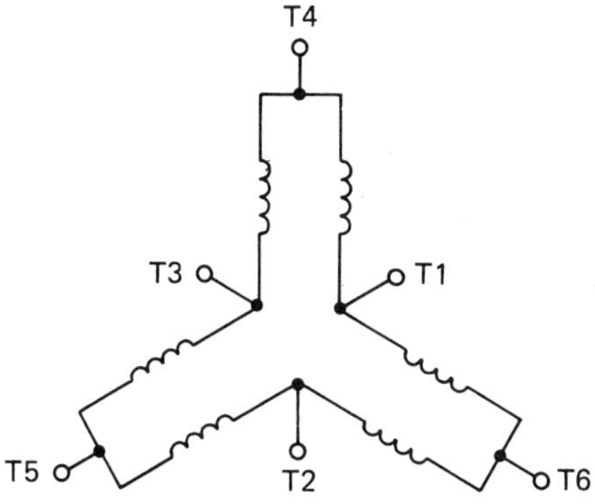

Speed	L1 L2 L3	Tie
Low	T1 T2 T3	
High	T6 T4 T5	T1-T2-T3

(a) Constant torque

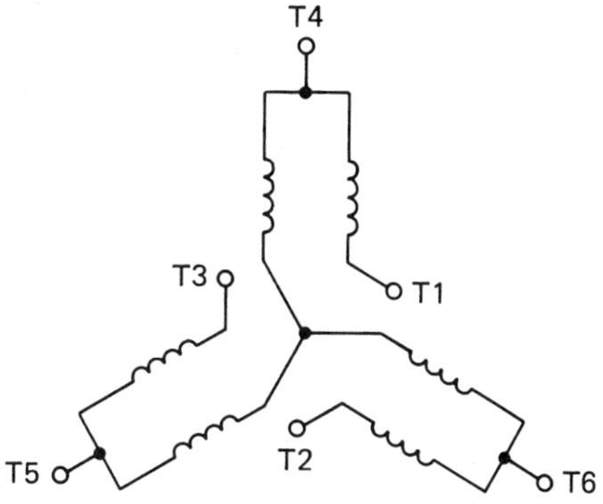

Speed	L1 L2 L3	Tie
Low	T1 T2 T3	
High	T6 T4 T5	T1-T2-T3

(b) Variable torque

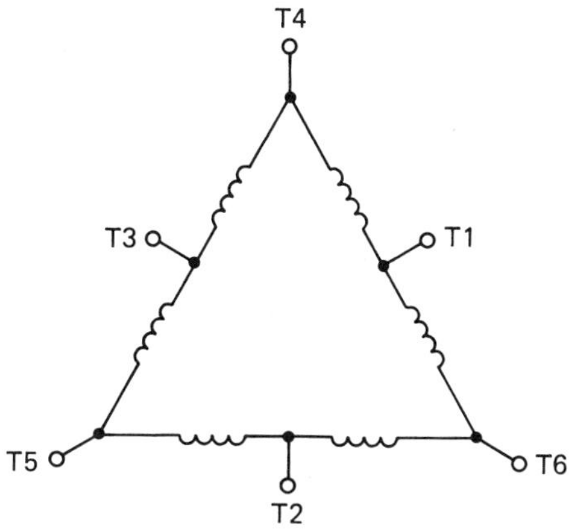

Speed	L1 L2 L3	Tie
Low	T1 T2 T3	T4-T5-T6
High	T6 T4 T5	

(c) Constant horsepower

Figure 6-17 Two-Speed Windings

switched to proper terminals in going from one to the other speed in order to avoid reversing the rotational direction of the synchronous flux. Unlike a wound rotor which must be physically wound for the same number of poles as the stator, a squirrel-cage rotor inherently reacts to match the stator poles. In converting from delta to wye, the rated power approximately doubles; with speed also doubled, torque is constant. This form of the motor is appropriate for friction-type loads for which torque is independent of speed.

The *variable-torque* arrangement is shown in Fig. 6-17(b). Low speed requires a series-wye, while high speed is parallel wye. The low-speed power input is appreciably less than that for constant torque, resulting in a low-speed torque of approximately half of the high-speed torque. Variable torque is appropriate for loads whose torque is generally proportional to speed.

Constant-horsepower characteristics result from the arrangement of Fig. 6-17(c), which employs parallel-wye at low and series-delta at high speed. This form is appropriate for loads in which load-power requirements are essentially independent of speed. The low-speed torque is approximately twice the high-speed torque.

Each of the foregoing methods offers a choice of two speeds. Greater options are provided by two or more independent windings in the stator, each of which can either provide one speed or be arranged for pole changing with two speeds. Thus, one winding might permit 4 or 8 poles (1800, 900 r/min) and the other 6 or 12 poles (1200, 600 r/min).

Frequency. Changing the *frequency* will change the synchronous speed and the operating speed of the motor. When frequencies are varied, machine reactances also change; to maintain rated current, voltage must be raised or lowered as the frequency increases or decreases. The principal limitation on this method of speed adjustment is the means for obtaining the controlled frequency. One method uses a woundrotor induction motor as a frequency converter; the stator is excited in the usual fashion, but the rotor is driven by a second machine whose speed is adjustable. Rotor voltages are extracted at the frequency $P(S_S - S_R)/120$ and applied to the controlled motor.

When the economics and performance of this system are compared with a system using a dc motor, the latter has clear advantages. However, the recent evolution of solid-state power electronics makes control through frequency changing more viable (Chapter Eight).

6-7. AUXILIARY SPEED-CONTROL METHODS

Because of the importance of motor speed control in industrial applications and the limitations on direct control of induction motor speed, auxiliary methods may be employed. Mechanical systems are used in which adjustable-pitch pulleys or speed-reducing gears couple the motor and the load. Adjustable-speed systems of this type are available in ratings up to about 25 hp, with speed ratios limited to approximately 6 : 1. Fixed-ratio speed-reducing gears range up to about 75 hp and permit load speeds as low as 84 r/min from a motor with an 1800-r/min synchronous speed. Economically, a low-speed motor is more expensive than a high-speed motor of comparable power rating inasmuch as the larger torque requires a heavier frame and the increase in number of poles requires a somewhat more complex electrical winding; at speeds below 750 r/min and power to 75 hp, a geared motor is attractive.

In the range to about 100 hp, regulated, adjustable speed is possible by means of eddy-current coupling. A simplified sketch of a coupling is shown in Fig. 6-18. Functionally, it is very similar to the induction motor. A solid iron drum is connected to the end of the motor shaft. A pole assembly is attached to the load shaft; this assembly may be likened to two clam shells with projecting fingers that cup a doughnut-shaped coil. Flux created by dc excitation in the coil enters one shell, continues to the fingers, crosses the air gap into the drum, then recrosses the air gap to enter the fingers of the second shell and complete the circuit. With relative motion between the pole assembly and the drum, eddy currents are created in the latter which produce a flux that reacts with the pole-assembly flux to transmit torque to the load shaft.

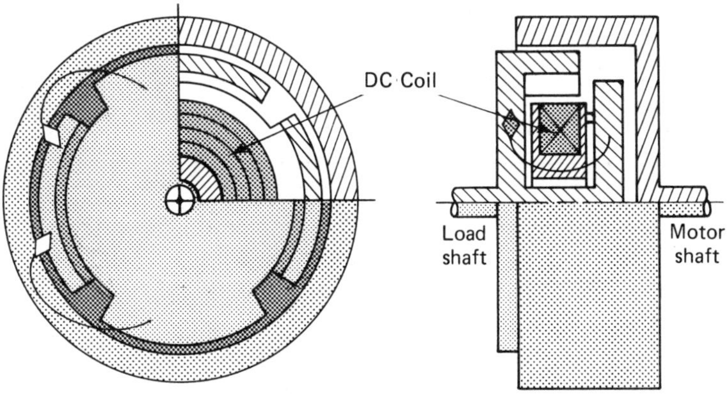

Figure 6-18 Eddy-Current Coupling

If the drum is conceived as a squirrel-cage rotor in which the number of bars has been increased to form a solid cage, the similarity of the coupling to an induction motor emerges. Like the induction motor, there must be slip between the pole assembly and drum for currents to be induced. The strength of the drum flux depends upon both the slip speed and the strength of the pole-assembly flux. The latter can be controlled by varying the current in the dc coil; by this means, the ratio of the load-shaft speed to the motor-shaft speed is continuously variable. If the dc excitation, in turn, is controlled by feedback, the speed may be regulated.

Speed ratios may range to about 16 : 1 in small-horsepower applications, but there is a price. Like the induction motor, the efficiency of the coupling is less than the slip; this coupling is not routinely applied, but where precise control of load speed is essential it performs well.

6-8. BRAKING AND GENERATION

As a motor, the slip range when started without load is from essentially zero to 1.0. Slips beyond this range are possible and useful. With the motor running in the conventional fashion, interchanging two of the three-phase lines reverses the synchronous flux and the slip becomes $(-S_S - S_R)/-S_S = (S_S + S_R)/S_S$, a slip that approaches 2.0. Also, if a second motor or an overhauling type of load drives the rotor in the direction of, but at a speed greater than, S_S, the slip becomes negative.

Equations 6-3 and 6-4 remain valid when the slip is outside the normal 0 to 1.0 range. In Fig. 6-19 the calculated torques of the motor used in the example problems are plotted for slips ranging from 2.0 to -1.0. In interpreting this curve, note that (1) the plotted torque is the electromagnetic torque developed by the motor—*not* the net torque acting on the rotor since the latter includes load, inertial, frictional, and applied torques, and (2) a torque is positive when, if no other torques were present, it would cause the rotor to follow in the direction of the synchronous flux.

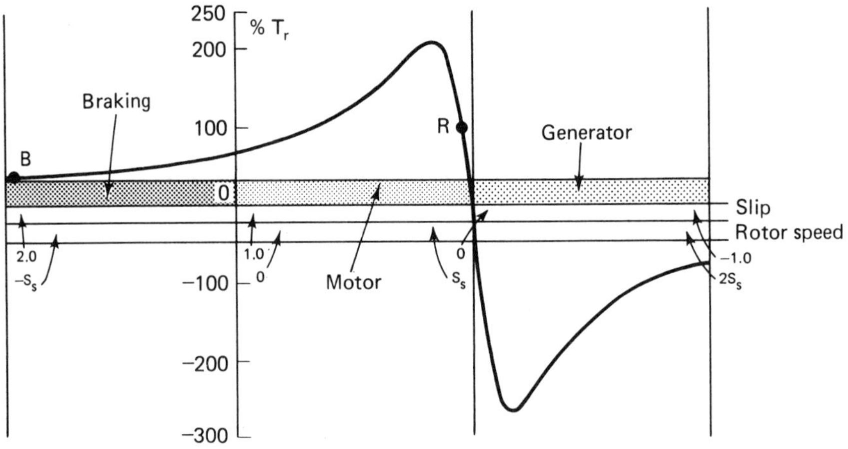

Figure 6-19 Torque-Slip Characteristics in Different Operational Modes

In the region $1 < s < 2.0$, torque is positive but the rotor speed is negative since it is revolving in a direction opposite that of the flux. Consequently, there is a braking torque; *plugging* is the term for stopping the motor by means of reversing two lines. As with the dc motor, when plugging is used for braking there must be provisions for disconnecting the line when the motor comes to rest in order to prevent subsequent acceleration in the reversed direction. While plugging is a common braking method, it does have some disadvantages. Rotor currents are large—for the example motor, somewhat greater than the starting rotor current. Second, if the motor is running at rated load corresponding to point R on the curve, immediately after plugging the operating point is at B; the braking torque is not impressive, although it increases as the motor slows down, and the deceleration period for a high inertial load is prolonged.

In the range $0 > s > -1$, the motor power, proportional to ST, is negative—the rotor is not absorbing power but returning it to the line. In physical terms, the counter emf, E_p, induced in the stator windings bears a relationship to V_p that causes currents to flow out of the stator windings when related to a positive reference input. In a sense, this is a corollary of a dc motor reverting to a dc generator under an overhauling load. The circumstance is created in the induction motor by an overhauling load; it also may be imposed by driving the rotor at greater-than-synchronous speed by a prime mover, and it briefly occurs in a pole-changing motor when, while running at high speed, the motor is switched to low speed.

In ranges up to about 500 hp, the induction motor may be used in special circumstances as an induction generator, but not as a primary power source. Generation is impossible without an available three-phase source; the source must provide the exciting current component (and thereby the synchronous flux), even though the superimposed induced currents cause delivery of power to the supply system.

With an overhauling load or in switching from high to low speed in a multi-speed motor, *regenerative braking* occurs, but the braking action is uncontrolled. Braking continues until the motor has decelerated to the speed required by the shaft torques for continuous rotation.

In addition to plugging and regenerative braking, *dynamic* braking may be applied to the induction motor. For dynamic braking, the ac lines are removed and dc is applied to the stator windings. The stator flux no longer rotates but is stationary; the flux–voltage relationships in the rotor are similar to those of a dc machine in which current flow in the rotor windings creates a quadrature rotor flux to react with the stator flux and decelerate the rotor. Compared to plugging, dynamic braking is advantageous in that it may be controlled by adjusting the applied dc voltage, results in smaller rotor currents and less heating, and cannot cause reversed motoring; however, it does require a dc source.

6-9. CONTROL CIRCUITS

The basic control functions for the induction motor are starting, reversing, braking, and speed selection.

The power circuit for reduced-voltage starting with autotransformers was illustrated in Fig. 6-12(b). For series-resistance or -reactance starting, the circuit is modified to provide short circuits across the starting impedance. Resistance can be stepped or stepless. The stepped resistance may employ tapped resistors; a timing circuit, energized at the moment of starting, subsequently eliminates an increment of resistance in each phase and later shunts the remainder, thereby providing three accelerating points. Stepless starting employs graphite resistors whose ohmic value is reduced when pressure on the resistor is increased; at starting, a solenoid is energized and operates against a delaying dashpot to gradually compress the resistors, followed by short circuiting after an appropriate time interval.

The induction motor rotates in the direction of the synchronous flux; since the latter will change direction if any two lines are interchanged, simple arrangements provide for two-directional running. The means for accomplishing the change in the power circuit is illustrated in Fig. 6-20(a); L2 is switched to T2 in both forward (F) and reverse (R), but the continuity of L1 and L3 is L1–T1 L3–T3 in forward and L1–T3 L3–T1 in reverse. A portion of the control circuit to permit

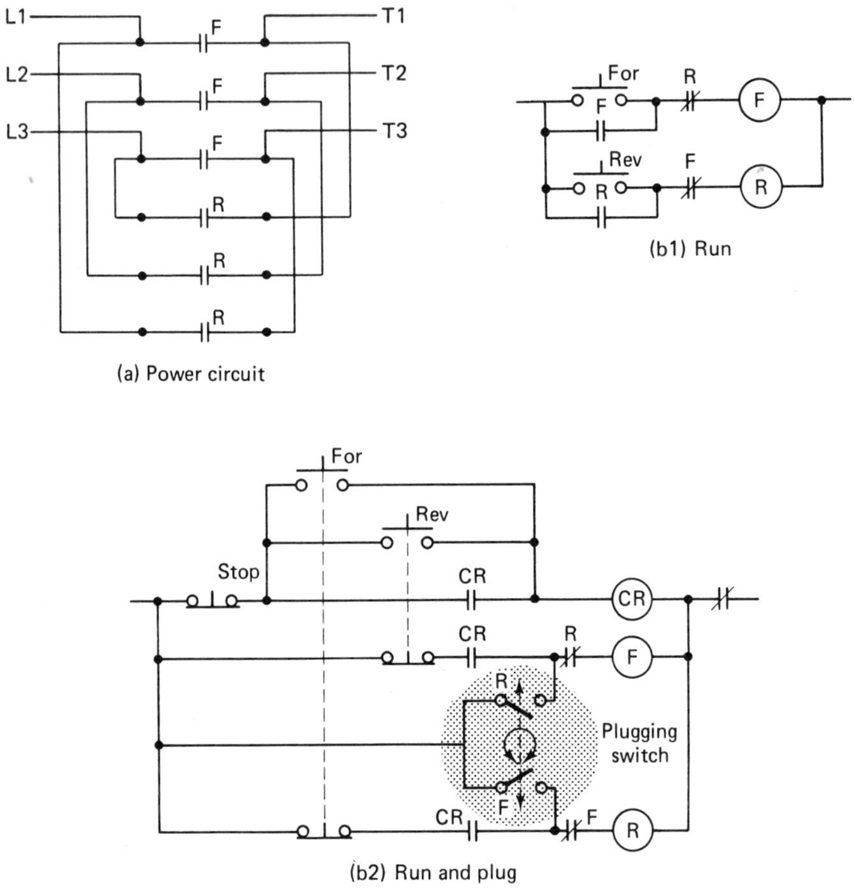

Figure 6-20 Bidirectional Rotation

this selection is shown in Fig. 6-20(b, 1); pressing FOR energizes the F coil, which seals the starting button, opens the normally closed interlock contacts F in series with the R coil, and closes the F contactors in the power circuit. Pressing REV duplicates the pattern with the R coil and contacts.

Since plugging also is accomplished by reversing two lines, a modification of the control circuit permits this form of braking with no change in the power circuit of Fig. 6-20(a). The control circuit is illustrated in Fig. 6-20(b, 2); the key addition is the plugging switch. This switch is open when the motor is at rest but closes to F when rotation is in the forward direction and R for reversed rotation. When the motor is started by pressing FOR, the CR coil is energized and CR contacts close; however, through the mechanical link to the FOR button the lower line of the control circuit is open when FOR is depressed. The CR contact closure energizes coil F, which opens the normally closed F contact in series with the R coil; with quick-acting CR and R relays, this opening occurs before the released

FOR button returns to its normal position. The motor now starts and runs in the forward direction, closing the plugging switch in the F direction but not establishing continuity through the R coil inasmuch as the intervening F contact is open. Later, when the STOP button is depressed, the CR and F relays drop out; the F contact in series with R recloses and coil R is energized, thereby plugging the motor. At zero speed, the plugging switch reopens and the control circuit is fully deenergized.

Circuitry for selection between two speeds in a constant- or variable-torque motor is illustrated in Fig. 6-21. The key change in this circuit is the provision of the tie point T1–T2–T3 for operation at the higher speed. At low speed, continuity is L1–T1, L2–T2, and L3–T3. At high speed, the H contactors in the shaded area close to form the wye and continuity is L1–T6, L2–T4, and L3–T5. In the control circuit, the normally closed H and L provide an electrical interlock that prevents simultaneous energization of the L and H coils. However, it is not necessary to stop the motor in order to change from one to the other speed; for example, when

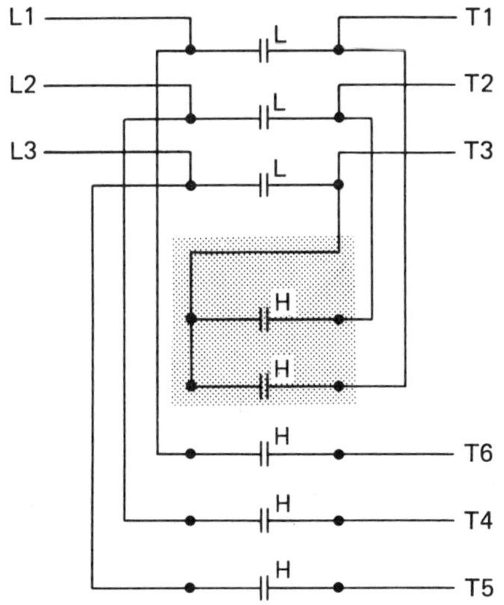

(a) Power circuit

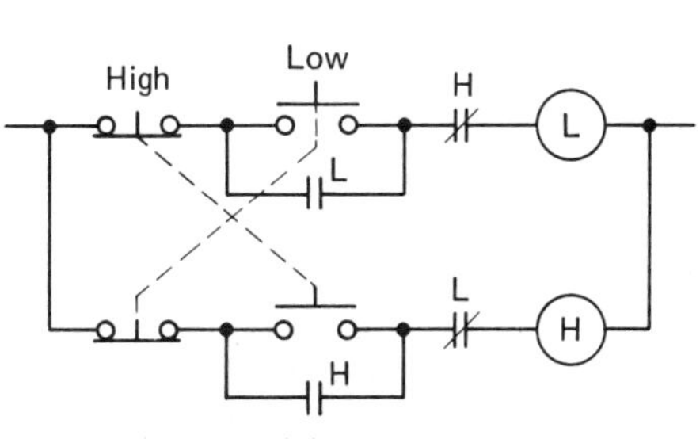

(b) Control

Figure 6-21 Multispeed Control

running LOW, the normally closed L contact in series with the H coil is open, but pressing the HIGH button interrupts continuity to the L coil, the L contact in the H coil circuit rapidly closes, and continuity is established through the H coil.

A combination starter and speed controller for a wound-rotor motor is illustrated in Fig. 6-22. Resistors in the rotor circuit serve the dual purpose of improving starting torque and providing a low-speed option. When started in low, neither the CR nor the A relay coil is energized; the A contactors in the rotor circuit remain open and the motor runs at reduced speed. When the HIGH button is depressed at the start, energization of CR provides continuity to the M relay coil and the motor starts with the resistors in the rotor circuit. After a preset delay, the M TC contacts in series with coil A close, energizing A and causing the A contactors to short-circuit the resistors.

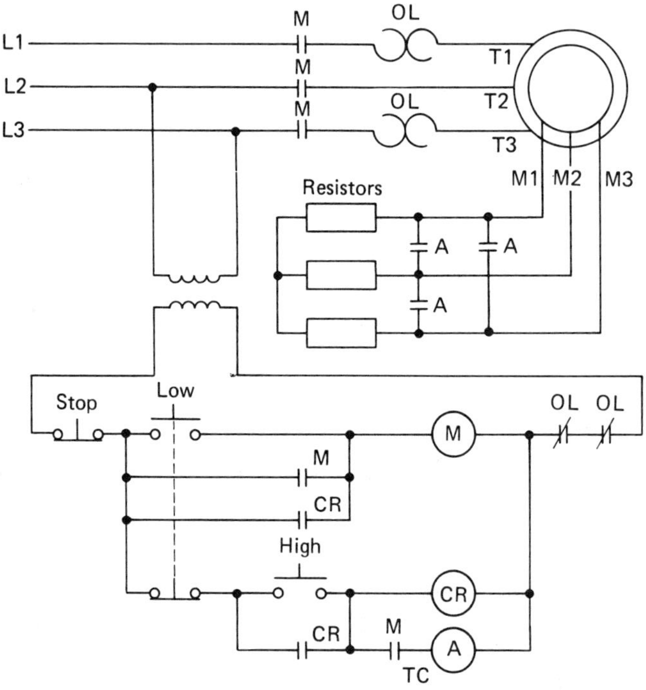

Figure 6-22 Wound-Rotor Starting and Speed Selection

In Fig. 6-22 a transformer is used to provide the control-circuit voltage. Controls are low-power circuits and, by their nature, relatively exposed to personnel. Ac controls normally are rated 120, 240, or 480 V.

To this point, control circuits for both ac and dc motors have been considered from the viewpoint of the motor only, with action initiated by an operator. But automatic initiation is common in industrial applications just as, in the home, a refrigerator or a heat-circulating pump motor automatically responds to temperature changes or a washing machine cycles under the control of a timed switch. Similar operation is possible by replacing the pushbuttons and/or contacts of a control circuit with switches responsive to temperature, pressure, elapsed time or other triggering.

Frequently, the sequential operation of two or more motors must be controlled. For example, one motor may drive a conveyor belt delivering production material to a second, separately driven belt—the second must be operating when the first is running to avoid pile-up at the transfer point. The following example illustrates this type of control circuit.

EXAMPLE 6-5: A system employs four motors. The running state of the motors must satisfy the following requirements:

1. M1 (motor 1) and M2 cannot run simultaneously.
2. M3 must run when either M1 or M2 is running, but it cannot run alone.
3. M4 starts only when M1 or M2, whichever is running, is stopped. M4 will run for a preset time, following which it will stop.

SOLUTION: (Fig. 6-23)

(a) The normally closed M1 and M2 contacts provide an interlock to preclude simultaneous operation of M1 and M2.

(b) The parallel combination of M1 and M2 contacts in series with relay coil M3 causes it to run when either M1 or M2 is running.

(c) Relay coil TR is energized when either M1 or M2 is running. (*Note:* The same objective could be achieved by a single M3 contact in series with the TR

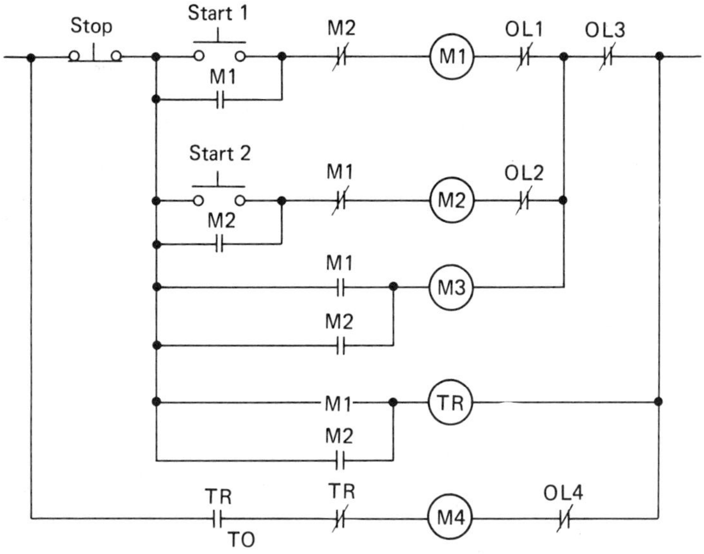

Figure 6-23 Example 6-5—Sequential Operation

coil.) When TR is energized, the two TR contacts in series with coil M4 reverse their state; there is no continuity through M4 when either M1 or M2 is running.

(d) When STOP is pressed, the normally closed TR contact in the bottom control line closes; however, the normally open TR contact remains closed for a preset time interval before opening. During this interval, M4 will run.

6-10. TESTING Test procedures that permit assessment of the performance of an induction motor are more elaborate than those required for the synchronous or dc motor. In particular, determination of stray-load losses and impedances require time and equipment that may not be available in the usual educational environment—for example, time to remove the rotor; a controlled ac voltage source at a frequency of about 15 Hz. The tests are specified in IEEE 112A; this discussion is limited to those that may be performed in a modest laboratory with limited time and is consistent with the method of input measurement, recognized in IEEE 112A, except that determination of stray-load losses is not included.

Stator Resistance. The measurements are made in the same manner as stator measurement for the synchronous machine (Paragraph 5-15). The resistance must be corrected to operating temperature.

No-load Losses. When running without load, the power input is P_{fwc} plus the stator I^2R loss. The power input is measured and the I^2R loss, computed for a resistance value corrected to the test temperature, is deducted to determine P_{fwc}. Friction, windage, and core losses are deemed constant at all loading levels.

Stray-load Losses. Determination of stray-load losses first requires removal of the rotor from the machine and measurement of power input to the stator for a series of stator currents, followed by driving the rotor at synchronous speed in a direction opposite to the synchronous flux and measuring input with and without power applied to the stator. The time and the effort in removing and replacing the rotor of a motor larger than a few horsepower precludes this testing in many educational labs. There is no recognized corrective factor comparable to the assumed 1% of power output for the dc machine which may be applied to the induction motor. Lacking inclusion of stray-load losses, the computed efficiency will be somewhat higher than actual.

Load Tests. Under load, measurements are made of the power input, line current, and slip at six load points between 25 and 150% rated. For slip, stroboscopic measurements are recommended; tachometer accuracy does not permit a meaningful measurement, since an error as small as 1% in measuring speed results in a far greater error in slip—for example, a motor with a synchronous speed of 1800 r/min and a slip range of 0.03 from no load to rated load has a speed differential of only about 54 r/min between no-load and full-load, and a 1% error in speed measurement is 18 r/min or 33% of the slip range.

Calculations from the measured data are as follows:

$$\text{pf:} \quad \text{pf} = \frac{\sqrt{3}V_L I_L}{P_{\text{in}}}$$

$$\%\eta: \quad P_0 = (1 - s)(P_{\text{in}} - 3I^2R_s) - P_{fwc}$$

$$\%\eta = \frac{100P_o}{P_{\text{in}}}$$

$$S_R: \quad S_R = (1 - s)S_S$$

$$T_L: \quad T_L = \frac{7.04P_o}{S_S}$$

6-11. RATINGS AND APPLICATION The voltage ratings of polyphase induction motors range from 115 V to 13.2 kV with horsepower ratings up to 100,000 hp. Power ratings of multispeed motors are based on the highest speed. Synchronous speeds range from 225 to 3600 r/min. Locked-rotor and pull-up torques are never less than 60% T_R and breakdown torques are at least 175% T_R.

Generally, in an emergency motors can withstand overspeeds without mechanical damage of 20% for synchronous speeds of 1800 r/min and less (50% for less than 200 hp at speeds below 1200 r/min).

Satisfactory performance with voltage and frequency variations are the same as for synchronous motors: maximum $\pm 10\%$ V_R, maximum $\pm 5\% f$, and combined V and f maximum $\pm 10\%$.

With the motor initially at ambient temperature, two starts, coasting to rest between starts, are possible; one start is possible at operating temperature.

A *general-purpose* motor is one rated 200 hp or less which is open construction, continuously rated, with a service factor in accordance with MG 1-12.47 (1.15 for integral horsepower motors other than design class D, somewhat higher in fractional horsepower motors) and with class B insulation for integral and class A for fractional horsepower motors. Ratings, operating characteristics, and mechanical construction are standardized for usual operating conditions. *Definite-purpose* motors are standardized for use under service conditions other than usual or use on a particular type of application.

Because of the variety of motors available, standard motors are not cataloged here. However, it is appropriate to clarify the definitions above in order to create an awareness of factors to be considered in selecting a motor for a particular task.

A machine with *open construction* has ventilating openings to permit cooling. Subcategories include classifications such as drip-proof (protected against liquids or particles dropping at an angle up to 15° from the vertical) and guarded (size of openings restricted to preclude accidental contact with hazardous parts). Distinct from open construction are *totally enclosed* motors, including air or water cooling, and *encapsulated* or *sealed* windings for severe environments.

Continuous rating defines the permissible load over a prolonged period. In contrast, short-time rating defines the load for a short, specified period.

The *service factor* is a constant multiplier that defines the permissible continuous load under usual service conditions at *rated voltage* and *frequency*. Since the motor performs satisfactorily with the earlier defined frequency and voltage variations, it is logical to assume a somewhat greater capability at rated levels; thus, with a service factor of 1.15, a 50-hp motor can continuously deliver (1.15 × 50 =) 57.5 hp at rated voltage and frequency.

The distinction between *fractional* and *integral* horsepower motors is not on the basis of whether the hp rating is greater or less than unity. The frame of a motor must be rugged enough to withstand the torques to which it will be exposed—these torques are greater in a 3/4-hp 900-r/min motor than in a 1.5-hp 3600-r/min motor. Generally, but with exceptions, a motor is fractional hp if, at 1800 r/min and the rated torque, its continuous rating would be less than 1 hp. Thus, with no change in torque, at 1800 r/min the 1.5-hp 3600-r/min motor would be rated 3/4 hp—a fractional-horsepower motor—but the 3/4-hp 900-r/min motor would deliver 1.5 hp at 1800 r/min and therefore is integral-horsepower.

Usual service conditions are an ambient temperature of 0 to 40°C, altitude of 3300 feet or less, rigid mounting, and unimpeded ventilation. If any of these elements vary, they must be considered in assessing the suitability of a particular motor.

Most of the information required to determine the suitability of a machine for a particular application will be found on the nameplate of the machine; however, much of it is in the form of code letters, which may be found in NEMA MG 1. For unusual circumstances, the manufacturer must be consulted.

PROBLEMS AND QUESTIONS

6-1. An eight-pole induction motor is supplied by a 60-Hz line and the shaft speed is 850 r/min. What is the (a) synchronous speed, (b) slip speed, and (c) slip?

6-2. Most induction motors have a full-load slip of 0.05 or less. The measured speed of a 60-Hz motor at rated load is 575 r/min. (a) How many poles does the motor have and what is its synchronous speed? (b) What is the full-load slip? (c) If the no-load slip is 0.01, what is the % speed regulation?

6-3. For the motor in Problem 6-2, what is the frequency of the rotor voltage at no-load, full-load, and the instant of starting?

PROBLEMS 6-4 to 6-11: Information on the motor to which these problems apply is as follows: (Note: Stray power loss is to be disregarded.)

Nameplate data: 100 hp 1160 r/min 60 Hz 3ϕ 117 A 460 V
Impedances (Ω): R_s 0.08, X_s 0.195, R_r 0.065, X_r 0.16, X_m 6.7
Test data:

	S_R (r/min)	I_L(A)	P_{in} (kW)
No load	~1200	38.7	3.7
Load	1145	176	128
	1152	156	114
	1166	117	83.6
	1175	91	62.7
	1183	69	44.0
	1190	51	25.6

6-4. From the no-load test data, determine the friction, windage, and core loss.

6-5. For each observation of the load test, find the following:
(a) Slip
(b) Power factor
(c) Stator copper loss
(d) Rotor power input
(e) Rotor copper loss
(f) Rotor power developed
(g) Power output (kW)
(h) Horsepower output
(i) Load torque
(j) Efficiency

6-6. From the phase equivalent circuit, for a slip of 0.0285, determine the following:
(a) Total circuit impedance
(b) Line current
(c) Power factor
(d) Power input
(e) Power developed (kW)

6-7. Convert the phase equivalent circuit to the Thévenin equivalent. From the Thévenin equivalent, determine the following:
(a) Maximum torque and the slip at which it occurs. Compare the latter with a frequently used approximation, $s_{mT} = R_r/(X_r + X_s)$.
(b) Maximum power and the slip at which it occurs.
(c) Rotor current, power developed (kW), and developed torque for a slip of 0.0285.

6-8. Assuming a wound rotor, what value of external resistance in series with each rotor phase would produce the maximum torque at a speed of 600 r/min?

6-9. What is the starting current and torque when the applied starting voltage is (a) rated voltage and (b) 80% rated voltage?

6-10. Assuming a wound rotor, what external resistance in series with the rotor phases will produce maximum starting torque? What is the starting current?

6-11. An autotransformer (open-delta) is used to obtain 80% rated voltage during starting. What are the line and motor currents and kVA?

6-12. Sketch a power control circuit (comparable to Fig. 6-12) for part-winding starting, showing line contactors, motor windings, and input line terminals.

6-13. A multispeed system of the type shown in Fig. 6-14 uses a 6-pole wound-rotor and 10-pole squirrel-cage motor. What are the synchronous speed options? Considering the general statement that efficiency is less than $1 - s$, discuss application of the statement to this system.

6-14. NEMA MG-1 states that two-pole, four-pole, six-pole, and eight-pole polyphase induction motors rated for 60 Hz are capable of satisfactory operation on 50-Hz circuits, with the following qualifications:
- Line voltage reduced to ($\frac{5}{6}$) rated voltage.
- Full-load horsepower reduced to ($\frac{5}{6}$) rated horsepower.
- Rated torque approximately the same for both frequencies.
- Locked rotor current at 50 Hz approximately 5% less than at 60 Hz.

Discuss these qualifications—that is, identify the rationale for each.

6-15. With the aid of Figs. 6-17 and 6-21, diagram the power and control circuits for two-speed (a) constant torque and (b) constant horsepower motors.

6-16. One routine method used by power companies to reduce the load on the distribution system when unusually high peak demands occur is to reduce the system voltages by 5%. Assuming only this voltage reduction causes changes in the motor operation, by what approximate percentage is the motor power reduced? What is the reduction if, in extreme emergencies, the system voltage is reduced by 10%?

6-17. The *linear induction motor* has a potential for numerous applications, particularly mass transit. The concept of the motor may be understood by visualizing a radial cut in Fig. 6-3 and an outward opening of the stator and rotor. Two forms of a transport motor may be conceived: (1) (Fig. 6-24) Armature windings in the car create a flux wave that repetitively travels the length of the car and reacts with an extended squirrel-cage track throughout the length of the run, or (2) a squirrel-cage winding in the car and an extended

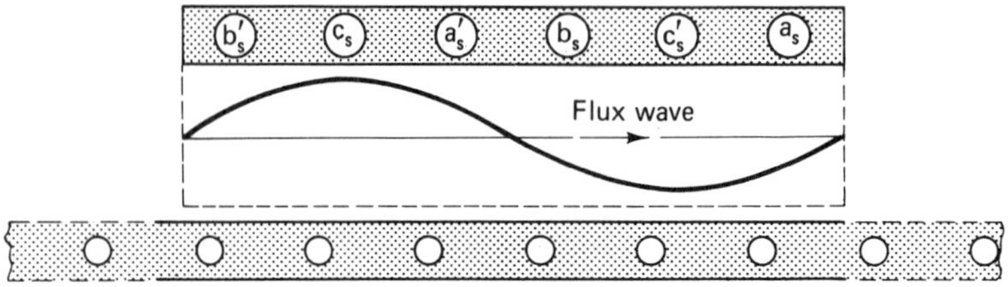

Figure 6-24 Problem 6-17

armature winding in the track, computer-controlled to energize coils in the proper sequence to create a traveling flux wave.

(a) Discuss advantages and disadvantages of the linear induction motor in comparison with contemporary systems that rely primarily upon dc motor drives.

(b) Suggest and discuss methods for controlling the speed of the motor.

(c) The synchronous speed of a conventional induction motor is $120f/P$. If d is the circumferential distance in feet between pole centers at the rotor surface, a fixed point on the surface of the rotor will travel Pd feet in one revolution. By projecting these relationships, show that the synchronous velocity of a linear induction motor is $1.364fd$ miles per hour (f, frequency; d, distance between pole centers in feet). Rationalize the nonappearance of P in this equation.

(d) What is the synchronous velocity of the car at 60 Hz if $d = 1.2$ ft? For a slip of 0.15 and a frequency of 50 Hz, what is the velocity of the car?

(e) If the car velocity is 80 mph and the frequency is 60 Hz, what is the slip?

(f) In what direction does the car of Fig. 6-24 move?

7

Single-Phase Motors and Servo Components

SECTION 7-1. SINGLE-PHASE MOTORS

Both in quantity and variety of types, single-phase motors are far more numerous than polyphase. In the home, washing machines, refrigerators, heat-circulating pumps, clocks, kitchen appliances, hand power tools, phonographs, and so on, all are driven by single-phase ac motors. Commercially and industrially, they are equally common.

Single-phase motors may be loosely grouped in one of three categories: induction, synchronous, and commutator. Each category embraces several types that, in turn, have numerous variations. This section considers basic types with no attempt to catalogue the many modifications that may be encountered.

7-1. CONSTRUCTION AND OPERATING PRINCIPLES—INDUCTION MOTORS

The construction of a two-pole single-phase stator is illustrated in Fig. 7-1(a). Windings are distributed much in the fashion of a polyphase motor. The positive reference direction of the currents is shown; these currents create a flux whose positive direction is upward on the vertical axis. With ac currents in the winding, the strength of the flux will be $\Phi_m \sin \omega t$. Unlike the polyphase winding, this winding does *not* produce a rotating flux; flux always lies on the vertical axis, directed upward during the positive and downward during the negative half of the cycle.

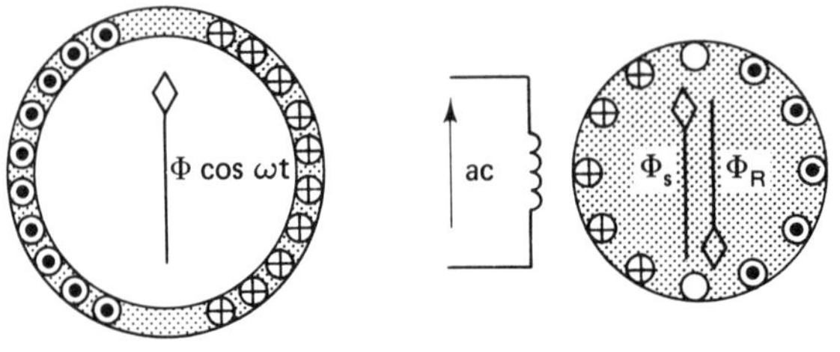

Figure 7-1 Construction of a Single-Phase Induction Motor

In Fig. 7-1(b), the stator is represented by a conventional coil symbol. A squirrel-cage rotor is inserted in the stator. With the rotor at rest and the stator excited, flux will induce voltages in the rotor; by Lenz's law, the direction of the flux established by the resulting currents opposes the stator flux. This rotor flux also lies on the vertical axis; consequently, there is no torque-producing angle and the rotor will not revolve.

Were the stator flux constant in direction and strength while the rotor revolved, voltages would be generated in the bars of the squirrel cage just as they are generated in the conductors of a dc armature. Voltages are generated with the rotor revolving in the sinusoidal flux, also, but several factors complicate the voltage–flux relationship: (1) Because the flux is time-variant, the voltage induced in a conductor is not the constant-amplitude voltage induced in the dc generator. (2) In the dc generator, the quadrature axis, perpendicular to the flux vector, separates conductor voltages of opposing polarities—that is, all those on one side of the axis have the same polarity while all those on the opposite side have opposing polarities. This principle remains valid with the squirrel-cage rotor revolving in the sinusoidal field. However, although the location of the quadrature axis remains fixed, the polarities of the induced voltages on either side change due to the reversal in direction of the flux. (3) The commutator of a dc machine assures a dc current in the windings. Except while commutating, there is no inductive delay and the magnitude of the current is constant. But in the short-circuited squirrel-cage bars, the currents are *not* dc but time-variant; inductance is a factor, and currents lag the induced voltages by nearly 90°.

The *cross-field* theory of the induction motor requires a detailed analysis of flux and current relationships; its conclusion is that, when rotating, there is a rotor flux on the quadrature axis that reacts with the stator flux to produce torque in a direction to sustain rotation. The torque angle is fixed at 90°, but with the fluxes time-variant, the instantaneous torque is not constant.

A second theory, *double revolving flux*, permits application of principles of the polyphase induction motor to the single-phase machine. In Fig. 7-2(a), two fluxes are shown, each with the magnitude $\Phi/2$ and rotating at angular velocity ω but in opposite directions. As phasors, each flux has instantaneous vertical and horizontal components as illustrated. The vertical components are additive and

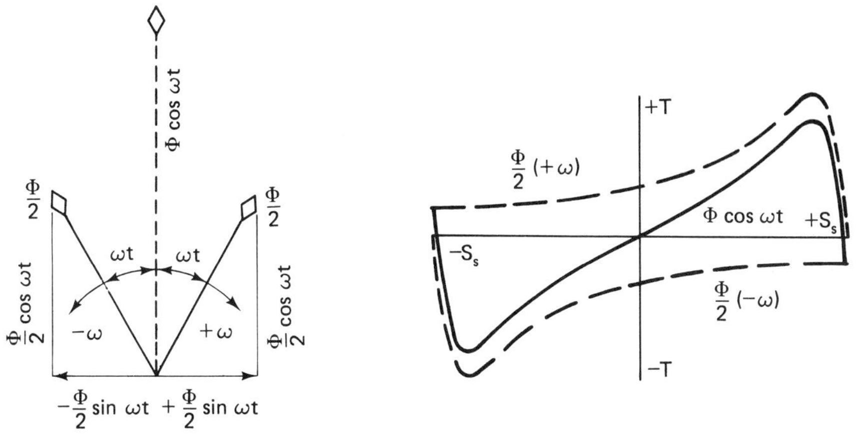

(a) Double revolving flux

(b) Torque-speed characteristic

Figure 7-2 Operational Theory

equal Φ cos *ωt*, but the horizontal components, equal in magnitude but opposite in direction, cancel. Thus, the double phasors are equivalent to the stator flux of Fig. 7-1.

Considering only the flux Φ/2 rotating at $+\omega$, the relationship between the flux and rotor is identical to that of a polyphase induction motor. Acting alone, this flux would produce the characteristic *T–S* curve labeled Φ/2 $(+\omega)$ in Fig. 7-2(b). (Note that this curve is plotted from $-S_S$ to $+S_S$.) Similarly, when Φ/2 rotating at $-\omega$ is separately considered, the *T–S* curve is the same except that both torque and speed are negative with respect to the assumed reference directions. This curve appears as Φ/2 $(-\omega)$.

The fluxes cannot be isolated—both are present at all times, and the *sum* of the curves is representative of the combined characteristics. This sum is the solid curve labeled Φ cos *ωt*.

While this curve correctly identifies the general nature of the *T–S* curve, it is not literally correct. When slip, induced currents, and counteremfs due to the rotor fluxes are considered, the strength of the flux rotating in the direction of the rotor increases with a concurrent decrease in the backward-rotating flux, and the maximum torque is substantially greater than indicated by the summation curve.

With two counter-revolving fluxes, it would appear that countertorques are inevitable, particularly as the fluxes revolve past one another—torques comparable to those in the synchronous motor if, at starting, both the stator and dc field are excited. Such torques do occur, causing pulsations at twice the line frequency. This phenomena is consistent with the discussion of single-phase power pulsations in Paragraph 5-2. The severity increases with loading and is noticeably audible particularly under heavy load.

From the characteristic curve of Fig. 7-2(b), it may be concluded that:

1. A single-phase motor develops no starting torque.

2. If started by auxiliary means, the motor will run in the direction in which started.

3. The motor is reversible if the auxiliary means provides an option for starting in either direction.

7-2. PHASE-SPLITTING INDUCTION MOTORS

A *two-phase* motor is illustrated in Fig. 7-3. The identical windings are positioned in the stator with a 90-electrical-degree separation. For a two-pole motor, the flux created by the current $I \sin \omega t$ always acts on the vertical axis and may be represented by $j\Phi \sin \omega t$. The

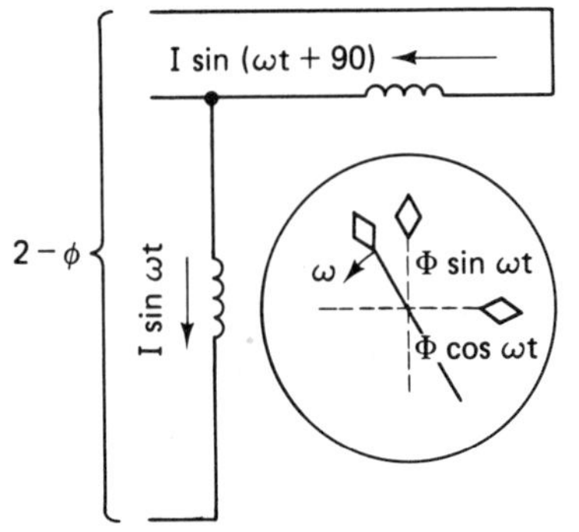

Figure 7-3 Two-Phase Induction Motor

current in the horizontal winding, $I \sin (\omega t + 90) = I \cos \omega t$, produces a horizontal flux $\Phi \cos \omega t$. These fluxes add to produce the stator flux, $\Phi \cos \omega t + j\Phi \sin \omega t = \Phi\underline{/\omega t}$. This is a rotating flux that differs from the three-phase revolving flux only in the magnitude of Φ (1.0Φ vs. 1.5Φ). Consequently, the principles of the three-phase motor are equally applicable to the two-phase machine. The former is self-starting, and so is the latter. This motor relies upon two voltages with a 90° phase separation; however, it is *not* voltage separation, *per se*, that assures the rotating flux but rather (1) the physical separation of the coils by 90°, and (2) a 90° phase difference between coil *currents*. If the coil locations are maintained, any means of separating the currents will accomplish the same purpose. One such means, using a single-phase source, is illustrated by the following example.

Example 7-1: The parallel network of Fig. 7-4(a) represents the windings of a two-pole single-phase motor. Physically, the windings are perpendicular. (a) What is the effective current in each winding when 115 V 60 Hz is applied? (b) If the flux created by each winding is ki, where i is the instantaneous current, what is the stator flux?

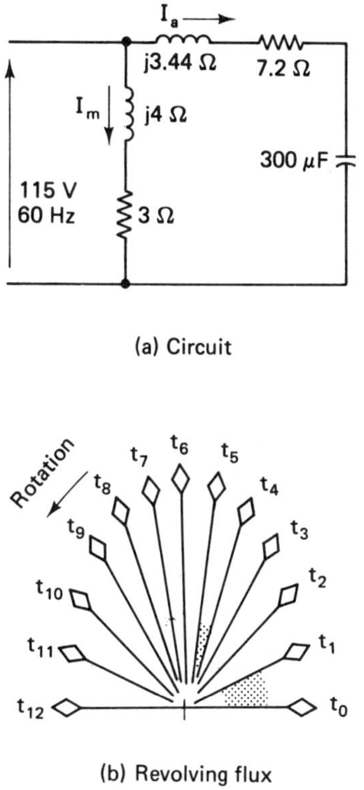

Figure 7-4 Example 7-1

SOLUTION:

(a) $Z_m = 3 + j4 = 5\,\Omega\,\underline{/53.1}$

$$I_m = \frac{115\underline{/0}}{5\underline{/53.1}} = 23\text{ A}\,\underline{/-53.1}$$

$$X_C = \frac{1}{2\pi f C} = \frac{1}{2\pi(60)(300 \times 10^{-6})} = 8.84\,\Omega$$

$$Z_a = R_a + j(X_a - X_c) = 7.2 + j(3.44 - 8.84) = 9\,\Omega\,\underline{/-36.9}$$

$$I_a = \frac{115\underline{/0}}{9\underline{/-36.9}} = 12.8\text{ A}\,\underline{/36.9}$$

The angular separation between I_m and I_a is 36.9 − (−53.1) = 90°.

(b) I_m and I_a are effective (rms) values. To express them as instantaneous currents, the maximum values are required ($I_{max} = \sqrt{2}\, I_{rms}$). The instantaneous currents are

$$i_m = \sqrt{2}(23) \sin \omega t = 32.5 \sin \omega t$$ (a phase angle of 0° is arbitrarily assigned to this current)

$$i_a = \sqrt{2}(12.8) \sin (\omega t + 90) = 18.1 \cos \omega t$$

Therefore, with the fluxes physically perpendicular,

$$\begin{aligned} \Phi &= k(i_a + ji_m) \\ &= k(18.1 \cos \omega t + j32.5 \sin \omega t) \end{aligned}$$

This flux is plotted for a half-cycle in Fig. 7-4(b). Each successive position corresponds to an added increment of $\omega t = 15°$—the time interval for rotation through shaded angles is identical. From the difference in angular movements, it is apparent that the instantaneous velocity of the flux is not constant; however, in one cycle the flux will complete one revolution and the *average* speed is $S_S = 120f/P$.

The strength of the flux also varies. Nonetheless, this rotating flux *will* induce voltages that produce starting and running torques.

For phase separations less than 90°, a rotating flux still is produced, again noting that the locations of the windings in the stator are separated by 90 electrical degrees. With only resistance and inductive reactance in each winding, each current lags the voltage by the angle $\tan^{-1} X/R$. If the X/R ratios of the windings differ, some separation is achieved; with a 25° separation, assuming for purposes of illustration that the maximum values of the two fluxes are identical, the resulting pattern is that of Fig. 7-5(a).

The resistance per unit of small-gauge wire is greater than that of large gauge. Inductance is proportional to N^2, where N is the number of winding turns. If one winding has fewer turns of smaller gauge wire than the second winding, its X/R ratio is smaller and its current is more nearly in phase with the applied voltage, leading the current in the second winding.

With capacitance added to increase the separation to 60°, still assuming equal maximum values of the two flux components, the strength and orientation of the flux with equal time increments is depicted in Fig. 7-5(b).

For a flux with a uniform strength and rotating at a uniform instantaneous velocity, the phase separation must be 90° and the maximum strengths must be identical. Departures from these ideals produce the elliptical patterns of the illustrations. But in all instances the *average* speed of the flux is S_S, based on the number of poles and frequency of the applied voltage.

Phase-splitting circuits can be used in one of two ways: (1) during the starting period, only, in order to provide initial rotation that then is sustained by the phenomena of the double-revolving flux, or (2) during starting *and* running by continuous simulation of the two-phase motor.

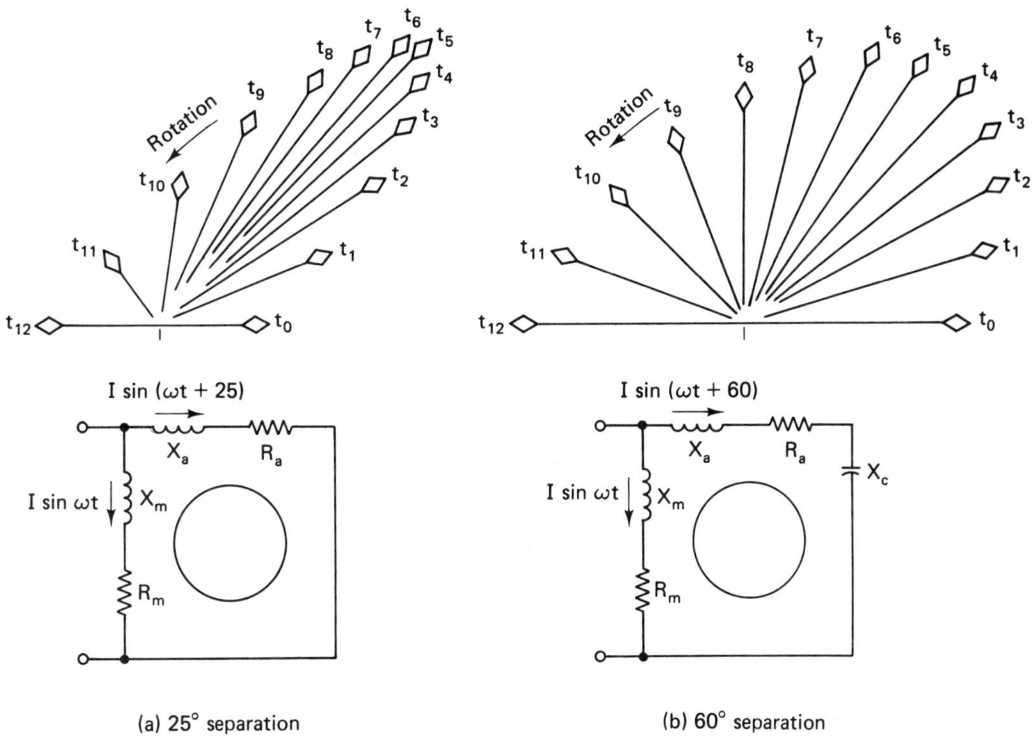

Figure 7-5 Phase-Splitting Windings

Split-phase Motor. A split-phase motor is illustrated in Fig. 7-6. The stator has two windings, a *main* and an *auxiliary*. The latter is energized only during the starting period; when the rotor has accelerated to a speed of about 75 to 80% S_S, a centrifugal switch opens to disconnect the auxiliary winding. Since the main winding is continuously energized while running, it is wound with heavy-gauge wire; further, the running torque is proportional to the flux, which, in turn, is proportional to the number of turns, and the main winding therefore has a relatively large number of turns. The impedance angle of the main is greater than that of the

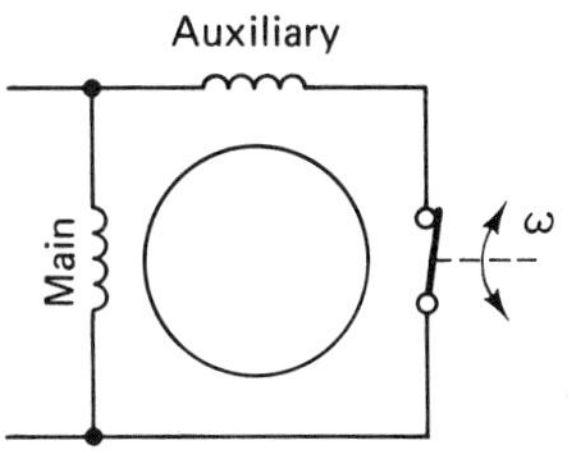

Figure 7-6 Split-Phase Motor

auxiliary, and the current in the latter leads the current in the main. The phase separation typically is on the order of 25°.

A *resistance-start* motor is one in which additional, external resistance is placed in series with the auxiliary to further increase the phase separation.

Starting currents are relatively large, 4 to 7 times rated current, but not large enough to disrupt the supply system as do the starting currents of large polyphase motors, and the motors are started across the line.

When a centrifugal switch is used to open the auxiliary circuit, it is located inside the frame; since it would be inaccessible for servicing in a hermetically sealed motor, the latter uses an external electromagnetic switch, with a sensing coil in series with the main winding to control normally open contacts in the auxiliary. The large starting current causes the contacts to close, but, as rotor speed builds and currents decay, the contacts reopen at a preset speed.

The rotational direction of a split-phase motor can be changed by reversing the line connections to the auxiliary winding; usually, this must be done at the terminal board inside the machine.

Breakdown torques range from about 185 to 250% rated torque, and starting torques are on the order of 100 to 200% (275% in high-torque motors).

The auxiliary windings are not constructed to withstand prolonged operation. If the switch fails to open when the motor is started, the resulting buildup of heat may cause the burnout of both windings.

Capacitor-start (Fig. 7-7). This motor has a capacitor on the order of 75 to 350 μF in series with the auxiliary winding. Starting torque is proportional to the product $I_m I_a \sin \alpha$, where α is the angular separation between the currents. Purely from the viewpoint of torque, it would appear that a 90° separation is optimal. However, when the size of the capacitor, magnitude of the starting currents, and economics are considered, a lesser separation is more attractive—the difference between 90 and 80 degrees, for example, reduces the starting torque by only 1.5% if all other factors are equal.

The capacitor used in a starting circuit is the electrolytic type, rated for short-

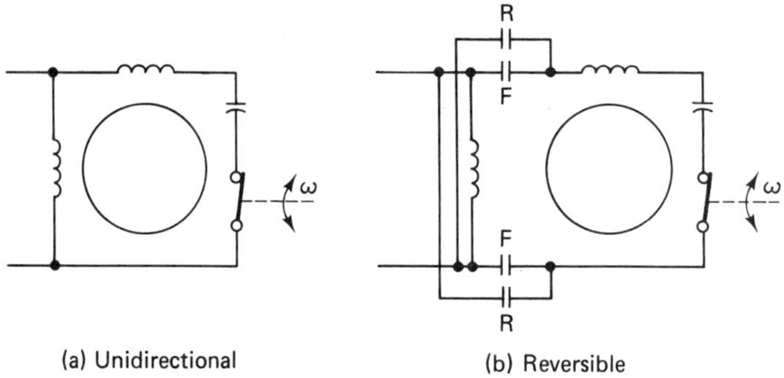

(a) Unidirectional (b) Reversible

Figure 7-7 Capacitor-Start Motor

time duty only (e.g., 20 operational periods of not more than 3 seconds duration per hour). Since it is possible for the voltage across the capacitor in an *RLC* circuit to be substantially greater than the applied voltage, precautions must ensure that destructive voltages will not occur.

Inasmuch as both split-phase and capacitor-start motors run on the main winding only, their running characteristics are the same. However, a comparison of Fig. 7-5(a) and (b) indicates a better flux in the capacitor-start circuit, which, in combination with the greater angle α, produces substantially greater starting torques. Starting, pull-up, and breakdown torques vary with speed and horsepower ratings, but the curves of Fig. 7-8 are representative. An additional torque parameter, *switching torque*, is the minimum developed torque as the motor accelerates through the switch-operating speed. The breakdown torque is the maximum developed torque with only the running winding excited.

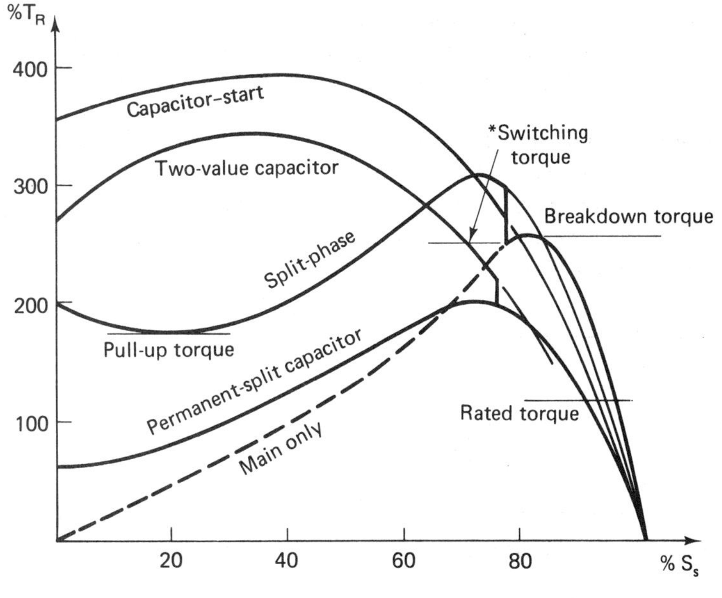

Figure 7-8 Characteristics of Phase-Splitting Induction Motors

A switch arrangement for reversing is illustrated in Fig. 7-7(b).

Permanent Capacitor. This motor runs on both windings. When running on a single winding, the power pulsations at twice the applied frequency result in a fairly noisy motor. On both windings with a 90° phase separation, power is smooth and the motor is quiet. Concurrently, the power factor and efficiency of the motor increase. With identical windings, a simple switch arrangement (Fig. 7-9) allows running in either direction.

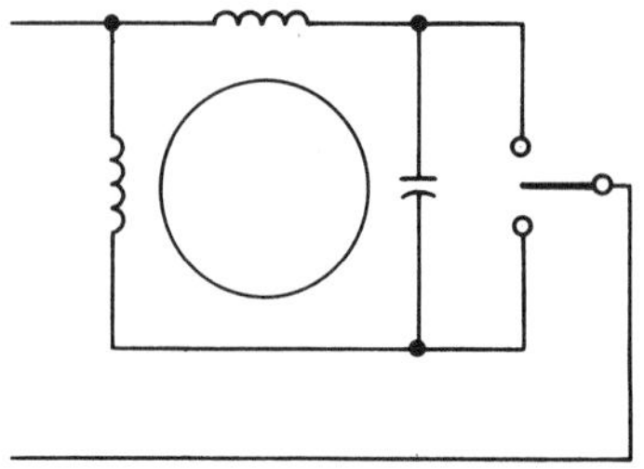

Figure 7-9 Reversible Permanent Capacitor

The speed of a permanent-capacitor motor may be controlled by voltage adjustment. Speed variations in motors with an auxiliary winding are restricted, since, in order to protect the auxiliary, the speed must be kept above the point where the centrifugal switch closes.

To optimize running characteristics, a relatively low capacitance is required. Low capacitance results in low starting torques, perhaps only 50% rated torque. The capacitor, rated for continuous duty, is the relatively expensive oil type.

Two-value Capacitor. To retain the running characteristics but improve the starting torque, a second capacitor may be paralleled with a permanent capacitor during the starting period, as shown in Fig. 7-10. The capacitance during starting may be 10 or more times as much as the running capacitance. The starting capacitor is rated for limited duty.

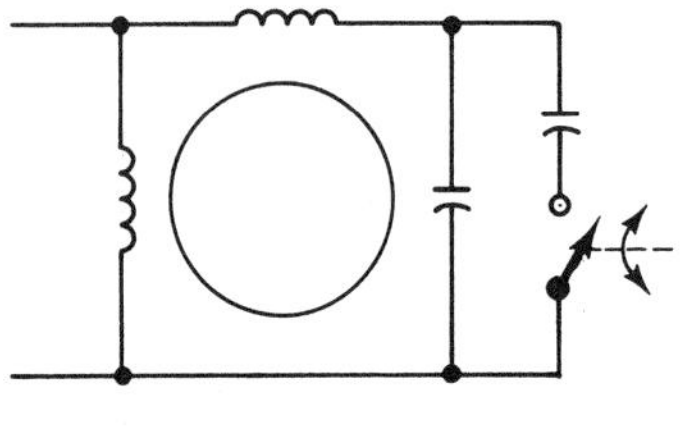

Figure 7-10 Two-Value Capacitor

The curves of Fig. 7-8 show the running speed with both windings in the circuit to be less than that with a single winding. Slip at rated load is 5% or less for split-phase and capacitor-start motors, whereas permanent-split motors have slips of about 10%.

7-3. SHADED-POLE INDUCTION MOTOR

The construction of a shaded-pole motor is illustrated in Fig. 7-11. The rotor is squirrel-cage. Unlike the distributed-winding poles of the phase-splitting motors, the stator has wound salient poles. A short-circuit band is placed around a *shaded* portion of each pole—in a small, cheap motor this may be done by merely drilling a hole, inserting a bare copper wire, and twisting the ends together to loop a portion of the pole.

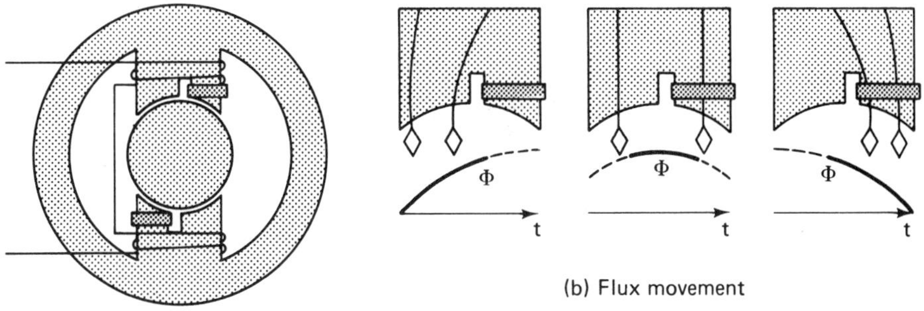

(a) Construction

(b) Flux movement

Figure 7-11 Shaded-Pole Motor

With a sinusoidal flux established by the pole windings, a voltage is induced in the copper band that, according to Lenz's law, opposes changes in flux through the band. Referring to Fig. 7-11(b), during the first portion of the sinusoidal buildup the rate of change is substantial, and the resulting short-circuit currents cause the flux to divert to the unshaded portion of the pole. When the flux is near its peak value, the rate of flux change and, thereby, induced voltage in the loop is small; the flux density is essentially the same in both the shaded and unshaded portions. Subsequently, when flux through the band is decreasing at a substantial rate, most of the pole flux is drawn through the shaded portion.

When both of the poles are considered, the resulting flux through the rotor swings across the face of the pole in a half-cycle, then repeats with a reversed general orientation during the second half-cycle. The flux is not constant, does not have a uniform velocity, and does not revolve completely around the stator, but the motion is sufficient to provide the necessary rotational–voltage component in the rotor to cause it to start. Flux movement continues during the running period, adding an additional torque component to that attributable to speed voltages.

The direction of movement of the flux and the rotor is from the unshaded to the shaded side of the pole. Consequently, if the permanently short-circuited band is replaced with independent loops around each half of the pole, with the ends of each loop connected to a switch, either half may be shaded to provide a choice of rotational direction.

The shaded-pole motor has a low power factor, is inefficient (30 to 40%), and has a low breakdown torque (about 125% rated torque). However, it is the cheapest motor to produce. In the range from $\frac{1}{20}$ to $\frac{1}{4}$ horsepower, where efficiency is a minor concern, it is employed in devices such as toys, fans, and phonograph turntables, with standard load speeds of approximately 1550, 1050, and 800 r/min.

With no centrifugal switches to limit speed options, the shaded pole can be constructed to permit selection of any of several speeds. This may be accomplished by providing a means of controlling the voltage across the full coil, but it commonly is done by applying line voltage to taps in the coil winding. When this method is used, the stator construction resembles that of Fig. 7-12. Regardless of the amount of the coil in the circuit, the induced voltage in the excited portion nearly equals the applied voltage. With relatively few turns, the voltage per turn (and coil current) is much larger than when the full coil is used. Since the voltage per turn is proportional to flux, the flux must be larger when only a fraction of the coil is used. With the larger flux, the motor runs at the highest speed, i.e., when the *smallest* portion of the coil is in the circuit.

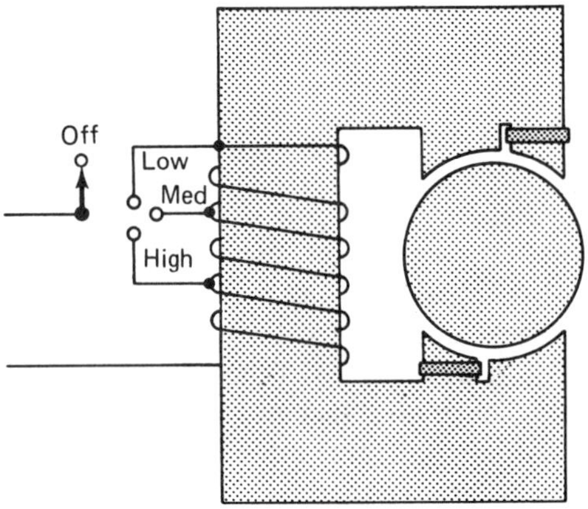

Figure 7-12 Multispeed Shaded-Pole Motor

7-4. RELUCTANCE MOTOR (SYNCHRONOUS)

The reluctance principle exploits the characteristic that magnetic flux will select the lowest-reluctance path in completing its circuit. A reluctance motor is not self-starting, but it is self-sustaining at synchronous speed provided that it is not loaded beyond the pull-out torque. Continuous rotation results from alternating inertial and electromagnetic torques.

The instantaneous magnetic torque on the rotor of Fig. 7-13 depends upon two factors: the instantaneous strength of the flux and the concurrent angular orientation of the rotor with respect to the axis of the flux. At t_0 the flux is zero; there is no magnetic torque, but inertial torque continues to drive a running motor. Between t_0 and t_1 the angular orientation of the rotor distorts the flux in a manner that provides a small torque in opposition to the inertial torque, T_j, but if the latter is dominant, the rotation continues. At t_1, the rotor is perpendicular to the flux axis and there is no defined, low-reluctance path to cause the flux to deviate; the magnetic torque, T_m, is zero. However, a moment later the rotor *will* be in a position to cause flux deviation and an accelerating torque is produced. Acceleration continues until the rotor is vertical at t_3. Beyond this point, the magnetic torque is in a decelerating direction but rapidly decays as the flux decreases; inertial torque sustains rotation. The pattern is repeated during the second half-cycle with the direction of the flux reversed.

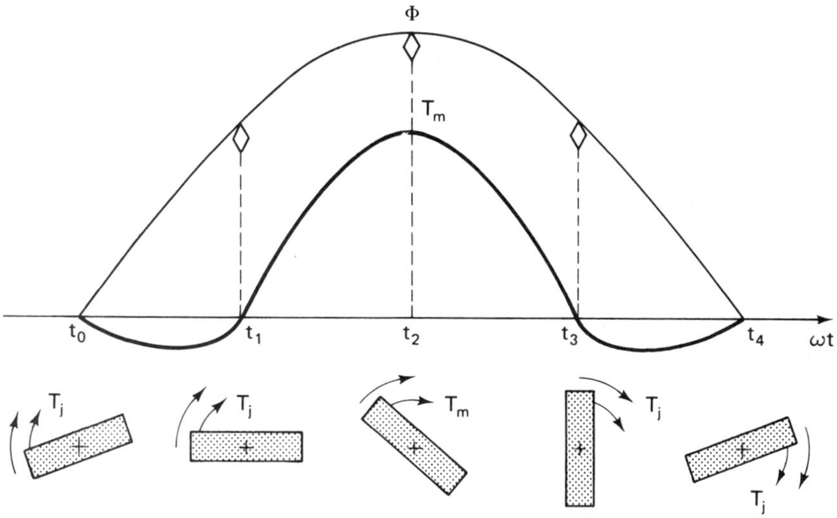

Figure 7-13 Reluctance Principle

The curve of T_m shows a positive torque that is substantially greater than the negative torque—a requisite for continued rotation. The precise curve primarily is dependent upon the load torque—with a large increase in the load, the accelerating period is shortened and the net torque may be inadequate to sustain rotation.

The rotor of a reluctance motor is basically a squirrel-cage with a number of teeth and conductors removed. A rotor for a four-pole motor is illustrated in Fig. 7-14. The stator winding is a phase-splitting type or shaded pole. To produce

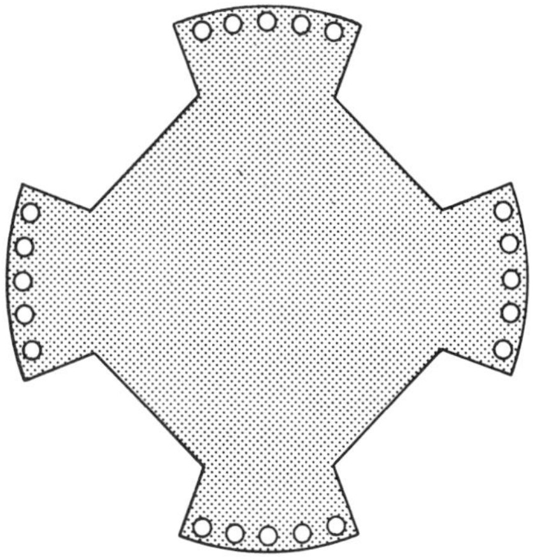

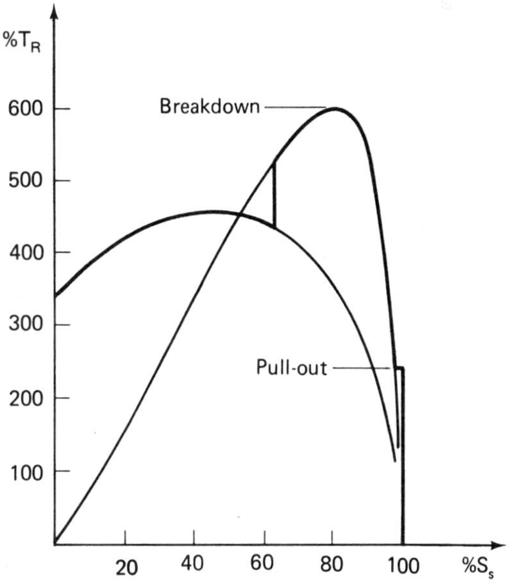

Figure 7-14 Reluctance Motor

the desired characteristics when running on the reluctance principle, the construction and design results in an exceptionally high induction-motor breakdown torque when running on one winding. The motor starts on the main and auxiliary, continues to accelerate when the auxiliary cuts out, and, at a speed near synchronous, snaps out of induction and into reluctance running. If loading is increased beyond the pull-out torque, the motor loses synchronism but continues to run as an induction motor.

7-5. HYSTERESIS MOTOR (SYNCHRONOUS)

In the discussion of hysteresis in Paragraph 1-8, illustrated in Fig. 1-15, it was noted that a substantial mmf is required to alter domain orientations in high-retentivity material. In Fig. 7-15 a smooth, cylindrical rotor is constructed from this type of material. If the rotor is locked and a constant-strength stator flux, Φ_s, is rotating at synchronous speed, the domains tend to align with this flux; however, work is performed in the continual realignment and the domain vectors lag the stator flux.

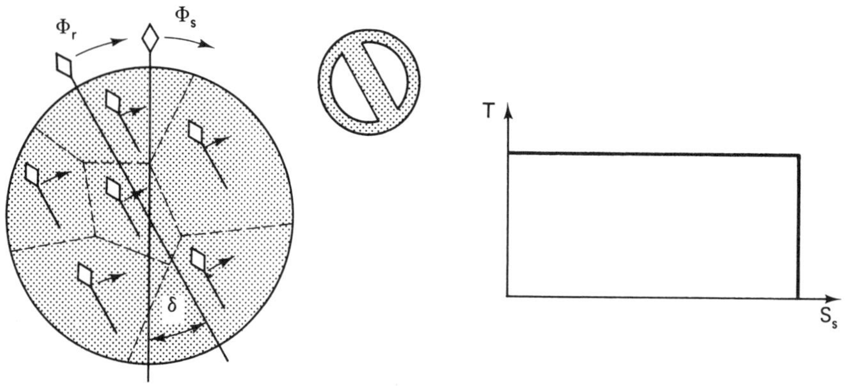

Figure 7-15 Hysteresis Motor

Their vector sum, the rotor flux Φ_r, therefore lags the stator flux by the hysteric angle δ. If the rotor now is unlocked, it will follow the stator flux, accelerating to synchronous speed. Throughout the accelerating period the lag angle, δ, remains constant and both Φ_s and Φ_r are constant; therefore, the torque also is constant. At synchronous speed, the lag angle decreases as the domain vectors more nearly align with Φ_s, but complete coincidence is not possible, since there then would be no torque angle to maintain rotation. In effect, at synchronous speed the rotor is permanently magnetized along a fixed axis which lags Φ_s by the torque angle required for the load. The effect may be strengthened by departing from the uniform cross section and providing a definite path for the flux, as shown by the smaller sketch in Fig. 7-15.

Unlike the reluctance motor, the hysteresis motor cannot operate at steady speeds less than synchronous. If the load torque increases, the motor will readjust the torque angle to provide equilibrium; loading beyond the maximum torque will stop the motor. Further, the motor will not operate with a stator flux pulsating on a fixed axis; either a permanent-capacitor or shaded-pole stator is required. With a permanent-capacitor stator, the hysteresis motor is the smoothest-running, quietest single-phase motor and therefore is preferred for quality sound-reproduction equipment. However, for identical horsepower ratings, it is more expensive than the reluctance motor.

A uniform-cross-section rotor inherently will match the number of stator poles. With provisions for pole changing in the stator, the motor is multispeed.

This motor is made in very small sizes only.

7-6. REPULSION MOTORS

A rotor with windings similar to the armature of a dc machine is represented in Fig. 7-16. The ends of the coils terminate at commutator segments. A distributed stator winding establishes a sinusoidally variant flux on the vertical axis, resulting in transformer voltages in the rotor. Voltages in the conductors have opposite polarities to the right and left of the direct axis.

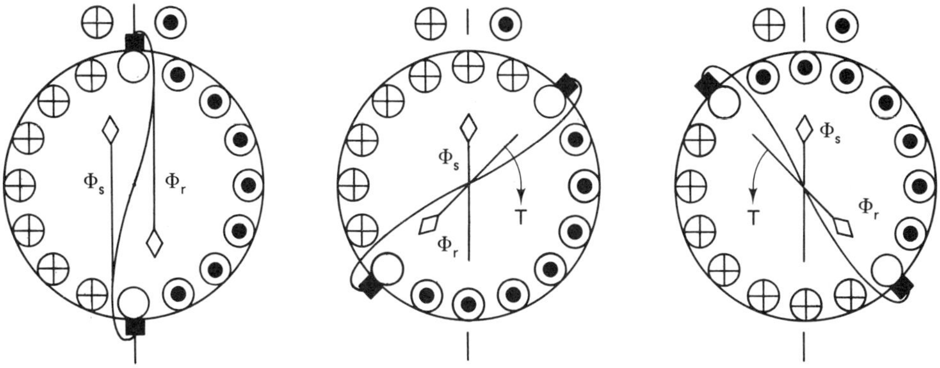

Figure 7-16 Reluctance Torque

If the brushes are electrically located on the direct axis and short-circuited, the armature currents produce a rotor flux coincident with but opposing the stator flux; no torque is produced. However, if the brushes are repositioned to the right or left of the direct axis, the net voltage in a series path creates coil currents whose polarity changes not with respect to the *direct* axis but rather the *brush* axis. The rotor flux no longer is coincident with the stator flux and a torque is produced. Rotation occurs because of the *repulsion* of like poles.

There can be no torque if the brushes are located on the quadrature axis—in this position the series-connected coils consist of matched pairs with equal but opposing voltages—the net voltage in a series path is zero and no current flows.

The repulsion principle is applied in three forms.

Repulsion Motor. The brushes are short-circuited during both starting and running. The torque–speed characteristics are shown in Fig. 7-17; starting torque is high but drops rapidly to a relatively low running torque as rotation produces a speed voltage in the armature. Speed is not frequency-sensitive as in the induction and synchronous motors but may be adjusted by shifting the position of the brushes. Note that the speed regulation is poor since a relatively small load change causes a comparatively large speed change.

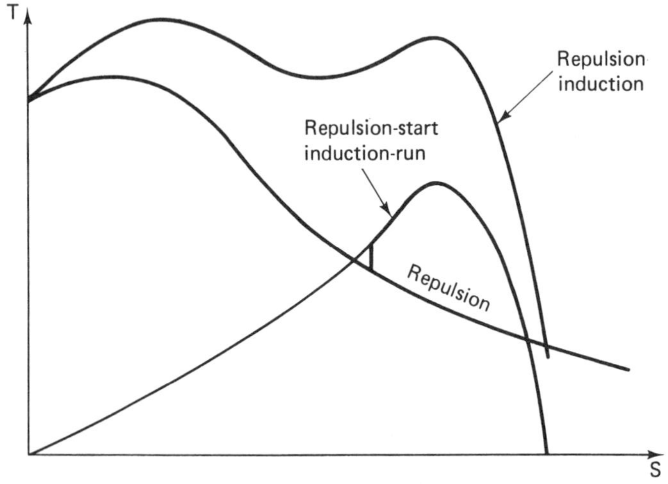

Figure 7-17 Repulsion Motors

Repulsion-start Induction-run Motor. If a short circuit is created across the commutator segments of the repulsion motor, the rotor is the electrical equivalent of a squirrel cage. In the repulsion-start motor, the brushes are short-circuited at starting but, as the speed builds, a complex mechanism raises the brushes and concurrently short-circuits the segments. Thus, the high starting torque of the repulsion motor is combined with the constant-speed overload-torque characteristics of the induction motor.

Repulsion-induction Motor. The rotor of this motor has two independent sets of windings, repulsion and conventional squirrel cage. Both are continuously operative, producing a characteristic curve in which the motor torque is the sum of the repulsion and induction torques. The large accelerating and breakdown torques of this motor are evident in Fig. 7-17.

All forms of the repulsion motor are expensive. They are employed only in the few instances where their unique characteristics are essential to the application. In size, they range up to 10 hp.

7-7. UNIVERSAL MOTOR

The universal motor is a series-wound motor designed to operate at approximately the same speed and output either on direct current or on single-phase alternating current with a frequency not greater than 60 Hz and an rms voltage approximately the same as the dc voltage.

From the discussion of the dc motor it is known that reversal of the directions of *both* the flux and the armature current does *not* change the direction of the torque. The direction of the flux depends upon the direction of the current through the coil. Since the coil of a series motor is excited by the armature current, both the current and the flux reverse simultaneously. The instantaneous torque always is in the rotating direction, but the ac torque will have the double-frequency pulsation common to single-phase power.

The torque–speed characteristics differ somewhat between ac and dc. With dc the counter emf is less than the applied voltage by $I_a(R_{ac} + R_{se})$, but with ac there is an added drop across the inductive reactances, resulting in a lesser generated countervoltage and a lower speed for the same torque. The ac rms voltage has a peak value of $\sqrt{2}$ V; saturation at the higher values results in a lesser effective flux than provided by dc.

In a dc machine, the stator and pole cores usually are solid castings inasmuch as the flux is constant; for ac operation, core losses would be high without lamination. Because armature reaction is more detrimental with ac, compensating windings are normal.

Rotor speeds of universal motors generally range from 5000 to 15,000 rpm, but often the rotor is geared down to a much lower speed at the load shaft. Speed is easily adjusted; although a number of methods have been used in the past, the current trend is toward solid-state speed control (Chapter Eight). Standard ratings are $\frac{1}{20}$ to 1 hp for speeds of 5000 r/min and above; at lesser speeds, there is a substantial disparity in performance between ac and dc operation.

SECTION 7-2. SERVO COMPONENTS

The principles of regulation and control were introduced in Chapter Three and illustrated with dc components. This section describes common ac devices found in data-transmission and automatic-control systems.

There are many requirements for visual display of the status of some aspects of a system—the amount of gasoline in the tank of an automobile, the operating status of generating capacity and feeder loads in a power distribution system, the many displays in the cockpit of a commercial jet. Requirements of this type can be met by electromagnetic devices: for example, the fuel gauge of the auto is a permanent-magnet rotor in a three-winding stator; winding currents, varied by a rheostat with a float-positioned arm, establish a net stator flux on an axis that varies with the liquid level and with which the permanent magnet aligns. The term *instrument servo* commonly is applied to indicators of this type.

In one sense, automatic control is simply an extension of data transmission; a visual display no longer is the end objective, but a similar sensing is compared with a reference to automatically cause the system to perform according to a preconceived plan. Often, comparison is accomplished in a computer which then dispatches correcting signals to power components.

The devices discussed in this section for the most part accept inputs in one form, electrical or mechanical, and provide outputs in the other form. The purpose is to illustrate capabilities of common electromagnetic transducers.

7-8. SYNCHROS Three identical windings, physically separated by 120° and wye-connected, are shown in Fig. 7-18(a). Terminal 2 is designated 0°; a positive angle is in a counterclockwise direction. Terminal 3 is oriented at 120° and terminal 1 at 240°.

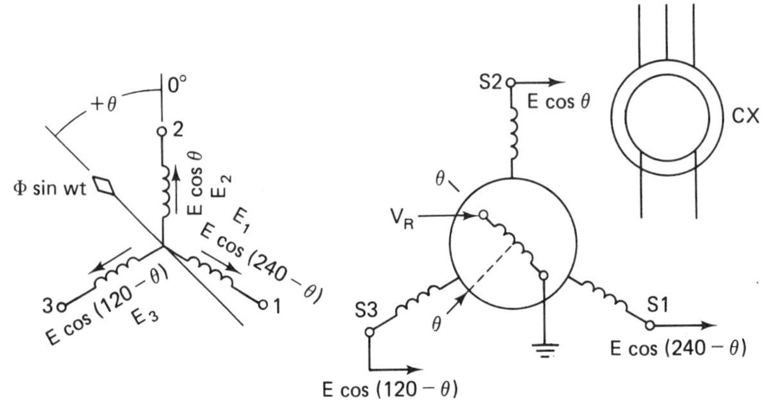

(a) Transformed voltages (b) Transmitter output

Figure 7-18 Synchro Transmitter

If a flux, $\Phi \sin \omega t$, is introduced on a fixed axis at $\theta°$, the magnitude of the voltage induced in each coil depends upon the angle between the flux and the winding axes. When a flux and winding axis coincide, the rms magnitude of the induced voltage is E; for example, if θ is 0°, E_2 has the magnitude E. However, when the axes are not coincident, the voltage is E times the cosine of the angle between them. Therefore, $E_2 = E \cos \theta$, $E_3 = E \cos (120 - \theta)$, and $E_1 = E \cos (240 - \theta)$, with the positive directions indicated by the voltage arrows.

The voltages induced in these windings *are not polyphase voltages*. Three-phase voltages are identical in magnitude but *phase-separated* by 120°; these voltages are *in phase* but differ in magnitude because of their physical orientation.

Transmitter. The windings may be placed in a stator or on a cylindrical rotor. They are distributed—*not* concentrated on salient poles. With the windings in the stator, a single rotor winding, excited by a constant ac reference voltage and with its axis oriented at $\theta°$, provides the sinusoidal flux as depicted in Fig. 7-18(b). A *mechanical input*, θ_i, positions the rotor and the output at the stator terminals is *electrical synchro data*, unique in magnitude and polarity for any angle θ.

This is a *synchro transmitter*; the standard symbol is shown.

Note that if the rotor is revolved through 180°, the magnitudes of E_1, E_2, and E_3 are unchanged but polarities are reversed. Synchro data validly reflect a full 360° range of input angles.

Example 7-2: $E = 20$ V. Find E_1, E_2, and E_3 for input angles of $+40°$, $-40°$, $(180 + 40)°$, and $(180 - 40)°$.

SOLUTION:

θ	E_2 $(20 \cos \theta)$	E_3 $[20 \cos (120 - \theta)]$	E_1 $[20 \cos (240 - \theta)]$
+40	15.3	3.5	−18.8
−40	15.3	−18.8	3.5
(180 + 40) 220°	−15.3	−3.5	18.8
(180 − 40) 140°	−15.3	18.8	−3.5

Each trio of voltages is unique for the corresponding θ.

Receiver. Suppose that the output of a synchro transmitter is connected to corresponding terminals of an identical stator. From awareness of reversible electromagnetic processes in devices such as the transformer and the synchronous motor, intuition suggests creation of a flux on the axis θ. For confirmation, consider Fig. 7-19(a).

If the ring forming the wye connection is at potential V, the sum of the currents flowing into the ring is

$$\frac{E \cos \theta - V}{Z} + \frac{E \cos (120 - \theta) - V}{Z} + \frac{E \cos (240 - \theta) - V}{Z} = 0$$

Therefore,

$$\begin{aligned} V &= \frac{E}{3}[\cos \theta + \cos (120 - \theta) + \cos (240 - \theta)] \\ &= \frac{E}{3}\left[\cos \theta + \left(-\frac{1}{2} \cos \theta + \frac{\sqrt{3}}{2} \sin \theta\right) + \left(-\frac{1}{2} \cos \theta - \frac{\sqrt{3}}{2} \sin \theta\right)\right] \\ &= 0 \end{aligned}$$

Consequently, letting $E/Z = I$, the winding currents are

$$\begin{aligned} I_1 &= I \cos (240 - \theta) \\ I_2 &= I \cos \theta \\ I_3 &= I \cos (120 - \theta) \end{aligned}$$

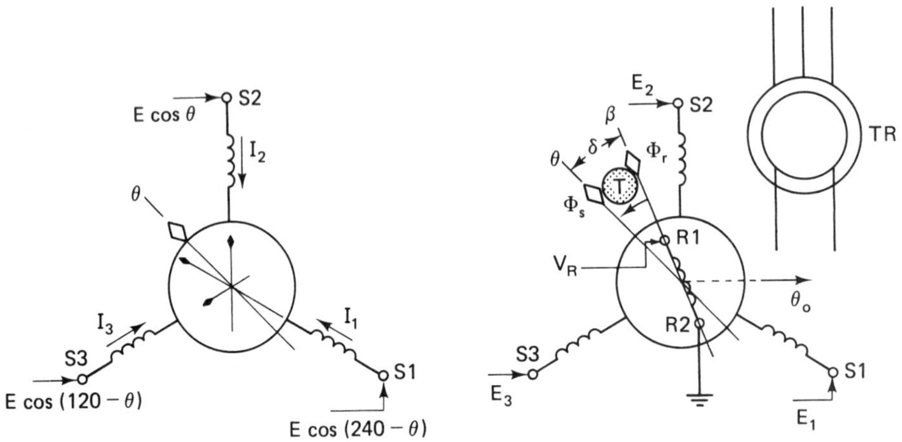

Figure 7-19 Synchro Receiver

Providing the iron is unsaturated, the flux produced by each current is directly proportional to the magnitude of the current. Therefore,

$$\Phi_T = \Phi \cos \theta \underline{/0^\circ} + \Phi \cos (240 - \theta)\underline{/240} + \Phi \cos (120 - \theta)\underline{/120}$$
$$= 1.5\, \Phi\underline{/\theta}$$

The orientation at angle θ is confirmed. Graphically, the component fluxes and the resultant are shown in Fig. 7-19 for $\theta = 40^\circ$.

If a rotor, identical to that of the synchro transmitter and excited by the reference voltage V_R but located at angle β, is inserted in a stator whose flux is at angle θ, a torque angle δ, equal to $\theta - \beta$, exists between the stator and rotor fluxes. The rotor turns until the fluxes are aligned. Thus, the electrical synchro data input to the stator produces a mechanical shaft-angle output. This is a *synchro receiver*, also referred to as a torque receiver. Generally, the receiver will respond to displacements on the order of 1°.

One common application of a transmitter–receiver pair is to instrument servos. For example, the rotor of a transmitter may be mechanically coupled to the elevator of an aircraft, continuously signaling the receiver (which is the movement of a cockpit panel meter) to provide a constant read-out of the elevator position.

Electrically, there is no difference between a transmitter and a receiver. However, the receiver requires mechanical damping in order to avoid substantial oscillation when responding to a new input. When $\theta_i = \theta_o$, there is no current flow between stators if the maximum coil voltages of the transmitter and receiver are identical.

The tie point of the wye of a winding seldom is accessible and test measurements must be made between the terminals. Considering E_{23}, the potential at S_2

with respect to S_3,

$$\begin{aligned} E_{23} &= -E_3 + E_2 \\ &= -E\cos(120 - \theta) + E\cos\theta \\ &= E\left[-\left(-\frac{1}{2}\cos\theta + \frac{\sqrt{3}}{2}\sin\theta\right) + \cos\theta\right. \\ &= \sqrt{3}\,E\left(\frac{\sqrt{3}}{2}\cos\theta - \frac{1}{2}\sin\theta\right) \\ &= \sqrt{3}\,E\sin(120 + \theta) \end{aligned}$$

Similarly,

$$\begin{aligned} E_{31} &= -E\cos(240 - \theta) + E\cos(120 - \theta) = \sqrt{3}\,E\sin\theta \\ E_{12} &= -E\cos\theta + E\cos(240 - \theta) = \sqrt{3}\,E\sin(240 + \theta) \end{aligned}$$

Example 7-3: $E = 20$ V. When $\theta = 30°$, what are the voltages between stator terminals?

SOLUTION:

$$\begin{aligned} E_{23} &= \sqrt{3}\,(20)\sin(120 + 30) = 17.32\text{ V} \\ E_{31} &= \sqrt{3}\,(20)\sin 30 = 17.32\text{ V} \\ E_{12} &= \sqrt{3}\,(20)\sin(240 + 30) = -34.64\text{ V} \end{aligned}$$

Example 7-4: The measured voltages between stator terminals are $E_{23} = 10.26$, $E_{31} = 19.28$, and $E_{12} = -29.54$. Find θ and E.

SOLUTION:

$$E_{23} = 10.26 = \sqrt{3}\,E\sin(120 + \theta) = \sqrt{3}\,E\left[\left(\frac{\sqrt{3}}{2}\right)\cos\theta - \left(\frac{1}{2}\right)\sin\theta\right]$$

$$E_{31} = 19.28 = \sqrt{3}\,E\sin\theta$$

Therefore,

$$\frac{10.26}{19.28} = \frac{(\sqrt{3}/2)\cos\theta - \frac{1}{2}\sin\theta}{\sin\theta} = \left(\frac{\sqrt{3}}{2}\right)\text{ctn}\,\theta - \frac{1}{2}$$

$$\theta = \text{ctn}^{-1}\frac{2}{\sqrt{3}}\left(\frac{10.26}{19.28} + 0.5\right) = 40°$$

$$E = \frac{E_{31}}{\sqrt{3}\,\sin\theta} = \frac{19.28}{\sqrt{3}\,\sin 40} = 17.3\text{ V}$$

Differential Transmitter. A rotor with three windings is shown in Fig. 7-20. The stator creates a flux at the angle θ and the rotor, whose reference axis is the R2 winding, is at angle ϕ with respect to the stator axis. Hence, the flux is at the angle

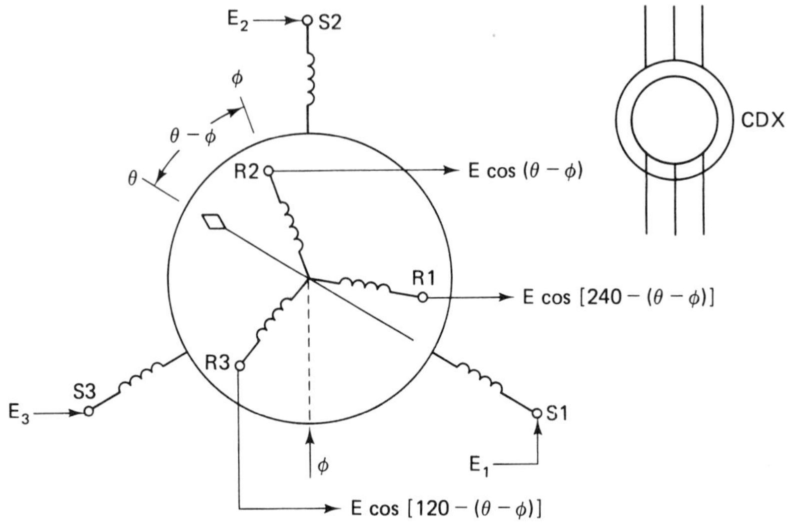

Figure 7-20 Synchro Differential Transmitter

$\theta - \phi$ with respect to the rotor, and the rotor voltages are electrical synchro data for the difference angle. This is a *synchro differential transmitter*. It has two inputs: electrical θ and mechanical ϕ; the output is electrical $\theta - \phi$. Coupled with a transmitter and receiver [Fig. 7-21(a)], a mechanical input to the transmitter (θ) is converted and applied to the stator of the differential transmitter; with a rotor input (ϕ), the electrical output of the differential transmitter is applied to the stator of the receiver to provide the mechanical output $\theta - \phi$.

If the S1–S3 connections to any winding are transposed, the winding flux is

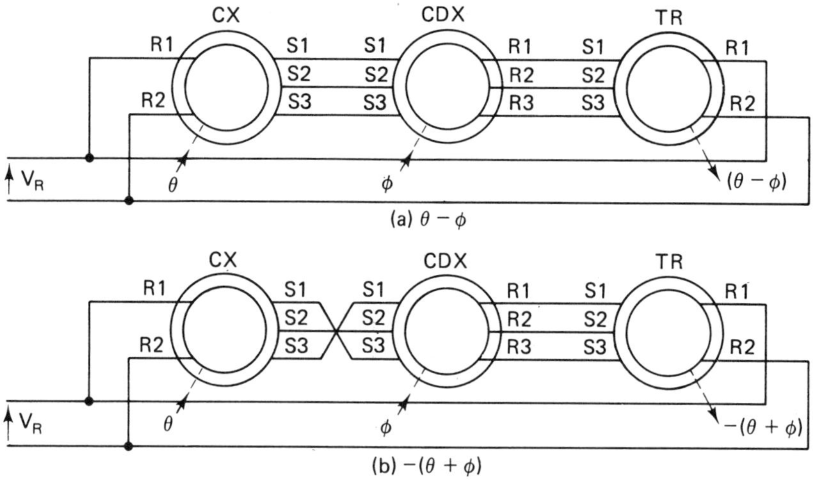

Figure 7-21 Data Transmission with the CDX

at the angle $-\theta$. With the transposition in Fig. 7-21(b), the electrical input to the CDX is $-\theta$ and the output is $-\theta - \phi = -(\theta + \phi)$.

Differential Receiver. In construction, the *synchro differential receiver* is the same as the differential transmitter; however, it accepts two electrical input angles and provides the difference angle as a mechanical output. In Fig. 7-22(a), two transmitters provide inputs to the differential receiver, one (θ) to the stator and the other (ϕ) to the rotor. The reaction of the receiver is illustrated in Fig. 7-22(b). The torque angle between Φ_s and Φ_r causes the rotor to turn until the fluxes are coincident. With both fluxes at θ, but the rotor axis at $-\phi$ with respect to Φ_r, the rotor is at the angle $\theta - \phi$ relative to the stator axis.

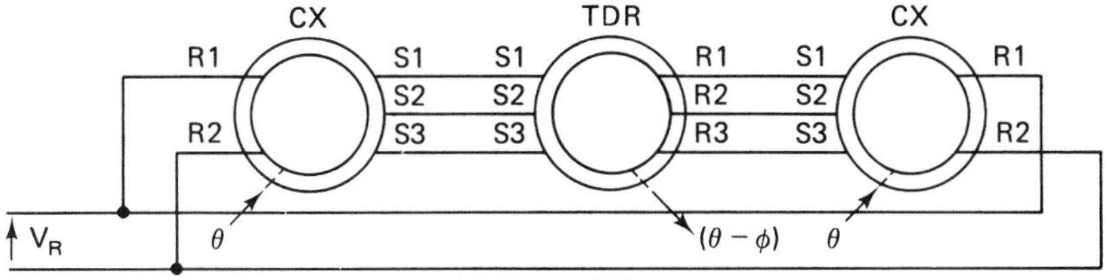

(a) Transmission

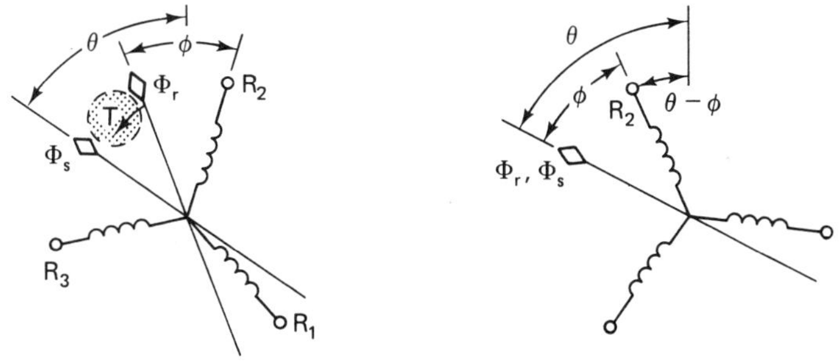

(b) Rotor torque and response

Figure 7-22 Synchro Differential Receiver

The synchro transmitter, receiver, differential transmitter, and differential receiver all have outputs—mechanical or electrical—that, by intent, remain fixed as long as there is no change in the input commands. They are data-transmission devices whose output can be introduced as a command to a feedback system in the same sense that the input potentiometer provided a command to the dc positioning system in Paragraph 3-3. The input potentiometer was not null-seeking but identified the null position for the system; similarly, these synchros are not inherently null-seeking.

Control Transformer. In contrast, the synchro control transformer is a device that *is* intended to force correspondence by providing an output signal only when a system response is in error. The construction of the control transformer is similar to the receiver but, unlike the latter, the rotor is not excited by a reference voltage.

In Fig. 7-23 the rotor is shown at four different locations with respect to a constant stator-flux axis. At position *a*, the coil is perpendicular to the flux and no voltage is induced. At position *b*, with R1 designated the positive reference terminal,

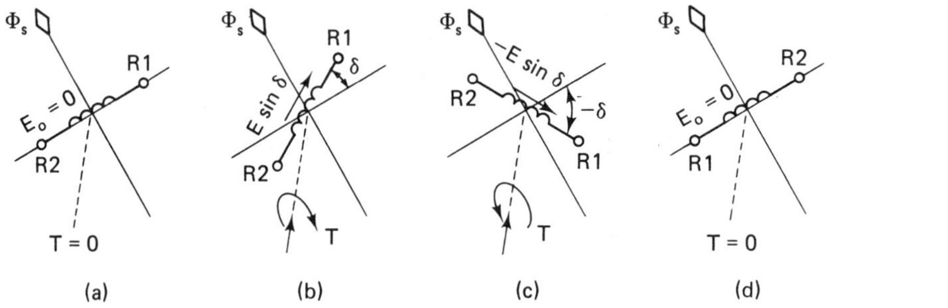

Figure 7-23 Induced Voltages in a Control Transformer

the induced voltage is $E \sin \theta$; this rotor voltage is applied as an error signal to a system that returns a clockwise torque to move the rotor to the null position. At position *c*, the induced voltage is $E \sin(-\theta) = -E \sin \theta$; with this polarity reversal, the system returns a counterclockwise torque. At position *d*, θ is 180° and the voltage again is zero; however, this is a highly unstable null that can be likened to a pendulum whose mass is delicately balanced directly above its pivot, where a slight dislocation causes it to swing to a hanging position. A slight displacement of the rotor in either direction results in a voltage that will drive it to the zero degree null position.

Note that the torques shown in positions *b* and *c* are *not* electromagnetic torques produced by stator–rotor fluxes but mechanical torques returned by a system such as that illustrated in Fig. 7-24; functionally, it is the equivalent of the dc system in Fig. 3-9. An input angle to the transmitter, θ_i, is transferred to the control transformer whose rotor, mechanically coupled to the output, is at θ_o. The voltage representing $\theta_i - \theta_o$ is amplified and applied to the motor, rotating it toward the null position. When $\theta_o = \theta_i$, the transformer voltage is zero and no signal is transmitted to the motor.

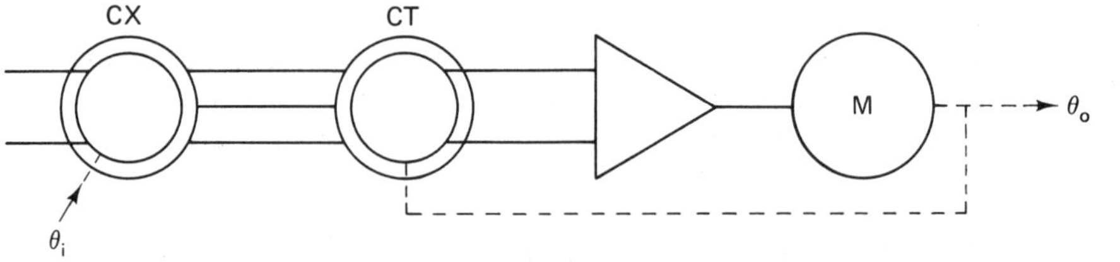

Figure 7-24 Control Transformer in Positioning System

7-9. RESOLVER In Fig. 7-25(a), a rotor has two independent, perpendicular windings, R1–R3 and R2–R4. The windings are not split coils but are sectionalized in the drawing for clarity and convenience in the following discussion.

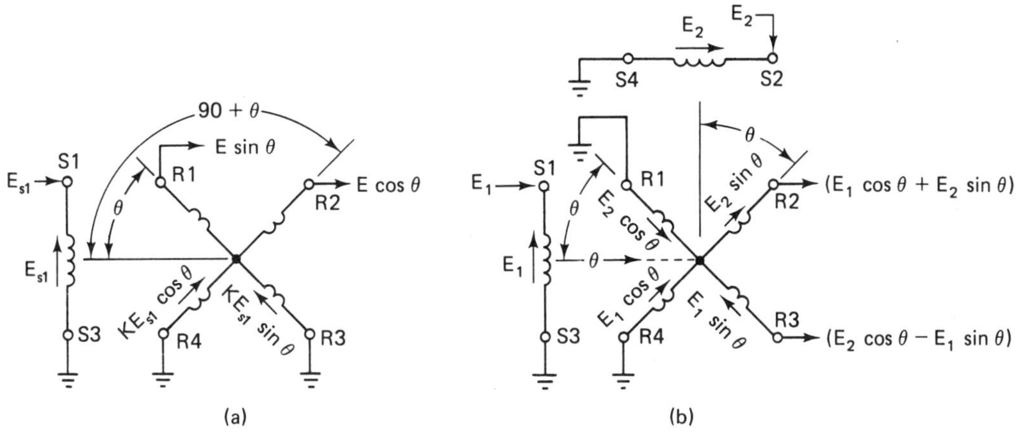

Figure 7-25 Resolver

S1–S3 is a stator winding. A displacement angle, θ, is defined as the angle between the R1–R3 winding axis and the *perpendicular* to the S1–S3 winding. The R2–R4 axis is at $90 + \theta$ with respect to S1–S3.

When $\theta = 0°$, R1–R3 is perpendicular to S1–S3; with voltage E_{S1} applied to the stator winding, no voltage is induced in R1–R3. But at $\theta = 90°$ the coils are aligned and the induced voltage is kE_{S1}, where k is the rotor-to-stator turns ratio. At any other angle, the R1–R3 voltage is $kE_{S1} \sin\theta$. Concurrently, the voltage induced in R2–R4 is $kE_{S1} \sin(90 + \theta) = kE_{S1} \cos\theta$.

If $k = 1$ and $E_{S1} = E$, at angle θ $E_{R1} = E \sin\theta$ and $E_{R2} = E\cos\theta$. This is a *polar-rectangular coordinate-conversion* system—that is, the polar coordinate $E\underline{/\theta}$ has been converted to outputs whose magnitudes are the X–Y coordinates.

A second stator winding, S2–S4 perpendicular to S1–S2, is added in Fig. 7-25(b). At input angle θ, R2–R4 is at θ and R1–R3 is at the angle $\theta - 90$ with respect to S2–S4. With $k = 1$, E_1 applied to S1 and E_2 to S2, each rotor winding has two induced voltage components that add arithmetically if E_1 and E_2 are in phase. In representing these components, voltages due to E_1 are shown in the lower half of the rotor coil and those due to E_2 in the upper half. The voltage arrows for rotor components have the same general direction as the corresponding stator arrows—components due to E_1 are oriented upward while those due to E_2 point generally to the right. The E_1 components are the same as those in Fig. 7-25(a), and the E_2 components are as shown. With R1 and R4 grounded, the potential at R2 is $(E_1 \cos\theta + E_2 \sin\theta)$ while that at R3 is $(E_2 \cos\theta - E_1 \sin\theta)$.

One use of this arrangement is conversion of *rectangular to polar* coordinates. In Fig. 7-26, E_x and E_y are voltages whose magnitudes represent the X and Y

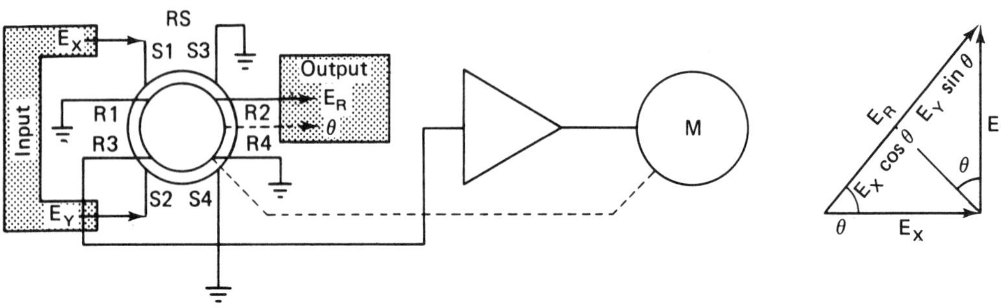

Figure 7-26 Rectangular-Polar Coordinate Transformation

coordinates, respectively. The rotor arrangements are the same as in Fig. 7-25; therefore, the output at R3 is $E_y \cos\theta - E_x \sin\theta$. This voltage is amplified and applied to a motor that rotates the resolver to the null position where the output at R3 is 0. At this position,

$$E_y \cos\theta = E_x \sin\theta$$

$$\frac{E_y}{E_x} = \frac{\sin\theta}{\cos\theta} = \tan\theta \quad \text{and} \quad \theta = \tan^{-1}\frac{E_y}{E_x}$$

Therefore, the mechanical orientation of the rotor is at the angle of the corresponding polar coordinate.

The potential at R2 is $E_x \cos\theta + E_y \sin\theta$. Referring to the triangle in Fig. 7-26, it is apparent that this is the hypotenuse E_R—that is, the magnitude of the voltage correctly represents the magnitude of the polar coordinate.

Coordinate conversions are particularly useful in navigational systems: for example, conversion of distance along a bearing referred to north (polar coordinate) to a map grid (rectangular coordinate).

Another use of the resolver is to supply the feedback signal in numerical control systems for machine tooling. The end objective of the system is precise positioning of a worktable along one or more axes in response to instructions fed to the system by a prepared tape. The movement along each axis results from motor-driven rotation of a lead screw to which the rotor of a resolver is coupled.

In this application reference voltages are applied to the stator windings of the resolver; the voltages are identical in amplitude and frequency but phase-separated by 90° ($V \sin \omega t$ and $V \cos \omega t$). Inasmuch as the stator windings are physically perpendicular, the stator condition is identical to that of the two-phase motor discussed in Paragraph 7-2 and illustrated in Fig. 7-3. The result is a flux rotating at the synchronous speed $\omega_s = 2\pi f_a$, where f_a is the frequency of the applied voltage.

Only one rotor winding is used in this application. The revolving flux generates a voltage in the winding; this voltage is the feedback signal for the system.

With the rotor at rest, the voltage for three different rotor positions is illustrated in Fig. 7-27. In Fig. 7-27(a) the rotor is at the reference position; the voltage

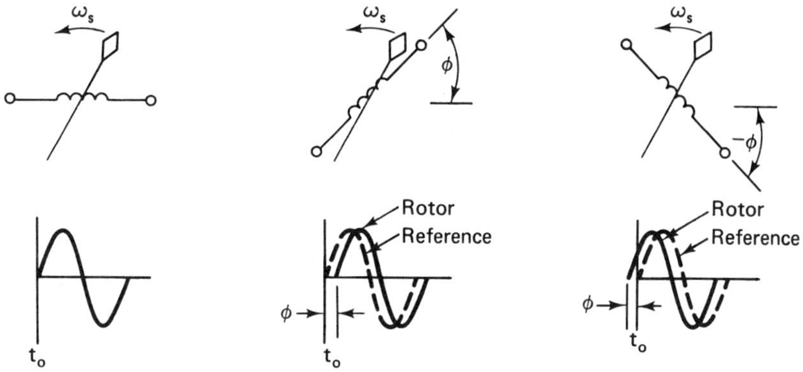

Figure 7-27 Phase Shift in Rotor Voltage

is $V \sin \omega_s t$ with a corresponding frequency f_a. With the rotor relocated at $\pm\phi$ (figures b and c) the magnitude and frequency of the rotor voltage are unchanged, but the *phase* is $\mp\phi$. This phase shift is the significant character of the feedback signal.

Now consider the circumstance of Fig. 7-28 in which the rotor is driven by an external motor. If the rotor is instantaneously accelerated to a speed of $\mp\omega_r$, the speed of the flux relative to the rotor becomes $\omega_s \pm \omega_r$. The frequency of the rotor voltage is therefore $(\omega_s \pm \omega_r)/2\pi = f_a \pm f_r$, where f_r is the *speed of the rotor* in revolutions per second.

Practically, the acceleration of the rotor from rest to a steady-state speed is *not* instantaneous. During the brief acceleration period, a lag angle ϕ accumulates as illustrated. At steady state, the resulting rotor voltage is $V \sin [(\omega_s \pm \omega_r)t \mp \phi]$.

The basic control loop of the system is illustrated in Fig. 7-29. A tape supplies the control unit with the distance (increments) and direction (sign) that the table is to be displaced along the screw axis. The control unit provides both (1) the refer-

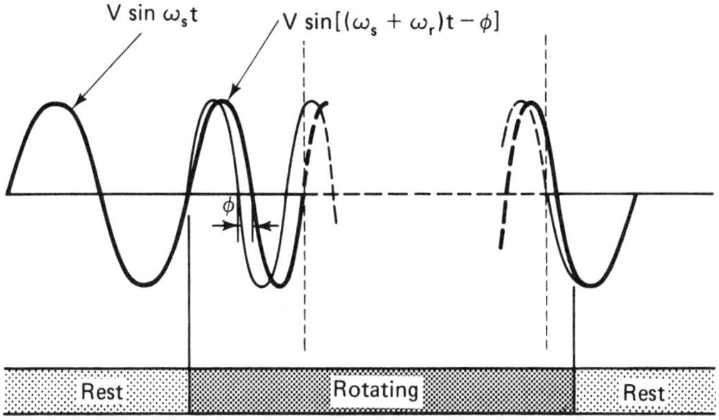

Figure 7-28 Resolver Response to Command Signal

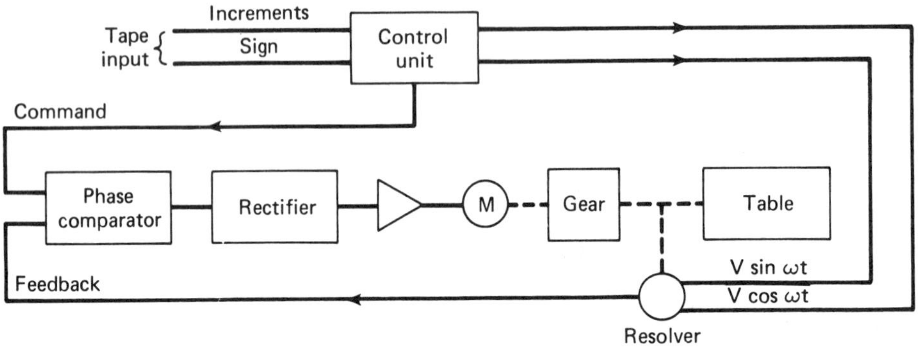

Figure 7-29 Numerical Control System

ence voltages for the resolver stator and (2) a command signal. The reference voltages never vary in frequency, amplitude, or phase, but the command signal is varied in frequency and phase in response to the tape instructions.

The command signal and the feedback signal (rotor voltage) feed into a phase comparator. If the command and feedback signals are identical and in phase, the output of the comparator is zero and the mechanical system is at rest. When the tape demands repositioning, the frequency of the command signal is changed to f_c, which may be greater or less than f_a. The comparator senses the deviation between the command and feedback signals and transmits an error voltage through a rectifier and amplifier to a motor. The motor drives the table and the coupled resolver in a direction and at a speed that causes the feedback signal to match the frequency of the command signal (i.e., $f_a \pm f_r = f_c$), but the phase separation (ϕ) developed during the acceleration period continues to cause transmission of an error signal by the comparator. The control unit emits the proper number of cycles at the control frequency to cause the required displacement, following which the command signal reverts to the reference frequency, f_a. The rotor then halts at the position where the rotor voltage is in phase with the at-rest control signal.

In a typical system, one revolution of the resolver corresponds to a linear table displacement of 0.1 in. The angular accuracy of the resolver is 0.001 revolution (0.36°), resulting in a linear positional accuracy of 0.0001 in. Typically, the reference frequency is 2.5 kHz and the control frequency may range from 2.475 to 2.525 kHz. The corresponding range of the rotor speed is ± 25 r/sec, with a displacement rate of 2.5 in/sec.

7-10. AC CONTROL MOTORS

Many ac control systems require low-power motors to respond to control signals. The two-phase induction motor discussed in Paragraph 7-1 and illustrated in Fig. 7-3 can fill the need. One winding is a reference which is constantly excited. The other is the control winding with appropriate circuitry to provide a 90° phase shift. The error signal of the system is applied to the control winding; with an error signal applied, the motor rotates but when the error is zero, the motor is at rest. The error signal may be positive or negative (i.e., 180° phase difference) and thereby produces rotation in the proper direction.

The torque–speed characteristics of the control motor depend upon the magnitude of the voltage applied to the control field and are typified in Fig. 7-30. Often they are linearized as indicated by the dashed line; the curves then have a character similar to the armature-controlled dc motor illustrated in Fig. 3-10.

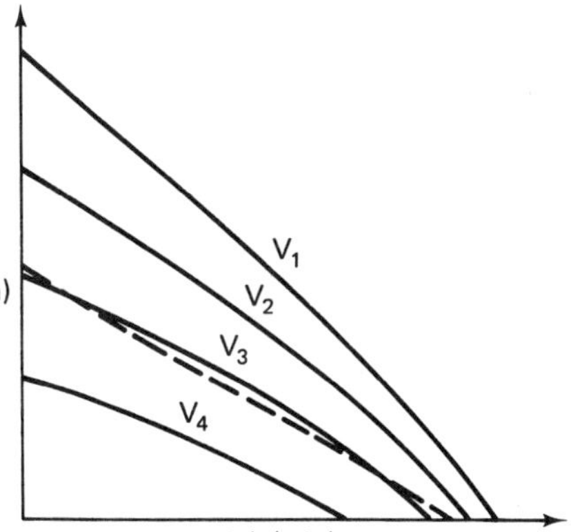

Figure 7-30 AC Control Motor

A single-phase induction motor will run on a single winding after it has been started. Application of a control signal to the two-phase control motor would initiate rotation that continued after removal of the signal if the motor were not properly designed. To preclude single-phasing, rotor resistance is large; this also produces maximum torque at zero speed rather than at a small slip, as is usual for induction motors (Fig. 6-9).

The control transformer and the resolver are among the possible sources for the signal input to the control winding.

7-11. AC TACHOMETER The basic construction of an ac tachometer duplicates that of the ac control motor. Again, the reference winding is excited by a constant ac voltage; however, the control winding is not excited but rather produces an output voltage that is proportional to the rotor speed. Either cross-field or double-revolving flux theory may be used to explain the functioning, the former being less complex. With the reference axis excited and the rotor revolving, a rotational voltage is induced in the conductors of the rotor, causing currents to flow whose polarities alternate with respect to the quadrature axis. Since the flux produced by the reference voltage has a fixed amplitude, the rotational voltage and resulting currents are proportional to the speed of the rotor. The currents produce an alternating flux on the quadrature axis that also is proportional to speed.

Inasmuch as the axis of the control winding also is on the quadrature axis, no voltage is induced by the reference flux, but the quadrature axis flux induces a voltage proportional to the rotor speed.

Frequently, the speed of response of a control motor must be monitored. For this purpose, a control motor and tachometer often are packaged together with both rotors mounted on a single shaft.

PROBLEMS AND QUESTIONS

7-1. At rated load the speed of a 40-millihorsepower 60-Hz permanent-split-capacitor motor is 1050 r/min. (a) What is the rated torque in oz-in? (b) If the full-load slip and torque are the same at 50 and 60 Hz, what is the rated speed and horsepower at 50 Hz?

7-2. The nameplate of a 3/4-hp capacitor-start motor is marked 115/230 V 60 Hz 3450 r/min 10/5 A. (a) To which voltage does the 3450 r/min apply? (b) At rated load and 230 V, what is the line current? (c) At rated load the measured power input is 980 W. What is the power factor and efficiency of the motor?

7-3. The impedance of the main and auxiliary windings of a single-phase induction motor are $6\underline{/40^\circ}$ and $10\underline{/20^\circ}$ Ω, respectively. (a) What capacitance in series with the auxiliary will produce a 90° phase separation in starting currents? (b) With this capacitance, what are the main, auxiliary, and line currents at starting?

7-4. Sketch a switching arrangement to allow a choice between forward and reverse rotation of a universal motor.

7-5. What is your choice of single-phase motors for the following applications? (a) Washing-machine motor (b) Air compressor (c) Wood lathe (d) Portable drill. State your reasons.

7-6. The discussion of the synchro receiver (Paragraph 7-8) showed that the electrical synchro data input created a flux of 1.5Φ on the axis θ. However, the development was based upon current flow resulting from the voltages induced in the transmitter coils only. Assuming identical electrical characteristics in both the transmitter and receiver, the current in each receiver winding is $(E_x - E_R)/2Z$, in which E_x and E_R are the rotor-induced voltages in corresponding phases of the transmitter and the receiver, respectively, and Z is the impedance of each winding. The resulting stator flux in the receiver is $\Phi_s\underline{/\phi} = 1.5\Phi\underline{/\theta} - 1.5\Phi\underline{/\beta}$. In the following table, ρ is the ratio $\Phi_s/1.5\Phi$ and $\Delta\theta$ is the angle by which angle ϕ leads angle θ; thus, for an angular difference $(\theta - \beta)$, $\Phi_s\underline{/\phi} = 1.5\rho\Phi\underline{/\theta + \Delta\theta}$.

$\theta - \beta$	ρ	$\Delta\theta$
0	0	—
30	0.518	75
60	1.0	60
90	1.414	45
120	1.732	30
150	1.932	15
180	2.0	0

(a) Verify that the torque on the rotor is in a direction to drive the receiver to angle θ.
(b) The torque acting on the rotor is $T_r = k\Phi_s\Phi_{rotor} \sin(\phi - \beta)$. Show that $T_r = k_T\rho \sin[(\theta - \beta) + \Delta\theta]$, in which $k_T = 1.5k\Phi\Phi_{rotor}$.
(c) Compute and plot the variable $\rho \sin[(\theta - \beta) + \Delta\theta]$ for the tabulated data. Discuss the plotted results.
(d) The discussion above focuses on the synchro receiver. Consider and discuss conditions in the transmitter.

7-7. A *stepping motor* is one for which a logic circuit controls switches in series with the stator windings in order to cause the rotor to move through a specified number of angular increments, thereby permitting controlled angular or linear displacement in automated manufacturing processes. Figure 7-31 illustrates a stepper motor. The rotor is a ferromagnetic or permanent-magnet bar. The repetitive sequence of switching is as follows:

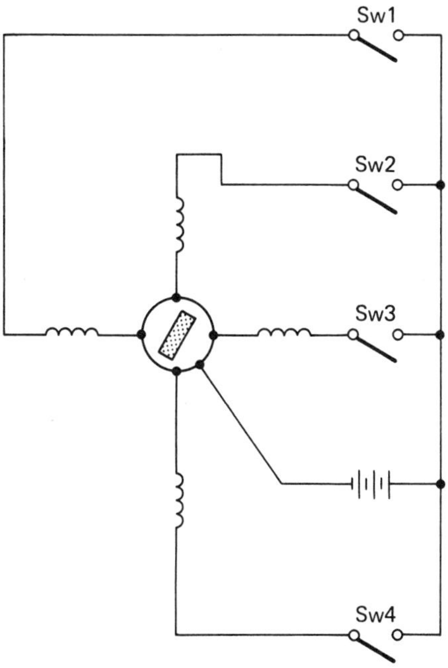

Figure 7-31 Problem 7-7

Step	*SW1*	*SW2*	*SW3*	*SW4*
1	ON	ON	OFF	OFF
2	ON	OFF	OFF	OFF
3	ON	OFF	OFF	ON
4	OFF	OFF	OFF	ON
5	OFF	OFF	ON	ON
6	OFF	OFF	ON	OFF
7	OFF	ON	ON	OFF
8	OFF	ON	OFF	OFF

(a) Explain the functioning of the motor.
(b) By what angular increment does the rotor advance for each step? If the rotor is geared to an output shaft by a ratio of 50 : 1 (i.e., 50 revolutions of the motor shaft produce one revolution of the output shaft), by what angular increment does the output shaft rotate for each step?
(c) If the switching sequence is changed to four-step by eliminating all even-numbered steps in the eight-step sequence, what is the angular increment per step? Is this sequence effective with either type of rotor (ferromagnetic bar or permanent magnet)?
(d) A ferromagnetic rotor may be designed to advance only a fraction of the angular increment of the stator flux on each step. If this fraction is 0.02, with a four-step sequence, how many steps are required per shaft revolution? For a stepping rate of 8000 steps per second, what is the rotor speed?

7-8. Find θ_o of the systems shown in Fig. 7-32.

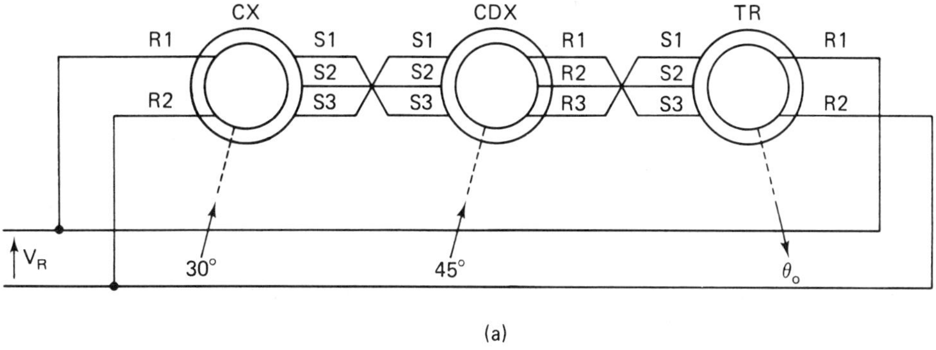

(a)

Figure 7-32 Problem 7-8

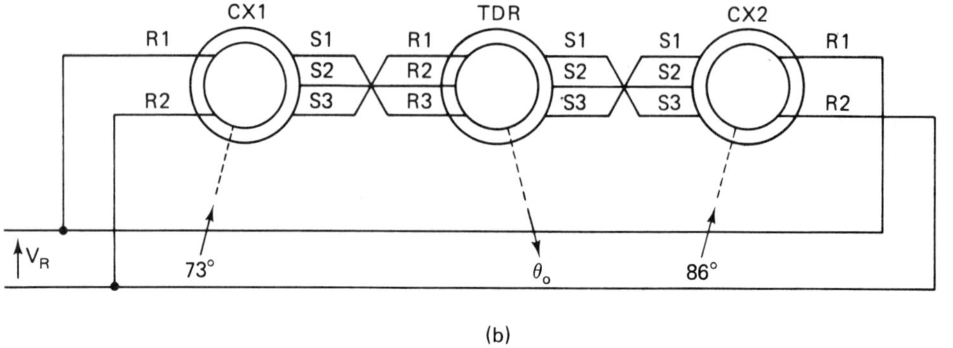

Figure 7-32 (Continued)

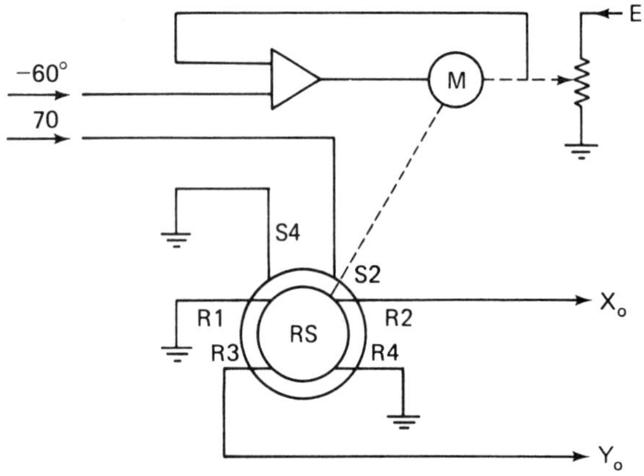

Figure 7-33 Problem 7-9

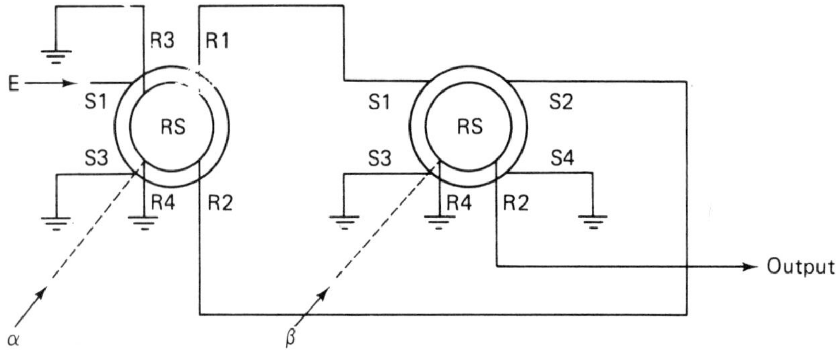

Figure 7-34 Problem 7-10

7-9. Find the outputs X_o and Y_o of Fig. 7-33.

7-10. In terms of E, α, and β, what is the output of the system of Fig. 7-34?

7-11. A ship is traveling on a bearing θ_B (Fig. 7-35). A radar whose zero orientation is the ship's bearing senses a target at range R_T and azimuth θ_T. The sensing is to be converted to the rectangular coordinates X_T, Y_T. Sketch a synchro system that accepts the input $R_T/\underline{\theta_T}$ and converts it to the output X_T, Y_T.

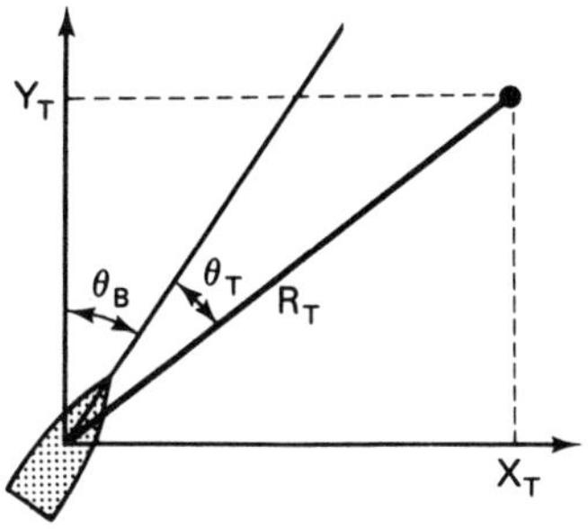

Figure 7-35 Problem 7-11

7-12. A number of transducers that sense linear or angular displacements are based upon the reluctance principle. A *microsen*, illustrated in Fig. 7-36, is extremely accurate in measuring small angular displacements in applications where the range of rotation is limited. An ac reference voltage, E_R, is applied to one set of windings and a second set provides the output signal E_o. A butterfly rotor is coupled to the element whose displacement is measured. Explain the functioning of the microsen and give the character of the output E_o when the rotor is located at 0, $+\theta$, and $-\theta$.

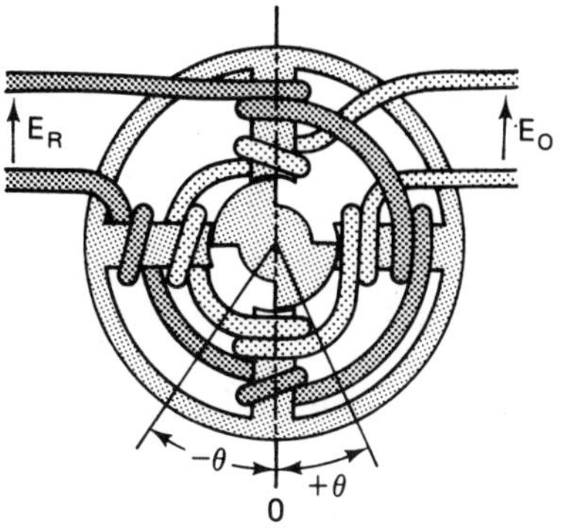

Figure 7-36 Problem 7-12

8

Introduction to Solid-State Power and Control

Although the vacuum tube is found in power and control circuits, development of solid-state devices has dramatically expanded the range of electronic applications in these fields. Many electromagnetic components (e.g., relays, rotating converters) can be supplanted with an accompanying increase in reliability and useful life, reduction in size, and often a greater economy. Some added performance options offered by solid state are not possible by other means.

Systems employing solid-state devices are extremely varied—for example, ANSI standards for rectifiers alone recognize more than 70 semiconductor circuits. This chapter considers fundamental principles and some illustrative applications. Pertinent characteristics of solid-state devices are reviewed only to the extent that the applications may be understood.

8-1. SWITCHING FUNCTION

In control and power applications, solid-state devices are used as switches to alter the circuitry in accordance with a preconceived pattern. Control of the ON or OFF status of the switch provides control of the system characteristics. For example, if a manual line switch to a motor was turned ON and OFF in the proper cadence, the alternating periods of acceleration and deceleration would produce a desired average speed. Similarly, half-wave rectification theoretically could be achieved by closing a mechanical switch at the beginning, and reopening at the ending, of each positive portion of the sinusoidal cycle.

Most solid-state switches are unilateral conductors. In the forward direction, resistance to current flow is very small and the voltage drop across the device is negligible. But in the reverse direction the resistance is very large; reverse current is negligible even though the voltage across the device is high. Idealized, the solid-state switch duplicates the ON and OFF characteristics of a mechanical switch. The principal devices are the diode, thyristor [silicon-controlled rectifier (SCR)], and transistor.

Switches are turned ON and OFF in the proper sequence and at the proper time to produce the desired output. In some instances, a change in the status of the switch occurs naturally as a result of inherent voltage and current variations in the controlled circuit, but in most cases switching control circuits are needed to implement the desired pattern. These control circuits are limited only by the imagination of a designer and range from relatively simple *RC* timing circuits to complex computer control. The need for switching control is recognized and the availability of appropriate means is assumed, but the text is concerned only with the performance of the power circuit when switching occurs.

8-2. CIRCUIT ANALYSIS

In the type of circuits to be discussed in this chapter, with few exceptions either the source or load voltage is sinusoidal. Although the sine wave is studied in introductory ac circuit courses, the usual steady-state ac circuit theory cannot be directly applied to solid-state switching circuits because of unique device characteristics. In this section, the sine wave is considered from the viewpoint of features essential to an understanding of the subsequent circuits. To the extent permitted by a mathematical limitation of algebra and trigonometry, several equations are developed, but the starting point in most instances is a general equation that is readily obtained by calculus methods only; these latter equations are given without proof but are traceable to application of basic laws (e.g., Kirchhoff, Ohm) to a circuit containing transients.

Waveforms. A basic *RL* circuit is shown in Fig. 8-1(a). In the solid-state circuits to be described, inductance usually is present due to (1) the load (e.g., motor), (2) a filter, or (3) a combination of (1) and (2); for this reason, the *RL* circuit requires particular emphasis. The input voltage has the rms value V and the instantaneous value $v = V_m \sin \omega t$, where $V_m = \sqrt{2}\ V$. This voltage, v, is the voltage across the *RL* branch—Kirchhoff's voltage law is inviolable.

In Fig. 8-1(b, 1), $\omega L = 0$ and the load is purely resistive. The impedance angle ($\tan^{-1} \omega L/R$) is 0°; current is in phase with the voltage. Instantaneous power is the product vi; the steady-state power over one cycle normally is determined from rms values—that is, $P = VI$ ($I = I_m/\sqrt{2}$). Because the polarities of the voltage and current always are identical, power always is positive—all input elec-

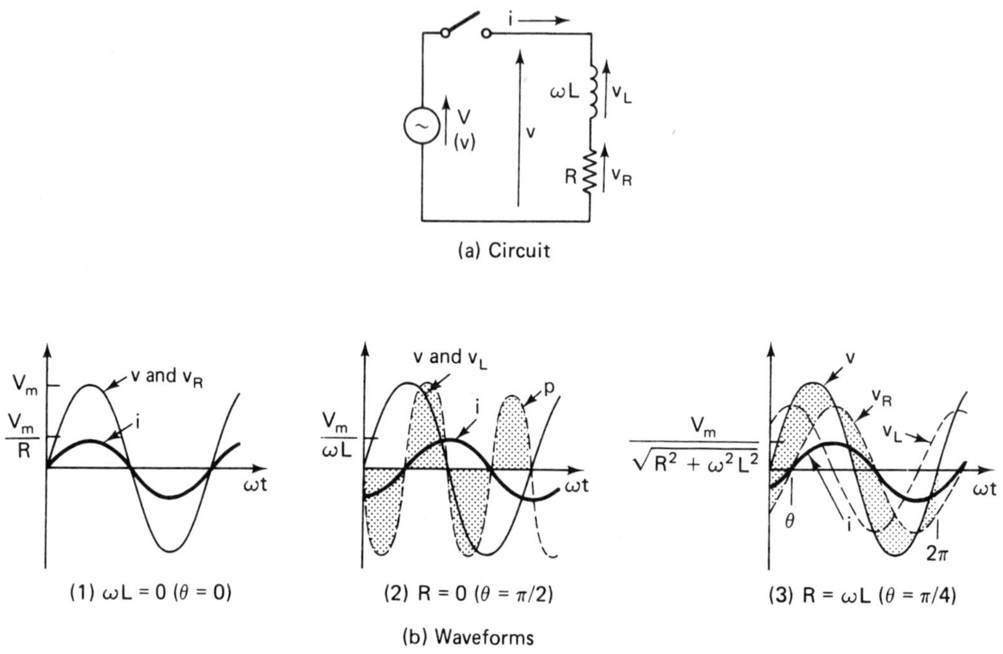

Figure 8-1 Steady-State Circuit Analysis

trical energy is converted to another form. Although average values of the voltage and current can be determined, generally they are meaningless in analyzing this type of circuit.

In Fig. 8-1(b, 2), $R = 0$ and the load is purely inductive. The impedance angle is $\tan^{-1} \omega L/0 = \pi/2$ rad (90°); consequently, current lags the voltage by 90°. The power curve is shown to emphasize an important point; the shaded areas under the power curve represent energy. From symmetry, it is evident that the net energy (and the power) over one cycle is zero. The physical interpretation is that energy is *stored in the inductor* during positive power intervals but *returned to the source* during negative intervals. Storage is in the form of the magnetic field surrounding the inductor. This field can exist only while the circuit is energized. If the switch is opened at $\omega t = \pi$ rad, the energy stored during the preceding quarter-cycle must be discharged; therefore, it will force a current across the air gap between the opening blades of the switch, usually forming a visible spark or arc. During the discharge, current retains the polarity it had immediately prior to opening the switch, but, inasmuch as energy is being discharged, the voltage polarity must be opposite that of the current. In the illustration, the energy discharge is forced to occur over the interval π to $3\pi/2$ rad since the inductor voltage must conform to the source voltage, but when a switch is opened there are no restraints except the time constant, $\tau = L/R_{air}$, in which R_{air} is the resistance of the air gap. Since R_{air} is very large, the duration of the discharge is brief and the accompanying voltage can be much greater than V_m. The term *discharge voltage* will be used for the inductor voltage under these circumstances.

In Fig. 8-1(b, 3), $R = \omega L$ and $\theta = \pi/4$ rad (45°). Consequently, the current lags the source voltage, v, by the angle $\pi/4$ rad. The voltage across the resistor, v_r, always must satisfy Ohm's law, and therefore is in phase with the current i, but the inductor voltage, $v_L = i(j\omega L)$, leads i by $\pi/2$ rad (90°). In the illustration, v_L is shown as a sinusoid; however, v_L also appears in the sketch in a second, unlabeled form. From Kirchhoff's voltage law, at all times $v_L + v_R = v$; in the shaded areas, the vertical difference, $v - v_R$, *also* is the instantaneous v_L. Note that v_L is positive when v lies above v_R and negative when v is below.

The shaded areas of Fig. 8-1(b, 3) frequently are referred to as volt–time areas and, as such, have a particular significance. From symmetry, it is apparent that, over a full cycle, the area *above* v_R equals the area *below* and the net volt–time area over one cycle is zero. This *always* must hold true regardless of distortion of the waveforms—in any periodic form the average voltage across an inductance is zero.

Returning to Fig. 8-1(b, 2), at $\omega t = 0$, note that $i = I_m \sin(\omega t - \pi/2) = -I_m \cos \omega t$. Also note that energy first is negative—that is, energy flow is from the load to the source. Neither of these circumstances is possible if the switch initially is closed at $t = 0$. However, this representation is common since it is consistent with steady-state performance of an inductor (1) following a transient period of perhaps several cycles immediately after energizing the circuit, and (2) with the understanding that the waveform represents the inductive component of an element that also contains some resistance, be it small—that is, the voltage across a practical conductor is $v_L + v_R$, where v_R is the voltage across the resistance inherent in the coil winding.

Another point to note with respect to all subfigures of Fig. 8-1(b) is that $v = 0$ at $t = 0$. But when the switch is closed the immediate instantaneous source voltage, v, is purely circumstantial—it may be at any point on the sine curve with any instantaneous magnitude between $\pm V_m$. In solid-state switching circuits, turn-on (and turn-off) may occur at different points on the curve, but the points of turn-on are not arbitrary and the ON (or conduction) period is of short duration, a fraction of a cycle. Therefore, in line with the reasoning of this and the preceding paragraph, *the steady-state principles applied to the circuits of Fig. 8-1 generally are invalid in solid-state switching circuits.* However, some principles *do* apply at all times, to wit:

1. **Kirchhoff's voltage and current laws.**
2. **Ohm's law.**
3. **The average voltage across an inductor over a full cycle is zero.**

If the transient period for the circuit of Fig. 8-1 is considered, the defining equation for i is

$$i = \frac{V_m}{\sqrt{R^2 + \omega^2 L^2}}[\sin(\omega t + \phi - \theta) - e^{-t/\tau} \sin(\phi - \theta)]$$

in which $\theta = \tan^{-1} \omega L/R$, ϕ is the angular location on the voltage sine curve at the instant the circuit is turned on, and $\tau = L/R = \omega L/\omega R$. Substituting this latter expression,

$$\boldsymbol{i = \frac{V_m}{\sqrt{R^2 + \omega^2 L^2}}[\sin(\omega t + \phi - \theta) - e^{-\omega t(R/\omega L)} \sin(\phi - \theta)]} \qquad (8\text{-}1)$$

(*Note*: **In this and ensuing discussions, all angles are in radians.**)

This equation has two components, one steady state and the other transient. Since neither θ nor ϕ are time-variant, $(V_m/\sqrt{R^2 + \omega^2 L^2}) \sin(\omega t + \phi - \theta)$ is the *steady-state component*; for $\phi = 0$ rad, this expression is consistent with the relationships shown in Fig. 8-1. The *transient component* is $(-V_m/\sqrt{R^2 + \omega^2 L^2})$ $[(e^{-\omega t(R/\omega L)} \sin(\phi - \theta)]$.

The key element is the exponential term; for pure resistance, it is zero and steady state is instantaneous at $t = 0$ with θ also equal to zero.

With pure inductance ($R = 0$) the value of the exponential expression is 1 and $\theta = \pi/2$ rad. The expression for i then becomes

$$i = \frac{V_m}{\omega L}\left\{\sin\left[(\omega t + \phi) - \frac{\pi}{2}\right] - \sin\left(\phi - \frac{\pi}{2}\right)\right\}$$

$$= \frac{V_m}{\omega L}[-\cos(\omega t + \phi) + \cos\phi]$$

For $\phi = 0$ rad, the expression is

$$i = \frac{V_m}{\omega L}(1 - \cos \omega t)$$

This curve is plotted in Fig. 8-2. Note that the current always is positive and has a peak value of $2V_m/\omega L$—this is *twice* the magnitude of the maximum current predicted by usual ac steady-state circuit analysis.

For $R = \omega L$ and $\phi = 0$ rad, the transient is $(V_m/\sqrt{2}R)[\sin(\omega t - \pi/4) + 0.707e^{-\omega t}]$. Practically, the transient decays to zero after five time constants, for which $\omega t = 5$. The steady-state component, transient component, and sum (i) are

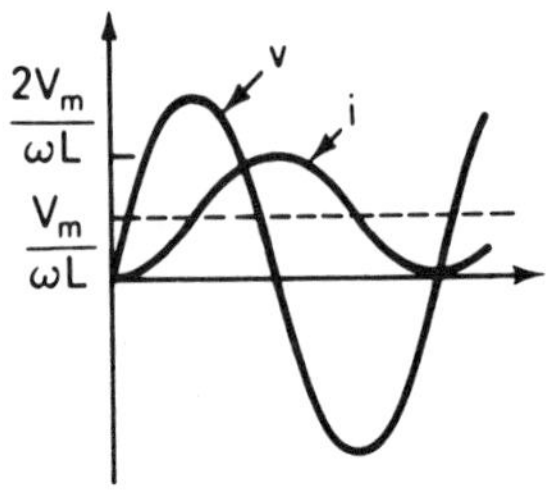

Figure 8-2 Inductance ($\phi = 0°$)

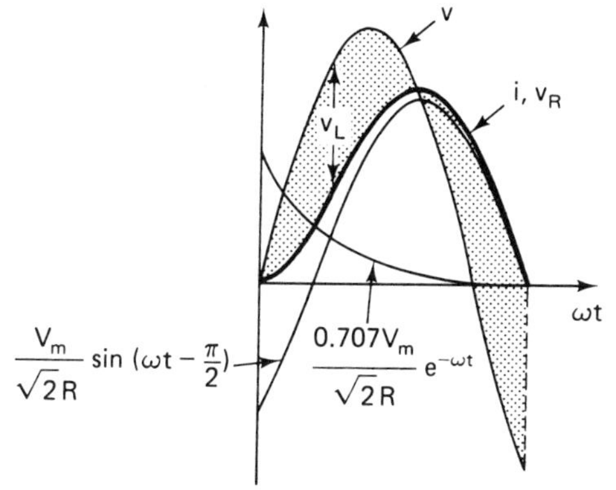

Figure 8-3 $R = \omega L(\phi = 0°)$

plotted in Fig. 8-3 for the interval during which i is positive. The instantaneous voltage v_R ($= iR$) is coincident with i and the instantaneous inductor voltage is $v - v_R$. Note that the positive (i.e., above v_R) volt–time area equals the negative volt–time area. In the negative volt–time area, that portion above the *horizontal axis* represents energy previously stored by the inductor which is absorbed (converted) by the resistance; the portion below the axis is energy returned to the source.

When ωL is much greater than R, the duration of the transient is prolonged well beyond the period of the steady-state component. Often, to simplify analysis of a rectifier circuit with a series-inductance filter, it is assumed that the inductance is infinite and causes the current to remain absolutely constant.

Average, rms, and Ripple Values. While $V_{\text{rms}} = V_m/\sqrt{2}$ and $V_{\text{av}} = 2V_m/\pi$ in steady-state circuits, these values are based on continuity of the sine wave for the full cycle. If the wave is discontinuous, the values will differ. By definition, V_{av} is the net area under the active segment of the $V_m \sin \omega t$ curve divided by the angular base; the base differs in various circuits and may be either the angular limits of the active segment, one-half cycle (π rad), or a complete cycle (2π rad). V_{rms} is the square root of the average value of the area under the $(V_m \sin \omega t)^2$ curve.

In Fig. 8-4 the active segment is from α to β. The shaded area applies for average values and the hachured area for rms values. The areas may be found by

$$(V_m \sin \omega t): \quad A_d = -V_m(\cos \beta - \cos \alpha) \tag{8-2}$$

$$(V_m \sin \omega t)^2: \quad A_r = \frac{V_m^2}{2}\left[(\beta - \alpha) - \frac{\sin 2\beta - \sin 2\alpha}{2}\right] \tag{8-3}$$

The average value, E_d, and rms value (E_r) are

$$E_d = \frac{A_d}{\text{base}} \tag{8-4}$$

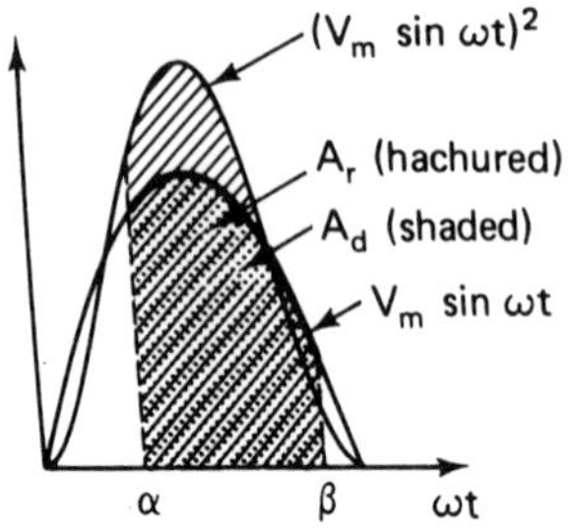

Figure 8-4 Average and Effective Values

$$E_r = \sqrt{\frac{A_r}{\text{base}}} \tag{8-5}$$

in which the base may be $(\beta - \alpha)$, π, or (infrequently) 2π, depending upon the particular circuit.

EXAMPLE 8-1: Find the average and effective (rms) voltage values of a half-wave rectifier.

SOLUTION: (Fig. 8-5). The instantaneous output voltage, v, of a half-wave rectifier follows the sine curve for the first half-cycle but is zero for the second half-cycle. Therefore, $\alpha = 0$, $\beta = \pi$ rad, and the base is 2π rad.

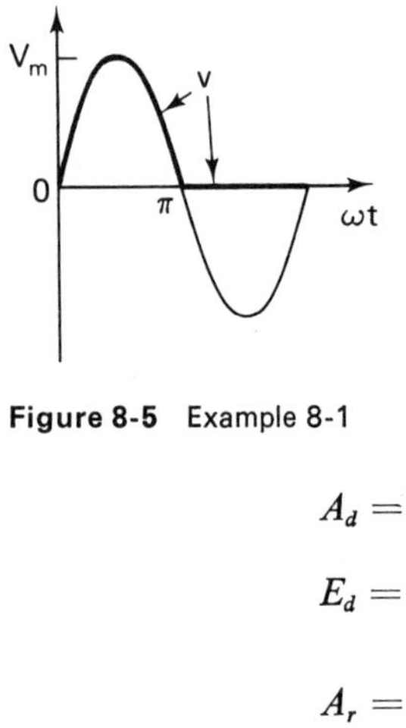

Figure 8-5 Example 8-1

$$A_d = -V_m(\cos \pi - \cos 0) = 2V_m$$

$$E_d = \frac{2V_m}{2\pi} = 0.318V_m$$

$$A_r = \frac{V_m^2}{2}\left[(\pi - 0) - \frac{\sin 2\pi - \sin 0}{2}\right] = \frac{V_m^2\pi}{2}$$

$$E_r = \sqrt{\frac{V_m^2\pi}{2(2\pi)}} = 0.5V_m$$

Note that E_d is one-half of the steady-state average for a full rectified sine wave ($0.636V_m$). However, the rms value is *not* one-half of the steady-state full-wave rms value ($0.707V_m$). [Recall that the reason for using rms values in ac circuits stems from the relationship $P = VI$. Thus, in a full-wave resistive circuit, $P = (0.707V_m)(0.707I_m) = 0.5V_mI_m$. In the half-wave rectifier, $P = (0.5V_m)(0.5I_m) = 0.25V_mI_m$—*one-half of the power* in the full-wave circuit.]

Example 8-2: Find the average and rms values for the voltage whose wave form is shown in Fig. 8-6.

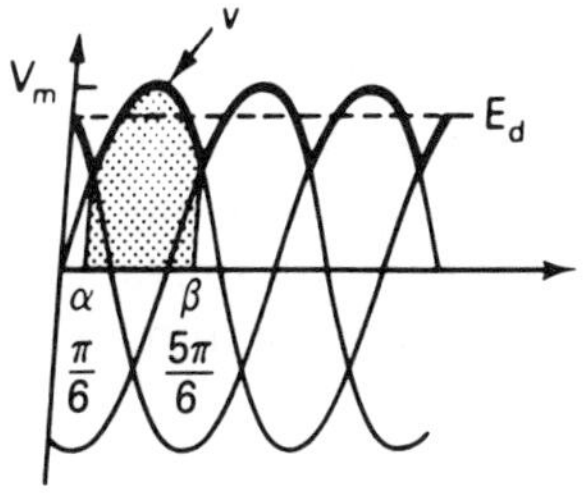

Figure 8-6 Example 8-2

SOLUTION: From the figure it is apparent that the output voltage is segments of three-phase voltages. The segments are uniformly repetitive; consequently, the average and rms values can be determined from the active segment of one phase, for which α and β are $\pi/6$ and $5\pi/6$ rad, respectively.

$$E_d = \frac{A_d}{\text{base}} = \frac{-V_m(\cos 5\pi/6 - \cos \pi/6)}{5\pi/6 - \pi/6} = 0.827V_m$$

$$E_r = \sqrt{\frac{A_r}{\text{base}}} = V_m\sqrt{\frac{\frac{1}{2}\left[\left(\frac{5\pi}{6} - \frac{\pi}{6}\right) - \frac{\sin 5\pi/3 - \sin \pi/3}{2}\right]}{5\pi/6 - \pi/6}}$$
$$= 0.841V_m$$

The voltage in Fig. 8-6 is dc although not constant. *Ripple voltage* is an ac component defined as the difference between the instantaneous voltage, v, and the average voltage, E_d. The *rms ripple*, E_a, is the rms value of the ripple voltage; $E_a = \sqrt{E_r^2 - E_d^2}$. *Percent ripple*, a characteristic that may determine whether the waveform is appropriate for a particular application, is defined as the ratio of the rms ripple to the direct (average) component, expressed as a percentage—that is, 100 E_a/E_d.

Example 8-3: Find the rms ripple and percent ripple for the waveform of Example 8-2.

SOLUTION:

$$E_a = \sqrt{E_r^2 - E_d^2} = V_m\sqrt{0.841^2 - 0.827^2} = 0.153V_m$$

$$\%\text{ ripple} = \frac{100(0.153V_m)}{0.827V_m} = 18.5\%$$

Pulse number is the ratio of the frequency of the ripple voltage to the frequency of the ac supply voltage. For purposes of this discussion it most readily is perceived as the number of waveform peaks in the interval equal to one cycle of the fundamental sine wave. In Example 8-1, the pulse number is one; in Example 8-2, in the interval corresponding to one full cycle of one of the three-phase sinusoids, the waveform peaks three times and thus the pulse number is three.

8-3. DIODE The symbol and essential characteristics of a diode are shown in Fig. 8-7. When the anode (A) is at a positive potential with respect to the cathode (K), the diode will conduct current with only a small voltage drop across the diode itself. But with the polarity reversed, providing the reverse breakdown voltage (V_{RB}) is not exceeded, a large voltage will produce only a very

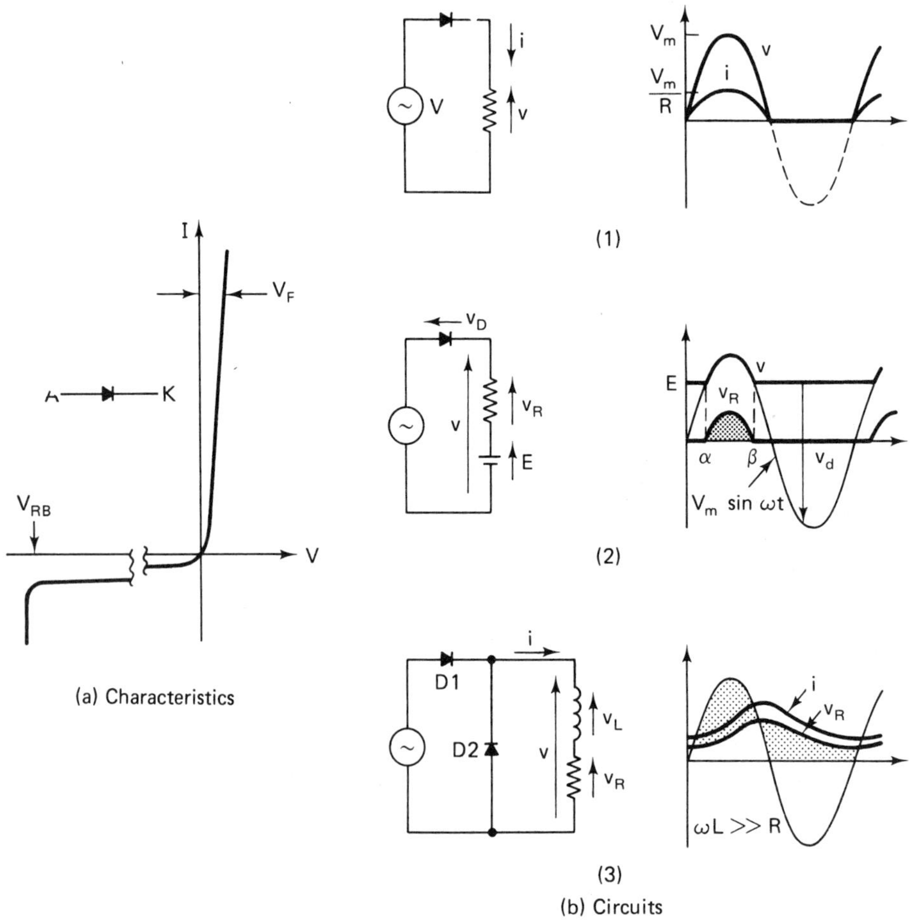

Figure 8-7 Diode Characteristics and Circuits

small leakage current. Idealized, the forward voltage drop across the diode ($V_F \simeq 1$ V) and the reverse leakage current ($\simeq 50$ mA in power diodes) are neglected. When forward-biased by circuit voltages, the current is determined by the load.

A half-wave diode rectifier with a resistive load is shown in Fig. 8-7(b, 1). Note that the waveform of the output voltage is that of Example 8-1 (Fig. 8-5), for which $E_d = 0.318V_m$. The instantaneous load voltage is $v = \sqrt{2}\,V \sin \omega t$, but only for the intervals $n\pi$ to $(n + 1)\pi$, where $n = 0, 2, 4, \ldots$; at other intervals $v = 0$.

In Fig. 8-7(b, 2), a constant voltage source, E, is in series with the load. During the interval 0 to α, the anode potential is positive, but the cathode potential, E, is larger, the diode is reverse-biased, and no current flows. With no current, $V_R = 0$; from Kirchhoff's voltage law, $v_d = V_m \sin \omega t - E$.

From α to β, the diode is forward-biased and conducting. Since $v_d = 0$, $v = V_m \sin \omega t = E + v_R$. This condition exists until, at $\beta(+)$, the diode again is reverse-biased and nonconducting. Thus, v_r is zero except during the interval α to β. Note that the peak reverse voltage across the diode is $E + V_m$, occurring at $3\pi/2$ rad (270°).

EXAMPLE 8-4: In Fig. 8-7(b, 2), $E = 0.67V_m$. Find the average voltage across the resistor.

SOLUTION: α and β occur at points where $V_m \sin \omega t = E = 0.67V_m$. Therefore,

$$\alpha = \sin^{-1} 0.67 = 0.734 \text{ rad} \qquad \beta = \pi - \alpha = 2.408 \text{ rad}$$

The area under the v_R curve is the area under the sine curve less the area under the E curve ($0.67V_m$) for the interval α to β. The area is

$$A = -V_m(\cos 2.408 - \cos 0.734) - (2.408 - 0.734)(0.67V_m) = 0.364V_m$$

$$E_d = \frac{0.364V_m}{2\pi} = 0.058V_m$$

If the load consists of a series inductor and resistor with $R = \omega L$, the waveform of v_R and i is that shown in Fig. 8-3. In that circuit, recall that part of the energy stored in the inductor was later returned to the source. If maximum power transfer is desired, one way to avoid return and dissipate all of the stored energy in the resistor is to add a *free-wheeling diode*. In Fig. 8-7(b, 3), ωL is substantially greater than R but otherwise differs from the circuit of Fig. 8-3 only in the addition of a diode in parallel with R and ωL. When $V_m \sin \omega t$ is greater than v, diode D1 is forward-biased and conducting while diode D2 is reverse-biased and OFF. But when the inductor discharges, D2 is forward-biased ($v_L = v_R + v_{D2}$) and D1 is reverse-biased. Current through R is continuous as a result of the discharge voltage of the inductor, but no reverse current flows through the source. Note that the positive volt–time area must equal the negative area.

The diode rectifiers of Fig. 8-7 all are half-wave. Two common schemes for full-wave rectifiers are illustrated in Fig. 8-8: center-tapped transformer and bridge. The term *single-way* identifies a circuit in which the current in a leg of the source is unidirectional; if current is bidirectional, it is *double-way*. Thus, the tapped-transformer circuit is single-way since current in each half of the secondary winding can flow in one direction only, but the bridge circuit is double-way. For both, the pulse number is two and, with V defined as shown on the schematics, the outputs are identical.

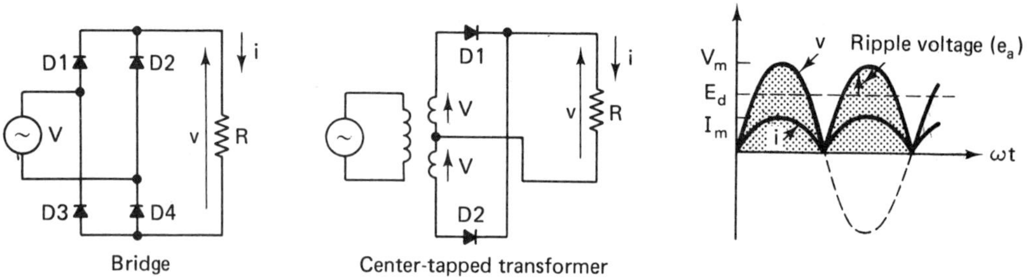

Figure 8-8 Full-Wave Rectification

A three-phase delta-wye, single-way, three-pulse rectifier circuit is shown in Fig. 8-9. A diode circuit is *uncontrolled*—that is, there is no way to fix the magnitude of E_d by controls imposed upon the diodes themselves. But by use of a transformer with the proper turns ratio, a desired E_d can be obtained from the available ac supply.

With respect to a neutral at the tie point of the wye, the instantaneous potential at the anode of diodes A, B, and C is the corresponding point on the sine curves A, B, and C, respectively. The potential of the positive output terminal is

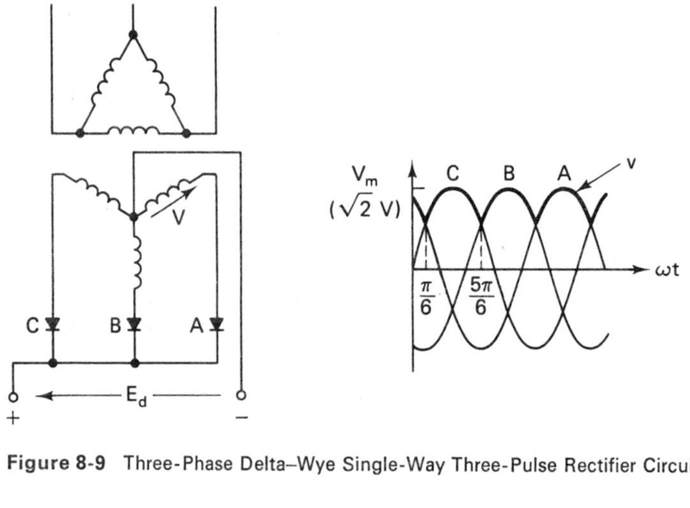

Figure 8-9 Three-Phase Delta–Wye Single-Way Three-Pulse Rectifier Circuit

that of the phase having the highest instantaneous potential. Thus, in the interval $\pi/6$ to $5\pi/6$ rad phase C has the highest potential and diode C conducts; although both phases A and B provide positive potentials to the anodes of their diodes for portions of this interval, the cathodes are at potential C and the diodes are reverse-biased. Therefore, the output voltage is the trace of the peaks of the phase curves over conduction intervals of $2\pi/3$ rad (120°). The waveform is recognizable as that of Example 8-2 (Fig. 8-6), for which the computed % ripple was 18.5%. If an inductive filter is connected in series with the load, the % ripple may be reduced, but the filter will not change the value of E_d.

A delta–wye, double-way, six-pulse circuit is shown in Fig. 8-10. The combination of diodes in each leg permits current flow in either direction through the secondary transformer windings. For analytical purposes, the tie point of the transformer windings again is considered to be at zero potential; therefore, determination of the conduction status of diodes A+, B+, and C+ is the same as for the single-way circuit of Fig. 8-9. During the negative portion of a phase, the point of connection of the phase terminal between diodes is at a negative potential; the most negative at any instant determines which of the A−, B−, and C− diodes is forward-biased and will conduct. Beneath the plotted curves is a key to diode conduction; in the interval 0 to $\pi/6$ rad, A+ is at the highest and B− the lowest potential—A+ and B− diodes will conduct. During this period, the voltage between the output terminals is the transformer line voltage V_{AB} with an rms value $\sqrt{3}\,V$ and peak value $\sqrt{3}(\sqrt{2}\,V) = \sqrt{6}\,V$ or $\sqrt{3}\,V_m$. With the horizontal axis

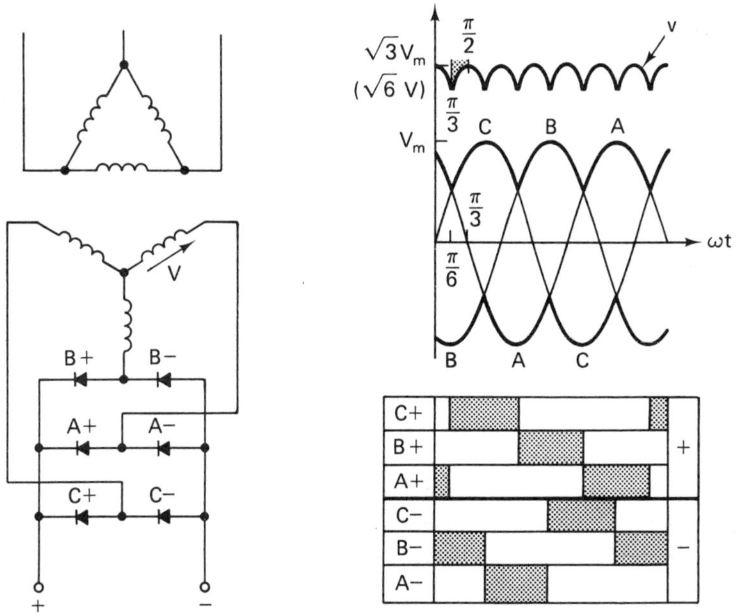

Figure 8-10 Three-Phase Delta–Wye Double-Way (Bridge) Six-Pulse Circuit

of the curves now regarded as zero potential at the *negative output terminal*, the potential at the positive terminal is the six-pulse curve with a peak value of $\sqrt{3}\,V_m$.

Recall that the phase angles of the *line* voltages in a wye connection lag the *phase* voltages by 30° ($\pi/6$ rad)—for example, $V_{BN} = V\underline{/30°}$ but $V_{BC} = \sqrt{3}\,V\underline{/0°}$. Therefore, an instantaneous angular position on the output voltage curve is not the angle ωt. The peak value of a pulse occurs at $\pi/2$ rad (90°) on a sine curve having the amplitude $\sqrt{3}\,V_m$, and the effective portion of the curve is from $\pi/3$ rad (60°) to $2\pi/3$ rad (120°).

EXAMPLE 8-5: (a) Find the average and rms values and the % ripple of the rectifier circuit of Fig. 8-10. (b) What is the peak reverse voltage across any diode?

SOLUTION: (a) Because of the repetitive character of the waveform, the required values can be obtained by computation for the interval $\pi/3$ to $\pi/2$ rad on the output waveform. Letting $V'_m = \sqrt{6}\,V$,

$$E_d = \frac{V'_m(-\cos\beta + \cos\alpha)}{\beta - \alpha} = \frac{V'_m(-\cos\pi/2 + \cos\pi/3)}{\pi/2 - \pi/3} = 0.955V'_m$$

$$E_r = V'_m\sqrt{\frac{1}{2} - \frac{\sin 2\beta - \sin 2\alpha}{4(\beta - \alpha)}} = V'_m\sqrt{\frac{1}{2} - \frac{\sin\pi - \sin 2\pi/3}{4(\pi/2 - \pi/3)}} = 0.956V'_m$$

$$E_a = \sqrt{E_r^2 - E_d^2} = V'_m\sqrt{0.956^2 - 0.955^2} = 0.044V'_m$$

$$\%\text{ ripple} = \frac{E_a}{E_d} \times 100 = \frac{0.044V'_m}{0.955V'_m} \times 100 = 4.6\%$$

[*Note*: With greater accuracy in the calculations (i.e., 10 significant figures rather than 3), the ripple is 4.2%; this is the value usually encountered in reference publications.]

(b) At $t = 0$, with diode B$-$ conducting, the anode of diode B$+$ is at the potential of the negative terminal while the cathode is at the potential of the positive dc terminal. The reverse voltage on diode B$+$ therefore is $\sqrt{6}\,V = 2.45V$.

If the transformer primary of Fig. 8-10 is eliminated and the secondary windings are replaced with the windings of a three-phase alternator, the resulting circuit is that usually found in the generating system of a modern automobile. Solid-state diodes made the alternator practical; without them, a substantially larger dc generator would be required in order to meet the demands of the numerous contemporary power accessories.

Diode rectifiers are used in applications where required control of the dc output can be provided by external means (e.g., transformer turns ratio). Among the uses are battery chargers, welding equipment, electrochemical processes, dc traction drives, radio and TV, and excitation systems for brushless ac generators.

When more finite control or adjustability of the dc output is required, the thyristor replaces some or all of the diodes in the rectifier circuit.

8-4. THYRISTOR The symbol and characteristics of the thyristor (silicon-controlled rectifier, SCR) are shown in Fig. 8-11. Like the diode, it can conduct only in the forward direction and with a forward voltage bias. However, it differs from the diode primarily in its ability to block forward voltages.

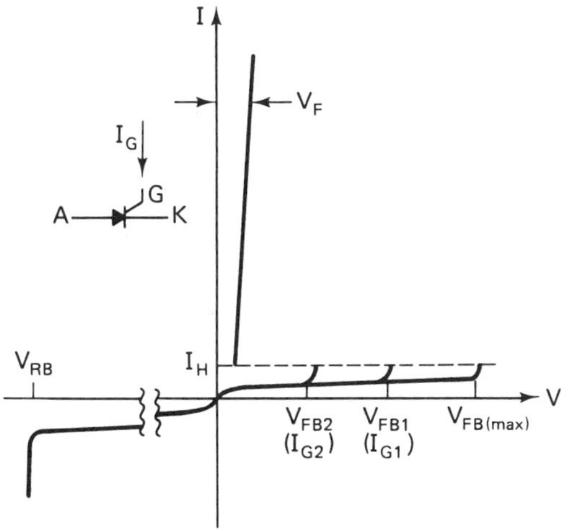

Figure 8-11 Thyristor Characteristics

In the illustration, if the anode–cathode voltage were increased to the maximum forward breakover voltage, $V_{FB(\max)}$, the thyristor would conduct; however, this is not the normal turn-on method and might be destructive to the device. Turn-on is accomplished by means of a *gate*—terminal G on the symbol. A current into the gate will decrease the forward breakover voltage of the thyristor. In the illustration, I_{G1} lowers the point to V_{FB1}; a larger gate current, I_{G2}, further reduces the breakover point to V_{FB2}. In each case, when breakover occurs the device turns ON and conducts a current whose magnitude is determined by the load, with only a small voltage drop ($V_F \simeq 1.5$ V) across the thyristor.

Although a gate current is required for turn-on, once conducting the gate has no further control; thus, only a short current pulse is required. When ON, the thyristor will continue to conduct as long as it remains forward-biased and the anode–cathode current does not drop below the small holding current, I_H.

Rectifier. In anlyzing rectifier circuits, usual assumptions are (1) negligible voltage drop when conducting and negligible leakage current (forward and reverse for the thyristor) when blocking voltage, (2) instantaneous turn-on and turn-off of the solid-state device, (3) sinusoidal ac line voltage, and (4) an ideal filter that eliminates ripple and maintains a constant, uninterrupted dc current. These assumptions apply unless specifically excepted.

In the rectifier circuits of Figs. 8-8 to 8-10, if the diodes are replaced with thyristors that turn on at the points where the diodes begin conducting, the performance of the thyristor circuits would duplicate that of the diode circuits. But with the added option of changing the point of turn-on, the rectifier can be controlled to vary the load voltage.

A single-phase bridge circuit is shown in Fig. 8-12(a). The delay angle, α, measured from the point where a diode would begin conduction, identifies the point where a pulse of gate current is applied to initiate conduction in the thyristor. For the positive half of the sine wave, CRs 1 and 4 are turned on at the delay angle α and conduct until, at $\pi + \alpha$ with the source voltage in the negative half-cycle, CRs 2 and 3 are turned on. Thus, the voltage across the filter and load resistance consists of repetitive segments of the sinusoidal input from α to $\pi + \alpha$. The conduction angle, θ_c, is the angular interval during which the thyristor conducts—for this circuit, $\theta_c = \pi$ rad (180°). Although v is at times negative, the inductor discharge voltage maintains both forward bias on the thyristor and current continuity.

Applying equations 8-2 and 8-4 with $\beta = \pi + \alpha$,

$$E_d = \frac{-V_m(\cos\beta - \cos\alpha)}{\pi} = \frac{-V_m[\cos(\pi + \alpha) - \cos\alpha]}{\pi}$$

$$= \frac{2V_m}{\pi}\cos\alpha$$

This voltage appears across the load resistor. Again note from the volt–time areas

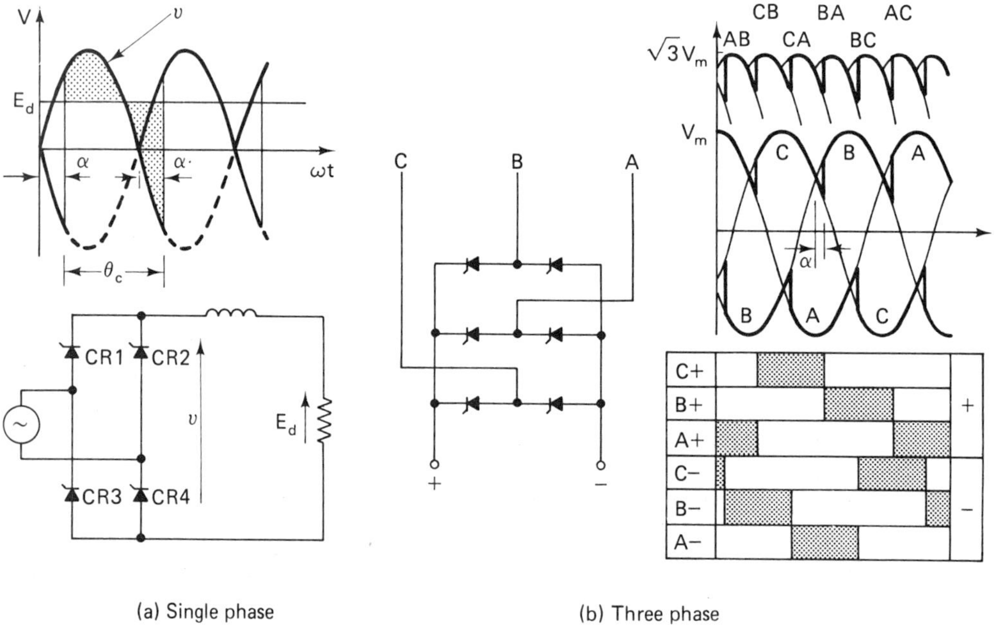

Figure 8-12 Full-Wave Controlled Rectifiers

that the average voltage across the inductive filter is zero, but the instantaneous inductor voltage is consistent with alternating storage and release of energy.

A full-wave, full-control, three-phase circuit is illustrated in Fig. 8-12(b). With thyristors between the source phase terminals and both the positive and the negative output terminals, the circuit is *full control*; diodes may replace the thyristors connected to the negative terminal and allow *half-control.* Note that α is measured from the point where a diode would begin conducting and is uniformly applied to all thyristors. With half-control, the principal change is that the negative trace of phases A, B, and C would appear as they do in Fig. 8-10 and E_d would be slightly greater for a given α than with full control.

Alpha may be varied from 0 to $2\pi/3$ rad with full control and from 0 to π rad with half control. Both permit reduction of the output voltage from maximum to zero.

The controlled rectifier output may be used in many applications where its output characteristics are required and suitable—for example, controlled lighting or heating; armature control of a dc shunt motor.

Chopper Control. The basic principles of a chopper circuit used to control the speed of a series motor powered by a battery (e.g., electric vehicles) are illustrated in Fig. 8-13. Assuming a constant torque load ($I_a = k$), speed may be varied by changing the average voltage, V, applied to the motor terminals. A thyristor is repetitively switched ON and OFF by a control circuit. Unlike ac, a battery voltage does not provide a natural means to turn off the thyristor; the dashed switch is a reminder that some provision must be made for turn-off but does not represent the actual method.

For low speeds the ON periods of the thyristor are of relatively short duration with substantial intervening OFF periods, as shown in Fig. 8-13(b, 1). The speed may be raised by increasing either the duration of the ON period or the frequency of the conduction intervals, as illustrated in Fig. 8-13(b, 2).

The resistance and inductance inherent in the field and armature of the motor impose a time constant on the current rise when the thyristor is ON and decay as the inductive voltage discharges through the motor and free-wheeling diode. The resulting form of the armature current is illustrated.

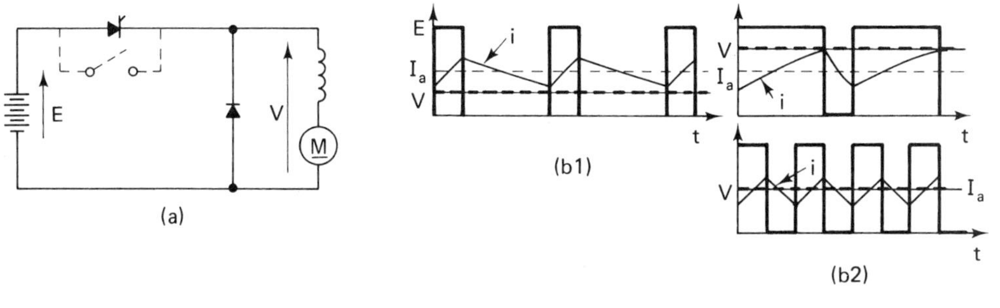

Figure 8-13 Chopper

Universal Motor Control. A popular, widespread use of the thyristor is to control the speed of universal motors. In domestic applications it appears in portable tools, appliances (e.g., blender, mixer, fan), toys, and so on. An elementary form of the control circuit is illustrated in Fig. 8-14.

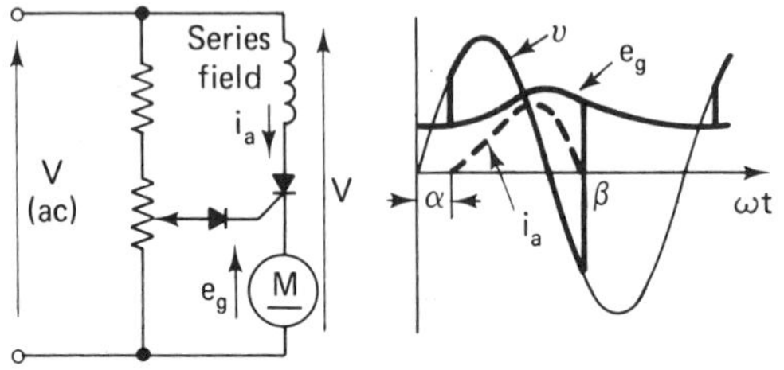

Figure 8-14 Universal Motor Control

The voltage applied to the gate of the thyristor is the difference between the variable potentiometer voltage and the counter emf, e_g, of the motor. This difference voltage determines the gate current and the point on the line voltage curve at which conduction begins. Varying the potentiometer setting allows external speed adjustment. At a fixed potentiometer setting, a decrease in e_g caused by increased loading and a drop in speed will cause conduction to begin at an earlier point, thereby providing some regulation of the speed.

Inverter. An inverter converts a dc input to an ac output.

The basic principle of the inverter is illustrated in Fig. 8-15. Ignoring for the moment the shaded components, if simultaneous conduction of CRs 1 and 4 were alternated with simultaneous conduction of CRs 2 and 3, the output would be square waves of alternating polarity. However, with CRs 1 and 4 conducting, there would be no means to turn them OFF, and turn-on of CRs 2 and 3 would create short-circuit paths through the thyristors. Further, a sinusoidal waveform normally is desired at the output. The *filter component* performs the functions of turn-off and wave shaping. A *choke* is added to prevent excessive current spikes when the thyristors are fired. Not shown is the firing circuit for turn-on of the thyristors. The complete circuit can provide a reasonably true sinusoidal output.

A basic single-phase parallel inverter is shown in Fig. 8-16. In analyzing the circuit, bear in mind that a voltage will be induced in the transformer secondary *only* if current in the primary winding is time-variant. Also note that a time-

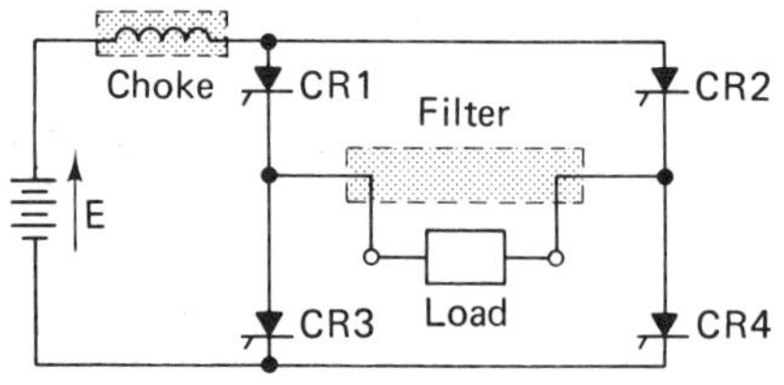

Figure 8-15 Basic Inverter

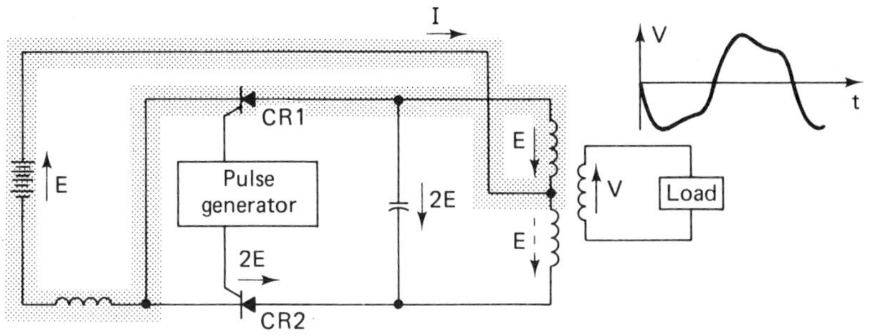

Figure 8-16 Parallel Inverter

variant current in one section of the primary induces a voltage not only in the current-carrying winding but also in the other half of the primary since the mutual flux is common to both. Finally, note that when switching occurs, the change in currents is governed by time constants due to capacitance and inductances in the circuit.

Assume that CR1 is conducting, the current is time-variant, and E is induced in the upper half of the primary; concurrently, E is induced in the lower half and the capacitor will therefore charge to $2E$. The current path for this condition is shaded.

CR2 is forward-biased at this time and will conduct when fired. The capacitor then applies a reverse voltage to CR1 to turn it off. Concurrently, two conducting paths are established: one duplicates the shaded path with CR2 and the lower half of the primary replacing CR1 and the upper winding. The second is a discharge path through the full primary and the capacitor. An interval follows during which the current in the upper winding decays to zero, current builds in the lower winding, and the capacitor recharges to $2E$ with the opposite polarity. The circuit now is set for turn-on of CR1 to repeat the process.

With the changing currents in the primary, a secondary voltage is induced with the general waveform as shown. If a truer sinusoid is desired, a filter may be added between the secondary winding and the load.

The obvious use of the inverter is to provide a required ac output when only a dc source is available. When combined with a rectifier, an uninterruptible power supply is possible; under normal operation, the input from an ac line is rectified and then inverted to an ac output. Between the rectifier and inverter is a battery that floats under normal operation but powers the inverter if the ac line input to the rectifier is interrupted. Continuity of output is assured—a critical requirement in many applications (e.g., a computer in which power failure not only would interrupt processing but would result in loss of stored data).

Dual Converter. The functioning of a *synchronous inverter* is illustrated in Fig. 8-17(a). Without the current source, the bridge circuit of Fig. 8-12(a) is duplicated, for which $E_d = (2V_m/\pi) \cos \alpha$. E_d is negative if α is greater than $\pi/2$ rad; with a passive load, only, this is not possible since (1) the inductor could not maintain

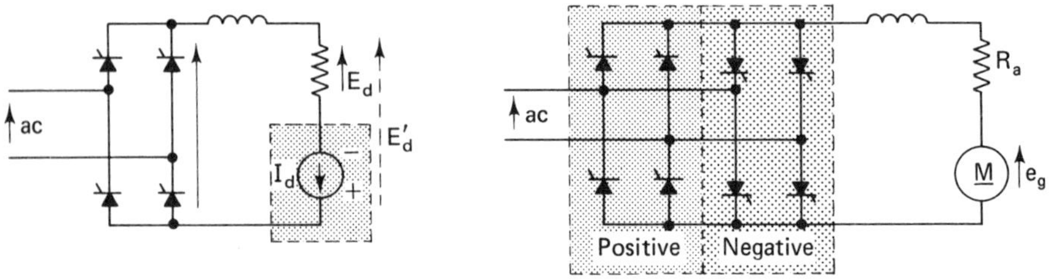

(a) Synchronous inverter (b) Dual converter

Figure 8-17 Inverters

current continuity, (2) a negative E_d would require a negative current which the thyristors cannot conduct, and (3) net power flow cannot be from the passive load to the source as would be required. However, with the addition of an active load component, represented by the current generator, current is continuous in the forward direction and E_d, the average voltage across the load resistance and current source (labeled E'_d), may be negative; thus, α may be continuously varied from 0 to π rad. From 0 to $\pi/2$ rad power flow is from the source to the load, but beyond $\pi/2$ rad power flow is into the source. This characteristic is useful in a link between high voltage dc and ac transmission lines since it permits changing the direction of power flow in response to changed system loads. The synchronous inversion also is useful in motor control.

A *dual converter* is illustrated in Fig. 8-17(b). The positive converter may be fired at delay angles between 0 and $\pi/2$ rad to control the speed of the shunt motor in one direction (a separately excited shunt field is not shown). With the motor running in a direction corresponding to a positive converter output, synchronous inversion cannot be accomplished by increasing the delay angle beyond 90° because e_g, the counter emf of the motor, has the wrong polarity for maintaining forward current in the thyristors. However, the negative converter *can* conduct with this polarity of e_g; by firing at angles greater than $\pi/2$ rad, controlled regenerative braking is achieved. As the motor approaches zero speed, the delay angle of the negative converter can be fired at less than $\pi/2$ rad to accelerate and drive the motor in the reverse rotational direction. Thus, the dual converter permits rotation and regenerative braking in either direction. Starting from rest is easily achieved without the usual starting resistors; the delay angle initially is large (approaching but less than $\pi/2$ rad) and then gradually reduced as the motor accelerates.

Frequency Conversion. One method of frequency conversion combines a rectifier on the input side and an inverter on the output side of the converter. The line input, probably at the standard 60 Hz, is converted to dc and then inverted to an ac output at a desired frequency, which may be greater or less than the input line frequency. One use of the frequency converter is control of the speed of ac synchronous and induction motors.

The *cycloconverter* is a frequency converter that cannot increase the input line frequency but is particularly effective for outputs of relatively low frequencies. The principle is illustrated with a three-phase dual converter. In Fig. 8-18(a), the positive converter is fired over a time period followed by an equal interval of firing the negative. By repetition of the sequence the output essentially is a series of square waves of alternating polarity. While this output conceivably could be filtered to a sine wave, it is doubtful if it would be economically feasible. The frequency of the output would be somewhat less than half of the input frequency.

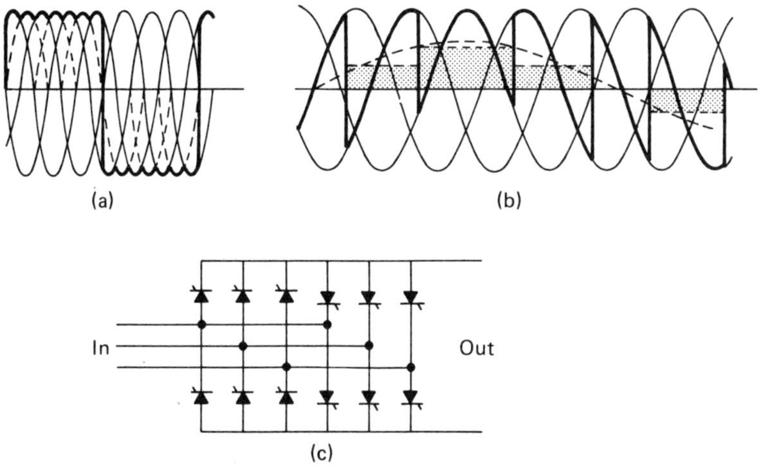

Figure 8-18 Cycloconverter

The unfiltered output can be made to more nearly conform to the desired output by proper combinations of delay angles and choice of positive or negative converter firing, as illustrated in Fig. 8-18(b). The switching pattern is designed to produce average voltages over conduction intervals as typified by the shaded areas, thereby providing a stepped approximation of the sine wave. Proper timing of the gating signals is influenced not only by the desired output voltage waveform but also by the current waveform and the load, and the converter performs synchronous inversion as well as rectification for portions of the cycle. Note that the input in the illustration is three-phase but the output is single-phase.

The cycloconverter primarily is used to control the speed of ac motors.

8-5. POWER TRANSISTOR

The characteristics and symbol of the transistor are shown in Fig. 8-19. The terminals of the transistor are the cathode (C), base (B), and emitter (E). When the transistor is ON, load current is conducted from the collector to the emitter; for this conduction to occur, there must be a current (I_B) between the base and the emitter and the collector must be at a positive potential with respect to the emitter. The shaded area of the characteristic curve is the *active region*; the relationship of I_C to V_{CE} is found by reference

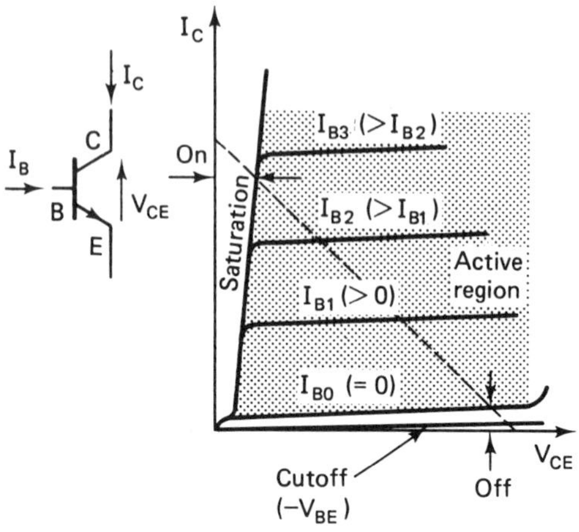

Figure 8-19 Transistor

to the curve for the existing base current, I_B. Numerous applications exploit the characteristics of the active region with varying base currents, but generally this area is not used in switching circuits and, by design, must be avoided to prevent overheating and probable failure of the transistor at typical operating power levels.

As a switch, the areas of interest are the *saturation* and *cutoff* regions. For illustration, an arbitrary dashed load line is shown. For this load, a base current equal to or greater than I_{B3} saturates the transistor, the voltage drop between the collector and emitter is negligibly small, the switch is ON, and the load current is determined by the load impedance and applied voltage. Similarly, a slight negative voltage applied between the base and the emitter turns the device OFF by driving it into the cutoff region. Between ON and OFF the active region is traversed, but the operating point is in this area for only a few microseconds.

The feature that distinguishes the transistor and thyristor switches is the control exercised by the base of the former and the gate of the latter. In the thyristor, a gate signal is required for turn-on but thereafter loses control, and conduction is continuous as long as the thyristor is forward-biased and the cathode–anode current is greater than the holding current of the device. But with the transistor, the base current must be continuous for the diode to conduct, and a slightly negative voltage pulse will cause turn-off even though the collector–emitter current is substantial. Thus, the transistor has the capabilities of the thyristor with the added feature of turn-off at any point without complex circuit arrangements.

Generally, the transistor can be used in all the thyristor applications discussed in Paragraph 8-4 but at much lower power levels, usually under 10 kVA for the transistor versus perhaps 1000 kVA for the thyristor.

An important use of transistors is in logic circuits to control turn-on and turn-off of power components. Transistor logic modules also can replace electromagnetic relays in much of the control circuitry for machines.

8-6. MOTOR APPLICATIONS The marriage of suitable power supplies and motors is highly influenced by the intended application, particularly the duty cycle. Similarly, transformer selection is complicated inasmuch as the current in a phase of the transformer commonly is an interrupted square wave rather than the continuous sine wave usually associated with ac power.

Standard ac polyphase motors are subject to the voltage and frequency variations discussed in Paragraph 5-18 (combined V and f deviation $\pm 10\%$). Since these limits do not permit substantial exploitation of converter capabilities, special design motors are built for this purpose.

The performance of a dc motor is influenced by the power supply—for example, an increase in rms current ripple will increase heating and an increase in peak-to-peak current ripple degrades commutator performance. The design and rating of standard motors intended for use with an electronic power supply considers both the source and the motor parameters. Ratings take into account the *form factor* of the current: the form factor (F) is the ratio of rms current to average current, $F = I_{rms}/I_{av}$. The factor is unique to a combination of source and motor and varies with speed and loading. Values range from about 1.2 to 2.0 for single-phase and 1.0 to 1.2 for three-phase source combinations.

Fractional horsepower motors are rated on the basis of the form factor. The motor is not intended for use at greater, but can be operated from supplies producing lesser, form factors.

Integral horsepower motor ratings specify a letter-class power supply for which the motor is designed. The classes, based on 60-Hz line input and no external inductance in series with the motor armature, are defined in terms of the following:

Class	*Type*	V_L *(ac Input)*
A	Dc generator or battery	
C	Three-phase, six-pulse, fully controlled bridge	230*, 460 V†
D	Three-phase, six-pulse, half-controlled bridge with free wheeling	230*, 460 V†
E	Three-phase, three-pulse, single-way, controlled	460 V*
K	Single-phase, two-pulse, controlled bridge with free wheeling	230‡, 115§

*Nominal dc voltage: 240 V.
†Nominal dc voltage: 500 V.
‡Nominal dc voltage: 180 V.
§Nominal dc voltage, 90 V.

These letter classifications are in the order of increasing ripple current. A motor may be operated from supplies with less ripple than its rated class. Thus, a motor rated for power supply D may also be used with classes A and C but *not* classes E and K.

Solid-state power and control circuitry is a developing field whose remarkable accomplishments are overshadowed by future expectations. Although it presages no revolution in machinery fundamentals, it is prompting innovation and modification. In automated systems, it provides speed of response, sensitivity of control, reliability, and efficiency beyond that previously possible.

PROBLEMS AND QUESTIONS

8-1. Find the average and rms values and the % ripple of the waveforms of Fig. 8-20.

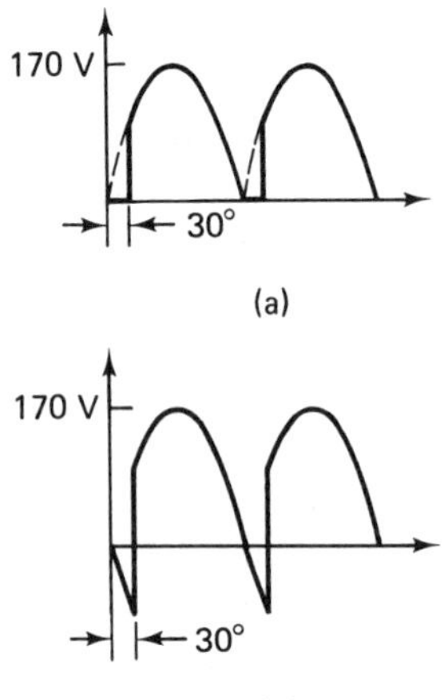

Figure 8-20 Problem 8-1

8-2. Find the average and rms values and the % ripple of the waveforms of Fig. 8-21.

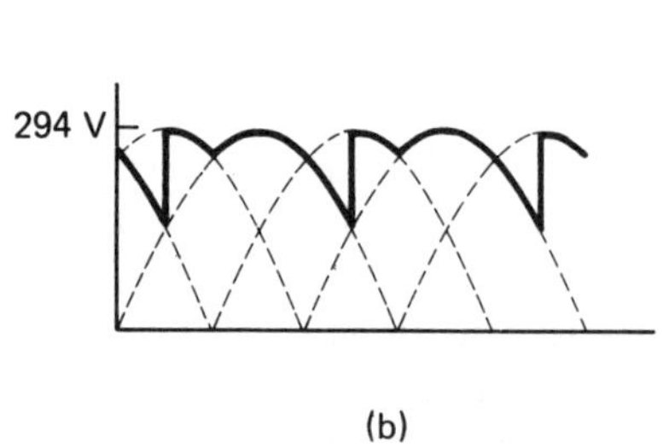

Figure 8-21 Problem 8-2

8-3. A small heater with a resistance of 10 Ω is supplied by a 120-V ac line. The heater power must be continuously variable between 100 and 1000 W.
(a) Using a series rheostat as the control means, what is the efficiency of the system for heater powers of 100, 500, and 1000 W?
(b) The series rheostat is replaced with a controlled bridge rectifier and series inductor that maintains an essentially constant dc current through the heater at any control setting. When conducting, the forward voltage drop across each thyristor is 1.6 V. What is the system efficiency for heater powers of 100, 500, and 1000 W?
(c) What are the delay angles, in the rectifier for the 100-, 500-, and 1000-W heater power?

8-4. In Fig. 8-12(b), the three thyristors connected to the negative output terminal are replaced with diodes. For a delay angle similar to that shown in the figure, sketch the output waveform.

8-5. The waveform for a controlled single-phase bridge rectifier is shown in Fig. 8-12(a); text discussion of the circuit shows that $E_d = (2V_m/\pi)\cos\alpha$. In Fig. 8-22 a free-wheeling diode has been added; sketch the resulting voltage waveform, v, and show that $E_d = (V_m/\pi)(1 + \cos\alpha)$.

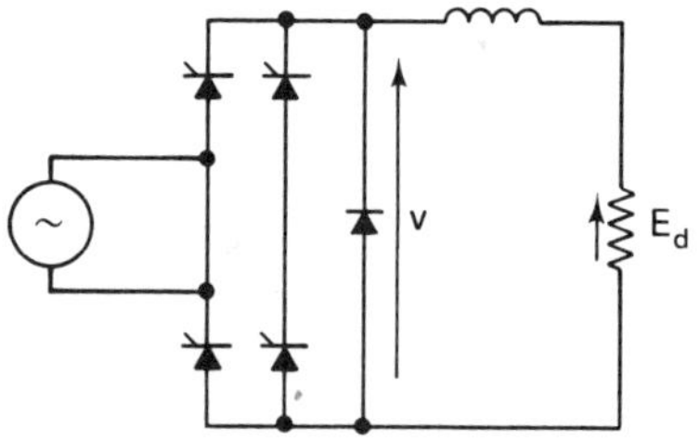

Figure 8-22 Problem 8-5

8-6. In Fig. 8-23(a) the switch closes to position A at t_0; at t_1 the voltage V_R is $E_B(1 - e^{-t_1/\tau})$.

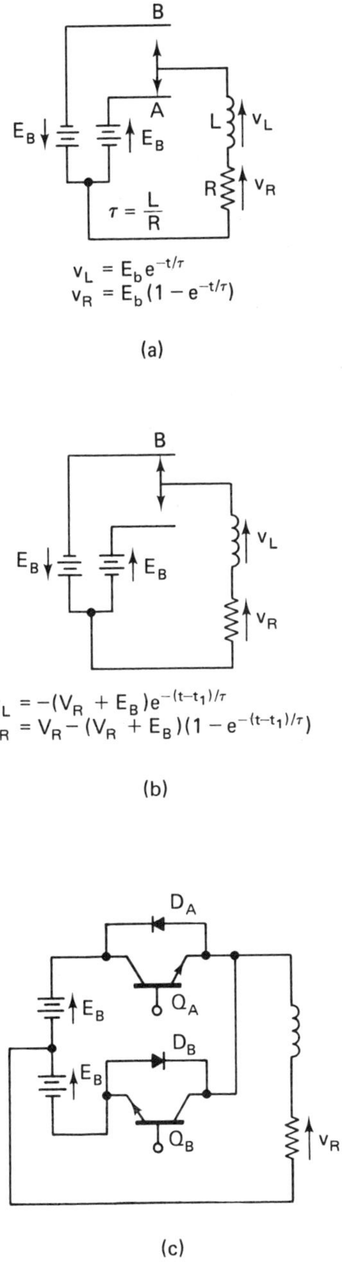

Figure 8-23 Problem 8-6

At t_1, the switch instantaneously shifts to position B, shown in Fig. 8-23(b); since current in an inductor cannot change instantaneously, the same current exists immediately prior to and immediately after switching. Consequently, $v_R = V_R$ immediately after switching and V_L is therefore $-(E_B + V_R)$. Thereafter, $v_R = V_R - (V_R + E_b)(1 - e^{-(t-t_1)/\tau})$. At t_2, v_R will be zero; this occurs when $-(t_2 - t_1) = \ln [1 - V_R/(V_R + E_B)]$.

(a) Plot E_B and v_R for the interval t_0 to t_6 with the switching sequence as follows:

Switch Position	Δt
A	$t_1 - t_0 = 2\tau$
B	$t_2 - t_1 = -\ln [1 - V_R/(V_R + E_B)]$ $= 0.623\tau$
B	$t_3 - t_2 = 2\tau$
A	$t_4 - t_3 = 0.623\tau$
A	$t_5 - t_4 = 2\tau$
B	$t_6 - t_5 = 0.623\tau$

(b) A transistor inverter is shown in Fig. 8-23(c). The ON period of transistors Q_A and Q_B correspond to switch positions A and B, respectively, of part (a). Show that the voltage waveforms of part (a) are duplicated. In particular, explain the circuit during the periods $(t_2 - t_1)$ and $(t_4 - t_3)$.

(c) If each diode–transistor combination is replaced with a thyristor, can the same results be obtained?

(d) Discuss the effect of impedances on the output.

8-7. A 240-V 20-hp dc shunt motor has a rated speed and armature current of 1150 r/min and 75 A. The machine resistances are $R_{ac} = 0.15\ \Omega$ and $R_f = 40\ \Omega$. At rated conditions, $V_F = 150$ V. The design of the motor permits speed variation with the following qualifications:

FIELD CONTROL: Maximum 2000 r/min. Armature current may not exceed rated.

ARMATURE CONTROL: Minimum 50% base speed with concurrent torque limited to 67% rated torque.

The control circuit is shown in Fig. 8-24. Analysis of the circuit is based upon the following assumptions:

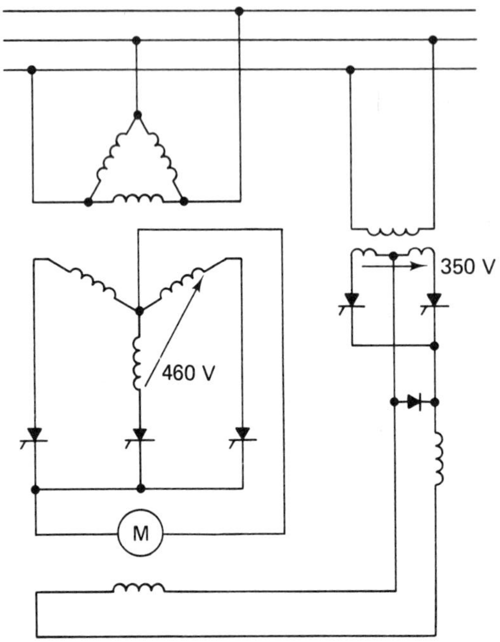

Figure 8-24 Problem 8-7

(1) The magnetic circuit does not saturate within the operating limits; therefore, $\Phi = kI_f$.

(2) Although I_a and I_f vary with the delay angle of the rectifier, at any constant delay angle the currents are essentially constant (i.e., ripple currents are negligible).

(a) What is the delay angle in each rectifier for (1) rated conditions, (2) maximum speed, and (3) minimum speed.

(b) Suppose that the motor was supplied from a constant-dc source of 240 V and control was exercised by variable resistors in the field and armature circuits. What would be the power loss in the rheostats for the permissible loads at base, maximum, and minimum speeds? Compare these losses with your estimate of losses in the rectifier circuits.

8-8. An ac switch is illustrated in Fig. 8-25.

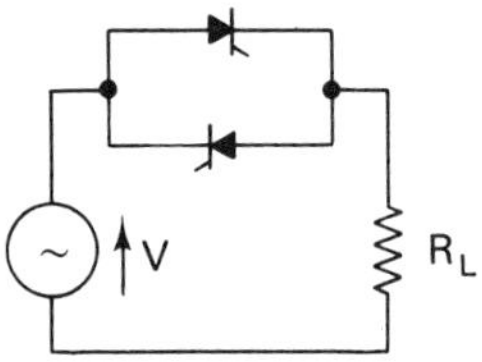

Figure 8-25 Problem 8-8

Because there is no smoothing inductor, the thyristor cannot conduct beyond the zero crossing of the voltage sinusoid, and the effective current is the rms value of the current waveform. Find the expression for the effective current as a function of V, R_L, and α.

A

Polyphase Circuits

A-1. INTRODUCTION Three systems are common in electrical power: single phase, three-wire single phase, and polyphase. The former is covered extensively in introductory circuits courses; this appendix distinguishes between the latter two and presents analytical methods for the solution of three-phase circuits.

A-2. THREE-WIRE SINGLE-PHASE AND POLYPHASE SYSTEMS A system is represented in the schematic of Fig. A-1(a) as two sources interconnected to provide three load terminals. The amplitude, frequency, and phase of the two source voltages are identical. Neglecting possible small differences in the voltage drops across internal impedances, the line voltages V_{AB} and V_{BC} also are identical; however, the voltage V_{AC}, although identical in frequency and phase to V_{AB} and V_{BC}, has a magnitude equal to their arithmetic sum. This is a *three-wire single-phase* system.

The common application of this system is to residential service; V_{AB} and V_{BC} have nominal effective values of 120 V and V_{AC} is 240 V. The center line is the neutral; it is physically grounded and, from a safety viewpoint, no line is at a potential greater than 120 V with respect to ground. With a reasonable balance in the division of outlets and installed equipment between the two 120-V circuits, the current in the neutral line is small, possibly zero.

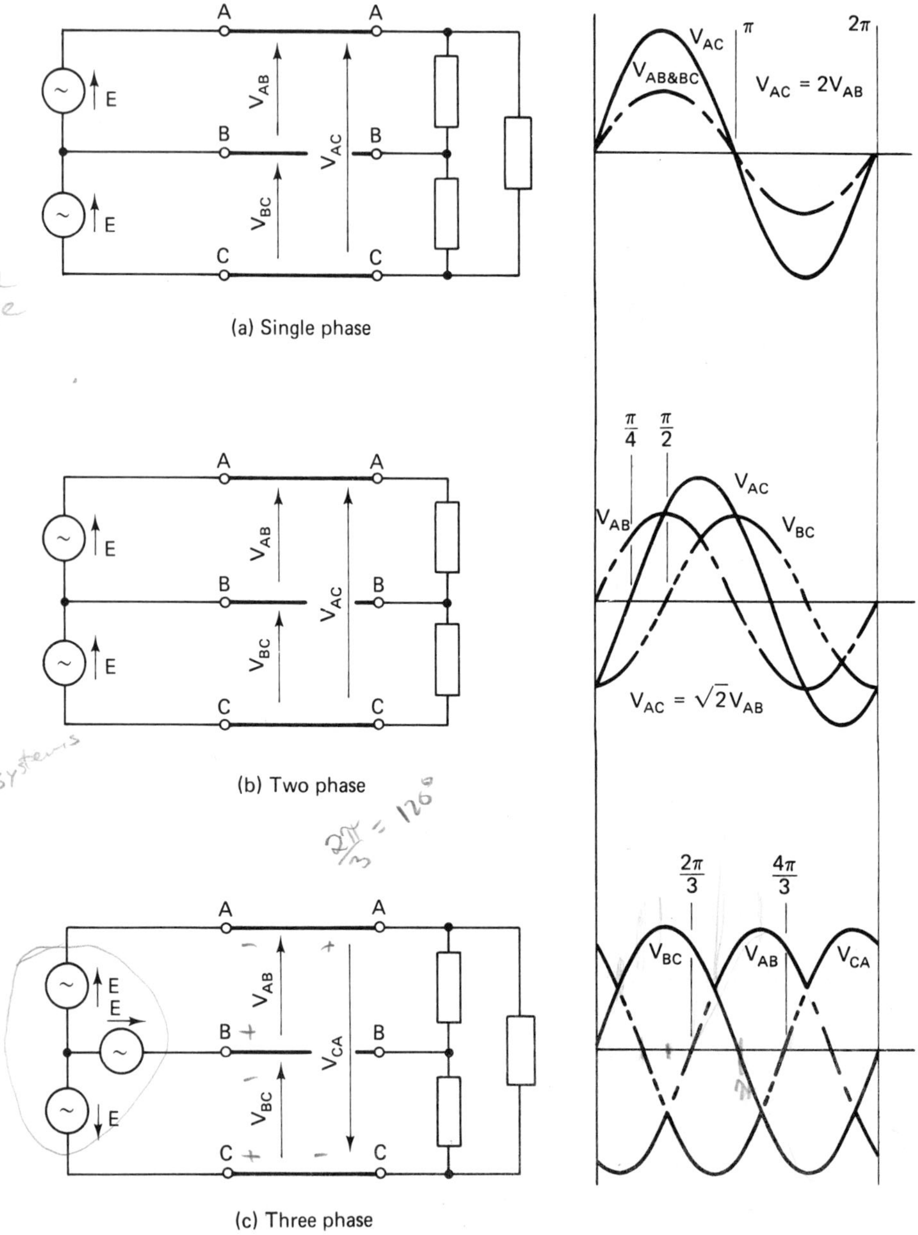

Figure A-1 Comparison of Three-Wire Systems

Two sources are retained in Fig. A-1(b), identical in amplitude anf frequency but phase-separated by 90°. The voltage V_{AC} is the vector sum of the source voltages—that is, $V_{AC} = V\underline{/0^\circ} + V\underline{/-90^\circ} = \sqrt{2}\,V\underline{/-45^\circ}$. This is a *two-phase three-wire* system, used primarily in control systems and approximated by phase-splitting circuitry in some forms of the single-phase motor.

The *three-wire three-phase* system provides three output voltages, identical in magnitude and frequency but phase-separated by 120°. Concentrating upon line characteristics and deferring consideration of the source and the load, the system is represented in Fig. A-1(c).

At this point a convention is introduced which always applies to three-phase circuitry in this text. Literally, any phase angles could be assigned to the line voltages providing the 120° separation is maintained. Also, the sequence in which positive peaks of the sinusoidal waves occur in lines *A*, *B*, and *C* are dependent upon the direction of rotation of the generator and may be *CBA*, as shown, or *ABC*. Phase sequence is significant—for example, it determines the direction of rotation of a polyphase motor—but, with few exceptions, the applications in this text result in voltage, current, and power relationships that are independent of the choice of base angles and sequence. This correlates to single-phase circuits in which there is no difference between $V\underline{/0^\circ}$ across $Z\underline{/\theta^\circ}$, resulting in $I\underline{/-\theta}$ and $V\underline{/40^\circ}$, $Z\underline{/\theta}$, and $I\underline{/40 - \theta}$. Little is gained by using a different combination of angles and sequence in each reference to a three-phase circuit. In this text, the *CBA* sequence and the voltage angles of Fig. A-3 always apply. Subscripting is patterned (Fig. A-2); V_{AB} is the potential at point *A* with respect to point *B*, I_{AB} is a current with positive flow from point *A* to point *B*, and I_A, I_B, and I_C are line currents with positive flow from the source to the load.

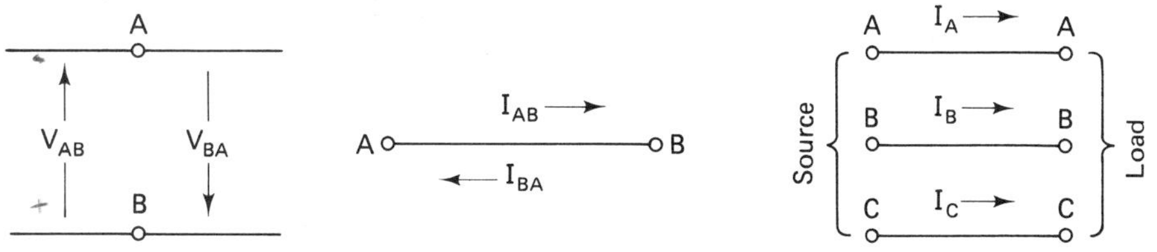

Figure A-2 Voltage and Current Subscripts

A-3. CBA SYSTEM The voltages in a three-phase system are generated not in three separate machines but rather in one polyphase alternator; however, for analytical purposes it is portrayed by three identical, interconnected, single-phase alternators. The term "phase" has several meanings; when referring to the voltage source, it denotes one single-phase branch. There are two methods of interconnecting the phases to create the three line terminals: wye (Y) and delta (Δ).

Y-connected (Fig. A-3), one side of each of the phases is tied to a common point, *N*. Point *N*, termed the neutral point, often is grounded and, in a four-wire

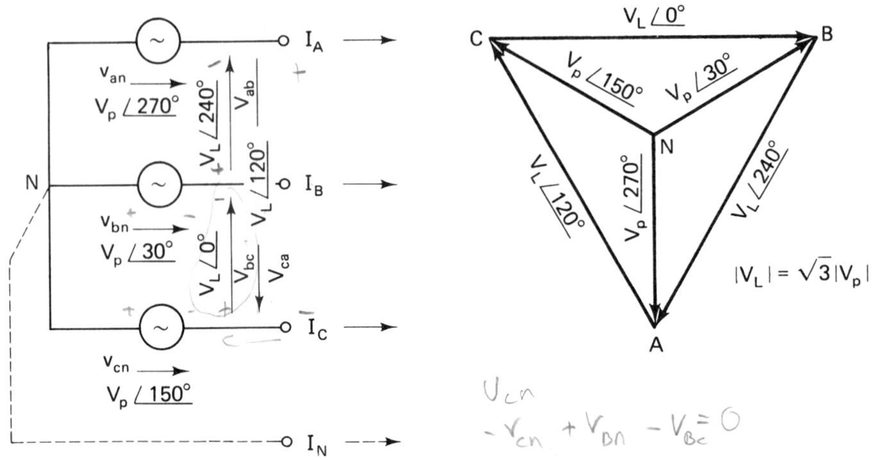

Figure A-3 Wye-Connected Source

system, is a terminal for a return path that usually carries little or no current; it is analogous to the neutral in the three-wire single-phase system.

The convention of this text employs the symbols V_p and V_L for the rms values of the phase and line voltages and assigns angles $V_{AN} = V_p\,\underline{/270°}$, $V_{BN} = V_p\,\underline{/30°}$, and $V_{CN} = V_p\,\underline{/150°}$. Tracking from terminal C to terminal B,

$$\begin{aligned}
V_{BC} &= -V_{CN} + V_{BN} \\
&= V_p(-1\,\underline{/150°} + 1\,\underline{/30°}) \\
&= V_p\{-(-\sqrt{3}/2 + j1/2) + (\sqrt{3}/2 + j1/2)\} \\
&= V_p(\sqrt{3} + j0) \\
&= \sqrt{3}\,V_p\underline{/0°} \\
&= V_L\underline{/0°}
\end{aligned}$$

Similarly, $V_{CA} = V_L\,\underline{/120°}$ and $V_{AB} = V_L\,\underline{/240°}$.

With few exceptions, loads considered in this text are balanced. In balanced loading, the equivalent load impedances across each of the three pairs of load terminals are identical both in magnitude and phase angle, and the phase currents I_{NA}, I_{NB}, and I_{NC} are identical in magnitude but separated in phase by 120°; the actual angles vary with the angle of the load impedance. Balanced loading is commonplace owing to the inherent characteristics of polyphase equipment and a conscious effort to distribute load components to maintain a balance.

The *CBA* triangle of Fig. A-3 conveniently relates phase and line voltages. The main triangle is equilateral, each side having the magnitude V_L. Phase voltage vectors originate at the center and terminate at the apexes of the triangle. With points A, B, C, and N identified, the geometry identifies the voltage angles. The magnitude of V_L always equals $\sqrt{3}\,V_p$.

In the wye connection, each phase of the source is in series with the corresponding line. Consequently, the phase current is the line current—for example, $I_{NA} = I_A$ in both magnitude and phase angle. The specific current angles are developed in the discussion of loading.

In a four-wire system, a line is connected to the neutral point. With four lines, loads may be connected across any of the line voltages (V_{AB}, V_{BC}, and V_{CA}) or phase voltages (V_{AN}, V_{BN}, V_{CN}). The neutral wire is of minor importance in balanced loading but essential to unbalanced loads.

The *delta* connection, shown in Fig. A-4, requires connection of the positive reference terminal of each phase to the negative reference terminal of another phase. The phases of the source form a closed loop, suggesting a circulating current; if there were differences in the magnitudes of the voltages or deviation from the 120° phase separation, circulating currents would occur, but a fundamental and important three-phase relationship is that the sum of three vectors, equal in magnitude but separated by 120°, is zero. Consequently, there is no net voltage and no circulating current in the loop formed by the sources.

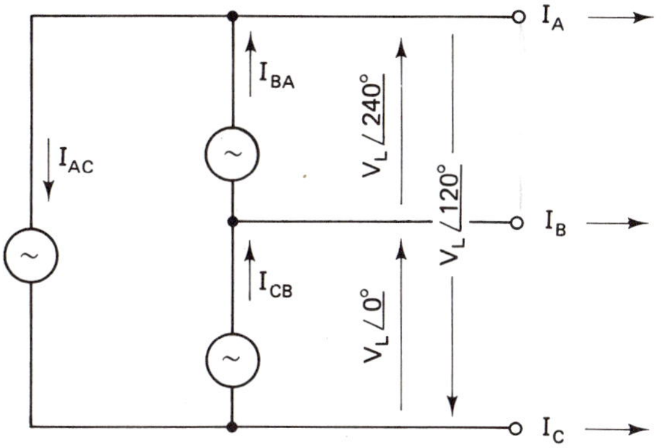

Figure A-4 Delta-Connected Source

In a delta, the phase voltages also are the line voltages; however, the line *current* is the vector sum of two phase currents. The magnitudes and phase angles of both the line and phase currents are functions of the magnitude and phase angle of the load impedances.

A-4. THREE-PHASE CURRENTS AND POWER

The Ohm's law relationship between voltage and impedance fixes the magnitude and phase angle of the load currents in the three-phase system. A three-phase load is identifiable as three two-terminal phase impedances, interconnected wye or delta to create three line terminals.

Wye Loading. A wye load is shown in Fig. A-5. Although connected to a wye source for analytical purposes, the load is affected only by the line voltages and the following relationships apply equally to a delta source.

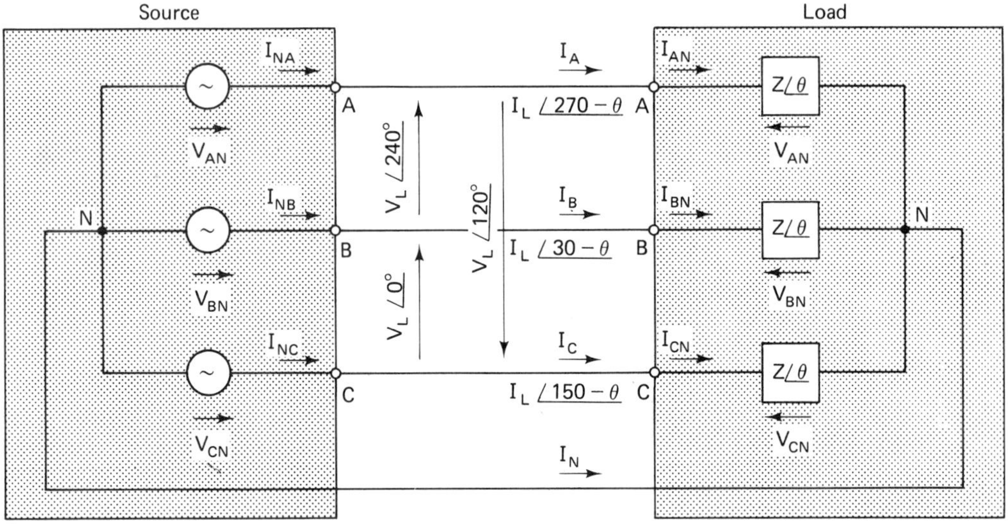

Figure A-5 Balanced Wye Load

With a connection between neutral points of the load and the source, application of Kirchhoff's voltage law to each of the three loops containing the neutral line identifies the voltage across each load phase as that of the corresponding source phase voltage. Hence, the voltage across Z_{AN}, extracted from the CBA voltage triangle, is $V_p\,\underline{/270°} = V_L/\sqrt{3}\,\underline{/270°}$. Applying Ohm's law, the current through Z_{AN} is $(V_L/\sqrt{3}\,Z)\,\underline{/270° - \theta}$; since Z_{AN} is in series with the line, this current is I_A. Similarly, $I_B = (V_L/\sqrt{3}\,Z)\underline{/30° - \theta}$ and $I_C = (V_L/\sqrt{3}\,Z)\underline{/150° - \theta}$.

At the neutral node, I_N must equal the sum of I_A, I_B, and I_C. These currents have a common magnitude I_L and are separated by 120°. Earlier, the statement was made that the sum of three vectors, identical in magnitude but separated by 120°, is zero; this is verified with line currents as follows. Applicable trigonometric identities are

$$\sin(\alpha \pm \beta) = \sin\alpha\cos\beta \pm \cos\alpha\sin\beta$$

and

$$\cos(\alpha \pm \beta) = \cos\alpha\ \cos\beta \mp \sin\alpha\sin\beta$$

$$\begin{aligned} I_N &= I_A + I_B + I_C \\ &= I_N\underline{/270° - \theta} + I_L\underline{/30° - \theta} + I_L\underline{/150° - \theta} \\ &= I_L\Big\{[(0 - \sin\theta) + j(-\cos\theta + 0)] \\ &\quad + \Big[\Big(\frac{\sqrt{3}}{2}\cos\theta + \frac{1}{2}\sin\theta\Big) + j\Big(\frac{1}{2}\cos\theta - \frac{\sqrt{3}}{2}\sin\theta\Big)\Big] \\ &\quad + \Big[\Big(-\frac{\sqrt{3}}{2}\cos\theta + \frac{1}{2}\sin\theta\Big) + j\Big(\frac{1}{2}\cos\theta + \frac{\sqrt{3}}{2}\sin\theta\Big)\Big]\Big\} \\ &= 0 \end{aligned}$$

With no current, the neutral line has no function and can be removed without altering the circuit. Henceforth, it is known that the voltage across an impedance of a *balanced* wye load is the corresponding phase voltage from the *CBA* triangle.

The total load power is the sum of the power in the three impedances. Since $|V_p|$, $|I_L|$, and θ are identical in the three phases,

$$P_t = 3|V_p||I_L|\cos\theta = 3\left(\frac{V_L}{\sqrt{3}}\right)I_L\cos\theta = \sqrt{3}\,V_L I_L\cos\theta$$

Delta Loading. In the delta load of Fig. A-6, the voltage across each impedance is a line voltage. The phase currents are $I_{AB} = I_p\,\underline{/240^\circ - \theta}$, $I_{BC} = I_p\,\underline{/-\theta}$, and $I_{CA} = I_p\,\underline{/120^\circ - \theta}$ with $|I_p| = |V_L/Z|$. Solving for the line current I_A by applying the current law to node A,

$$\begin{aligned} I_A &= I_{AB} - I_{CA} \\ &= I_p\underline{/240^\circ - \theta} - I_p\underline{/120^\circ - \theta} \\ &= I_p\left\{\left[\left(-\frac{1}{2}\cos\theta - \frac{\sqrt{3}}{2}\sin\theta\right) + j\left(-\frac{\sqrt{3}}{2}\cos\theta + \frac{1}{2}\sin\theta\right)\right]\right. \\ &\qquad \left. -\left[\left(-\frac{1}{2}\cos\theta + \frac{\sqrt{3}}{2}\sin\theta\right) + j\left(\frac{\sqrt{3}}{2}\cos\theta + \frac{1}{2}\sin\theta\right)\right]\right\} \\ &= I_p(-\sqrt{3}\sin\theta - j\sqrt{3}\cos\theta) \\ &= \sqrt{3}\,I_p\underline{/270^\circ - \theta} \end{aligned}$$

Similarly, $I_B = \sqrt{3}\,I_p\underline{/30^\circ - \theta}$ and $I_C = \sqrt{3}\,I_p\underline{/150^\circ - \theta}$.

Comparing the delta and wye line currents, assuming the line voltages and phase impedances are identical,

$$\frac{I_Y}{I_\Delta} = \frac{V_L/\sqrt{3}Z}{\sqrt{3}\,V_L/Z} = \frac{1}{3}; \qquad \text{therefore, } I_\Delta = 3I_Y$$

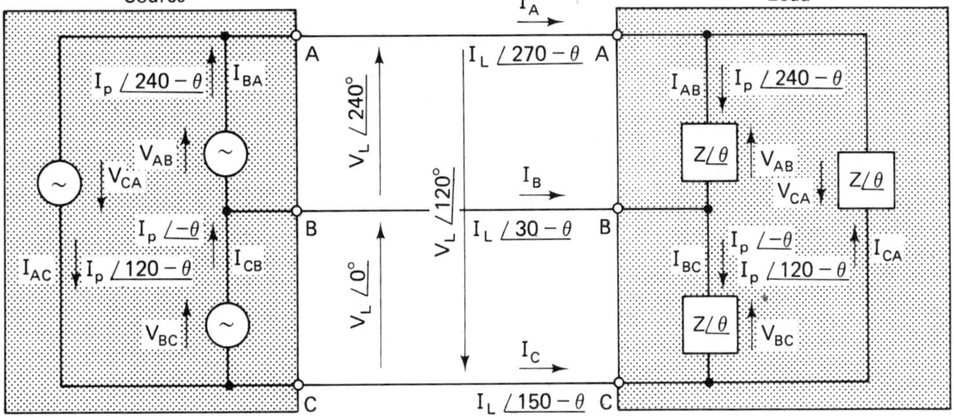

Figure A-6 Balanced Delta Load

Note that the phase angles of the line currents are identical in both wye and delta loading (e.g., $I_A = I_L\underline{/270° - \theta}$). Further, the constants in the expressions for phase angles are the angles of corresponding phase voltages in the standard *CBA* triangle.

The total power in the delta load is

$$P_t = P_A + P_B + P_C = 3V_L I_p \cos\theta = 3V_L\left(\frac{I_L}{\sqrt{3}}\right)\cos\theta$$

$$= \sqrt{3}\,V_L I_L \cos\theta \qquad \text{the same as for wye loading}$$

Example A-1: A 208-V *CBA* system supplies a balanced wye-load with a phase impedance of $10\ \Omega\ \underline{/40°}$. Find the phase voltages and the line currents. What is the total load power?

SOLUTION:

$$|V_p| = \frac{|V_L|}{\sqrt{3}} = \frac{208}{\sqrt{3}} = 120\text{ V}$$

$$V_{AN} = 120\text{ V }\underline{/270°}$$

$$V_{BN} = 120\text{ V }\underline{/30°}$$

$$V_{CN} = 120\text{ V }\underline{/150°}$$

$$I_A = \left(\frac{V_L}{\sqrt{3}\,Z}\right)\underline{/270° - \theta} = \left(\frac{208}{10\sqrt{3}}\right)\underline{/270° - 40}$$

$$= 12\text{ A }\underline{/230°}$$

$$I_B = 12\text{ A }\underline{/-10°}$$

$$I_C = 12\text{ A }\underline{/110°}$$

$$P_t = \sqrt{3}\,V_L I_L \cos\theta$$

$$= (\sqrt{3})(208)(12)(\cos 40°)$$

$$= 3.31\text{ kW}$$

Example A-2: A 4160-V *CBA* system supplies 100 kW to a balanced delta load at a power factor of 0.85. Find I_{AB} and I_A.

SOLUTION: The power factor is the cosine of the impedance angle. Therefore,

$$\theta = \cos^{-1} 0.85 = 31.8°$$

$$P = \sqrt{3}\,V_L I_L \cos\theta$$

$$|I_L| = \frac{P}{\sqrt{3}\,V_L \cos\theta}$$

$$= \frac{100 \times 10^3}{(\sqrt{3})(4160)(0.85)} = 16.33\text{ A}$$

$$I_A = I_L\,\underline{/270° - \theta} = 16.33\text{ A }\underline{/238.2°}$$

$$I_{AB} = \frac{I_L}{\sqrt{3}} \underline{/240° - \theta}$$

$$= \frac{16.33}{\sqrt{3}} \underline{/240° - 31.8} = 9.43 \text{ A } \underline{/208.2}$$

A-5. MEASUREMENT OF THREE-PHASE POWER

There are several ways to insert wattmeters in order to determine the total power in a three-phase system. Since the total power is the sum of the power in the three impedances, by inserting a wattmeter to measure the power in each phase the total power can be determined. However, for a wye load this approach requires access to, or creation of, the neutral point; for the delta load it requires accessibility of the phase current. This method may be impractical because of the physical structure of the load. But total power can be measured with access to line voltages and currents only, using two wattmeters. For the moment, only balanced loading is considered.

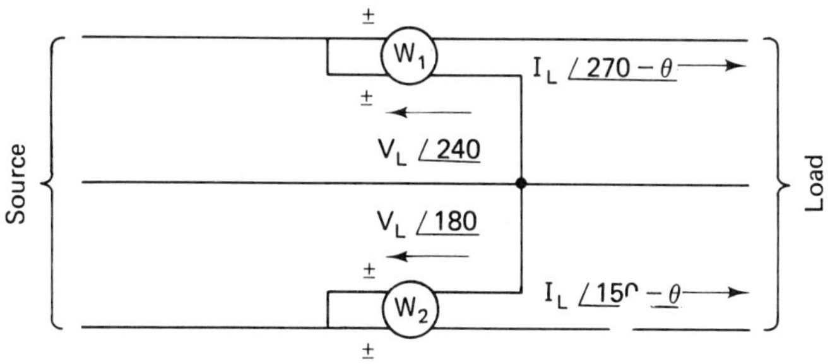

Figure A-7 Two-Wattmeter Power Measurement of Balanced Loads

The wattmeter connections are shown in Fig. A-7. The polarities of the connections are important and must be observed. In W1, the voltage is $V_L\underline{/240°}$, the current is $I_L\underline{/270° - \theta}$, and the power is

$$V_L I_L \cos[240° - (270° - \theta)] = V_L I_L \cos(\theta - 30)$$

In W2,

$$V = V_L\underline{/180°}, I = I_L\underline{/150° - \theta},$$

and

$$P = V_L I_L \cos[180° - (150° - \theta)] = V_L I_L \cos(30 + \theta).$$

Summing the wattmeter readings,

$$W1 + W2 = V_L I_L \cos(\theta - 30) + V_L I_L \cos(30 + \theta)$$

$$= V_L I_L \left[\left(\frac{\sqrt{3}}{2}\cos\theta + \frac{1}{2}\sin\theta\right) + \left(\frac{\sqrt{3}}{2}\cos\theta - \frac{1}{2}\sin\theta\right)\right]$$

$$= \sqrt{3}\, V_L I_L \cos\theta$$

Thus, the sum of the readings is the total three-phase power.

One problem encountered in this method occurs when the power factor is less than 0.5 ($|\theta| > 60°$). If θ is greater than 60°, cos $(30 + \theta)$ is negative, and W2 will deflect down-scale. Similarly, if θ is less than $-60°$, W1 will move down-scale. Should this occur, interchanging the current leads of the down-scale wattmeter will cause it to read up-scale, and the total power is the difference between the wattmeter readings. This circumstance is common with induction motors; running at no load, θ is greater than $+60°$, but as the load increases, θ decreases. In graduated, increasing loading, restoration of the original polarity and reversion to the sum, W1 + W2, is required at some point.

A-6. UNBALANCED LOADING

If at least one phase impedance differs from the others in either magnitude or phase angle, the load is unbalanced. For unbalanced loads, the relationships developed in earlier paragraphs are invalid, and differing analytical techniques must be applied. Circuit solutions are more laborious but use techniques common to single-phase circuitry.

Delta Load. The solution of delta loading is direct inasmuch as the voltage across each phase impedance immediately is known. Phase currents are determined by application of Ohm's law to each phase, and the line currents are found by applying Kirchhoff's current law at each of the three nodes. Power is computed by summing the power of the three phases. (See Example A-3a for an illustrated solution.)

Wye Load. Solution of unbalanced wye loading is more involved than that for the delta because the voltage across each phase initially is unknown. Only in a balanced wye or four-wire system do V_{AN}, V_{BN}, and V_{CN} occur across the phase impedances; unbalanced, the magnitudes and angles depend upon the unique impedance combination. In Fig. A-8, one of numerous possible displacements of the common tie from the neutral point N to a variable point O is illustrated, resulting in phase voltages V_{AO}, V_{BO}, and V_{CO}; the line voltage vectors are inflexible.

Solution of an unbalanced wye load is keyed to determination of either the phase voltages or line currents. Three methods are presented here.

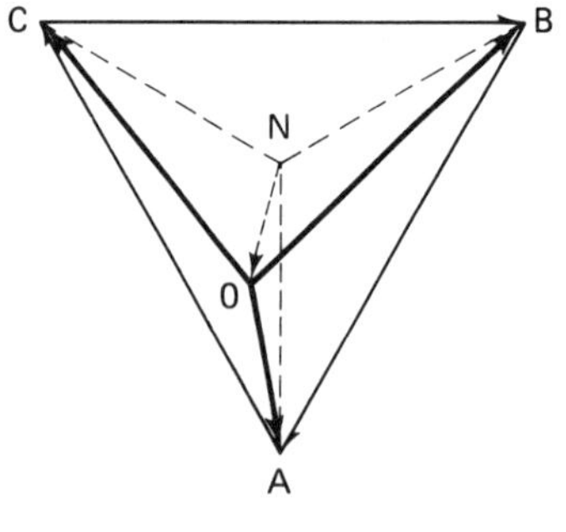

Figure A-8 Displaced Neutral

Equivalent Delta. The relationships for interconversion of wye and delta networks are given in Fig. A-9. Solution of the equivalent delta yields results for the line currents and total power. If required, phase voltages can be obtained by multiplying each line current by the corresponding phase impedance of the original wye. [See Example A-3(a).]

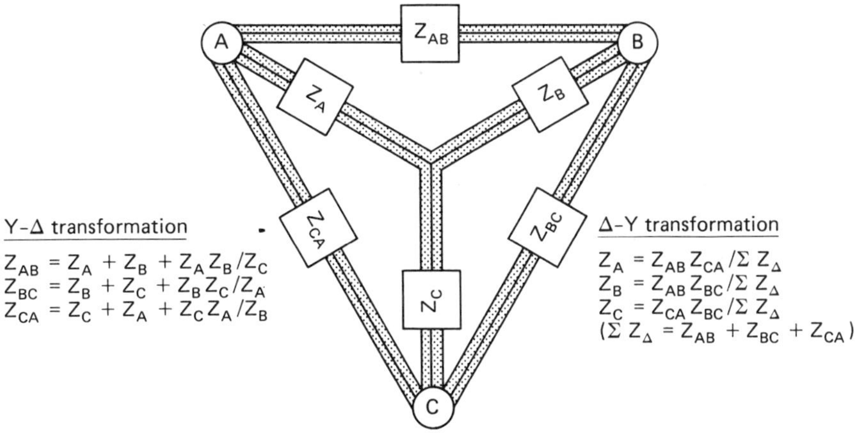

Figure A-9 *Y*–Δ and Δ–*Y* Transformation

Nodal Analysis. In nodal analysis, a ground is assumed at one line terminal and the potentials at the remaining line terminals are defined. A nodal equation, written at the impedance tie point O, is solved for potential V_O. The phase voltages are $V_{AO} = V_A - V_O$, $V_{BO} = V_B - V_O$, and $V_{CO} = V_C - V_O$. [See Example A-3(b).]

Loop Analysis. Loop equations are written for two line currents and solved simultaneously. The third line current is the negative of the sum of these two-currents [See Example A-3(c).]

Example A-3: Solve the circuit of Fig. A-10 by the following methods:

(a) I_A, V_{AO}, and P_A using Y–Δ conversion.
(b) I_B, V_{BO}, and P_B using nodal analysis.
(c) I_C, V_{CO}, and P_C using loop analysis.
(d) $P_t = \sum P_\Phi$.

SOLUTION:

$$\text{(a)}\ Z_{AB} = Z_A + Z_B + \frac{Z_A Z_B}{Z_C} = 10\angle{-30} + 15\angle{30} + \frac{(10\angle{-30})(15\angle{30})}{10\angle{0}}$$

$$= 36.7\ \Omega\angle{3.9^\circ}$$

$$Z_{CA} = Z_C + Z_A + \frac{Z_C Z_A}{Z_B} = 10\angle{0} + 10\angle{-30} + \frac{(10\angle{0})(10\angle{-30})}{15\angle{30}}$$

$$= 24.5\ \Omega\angle{-26.1^\circ}$$

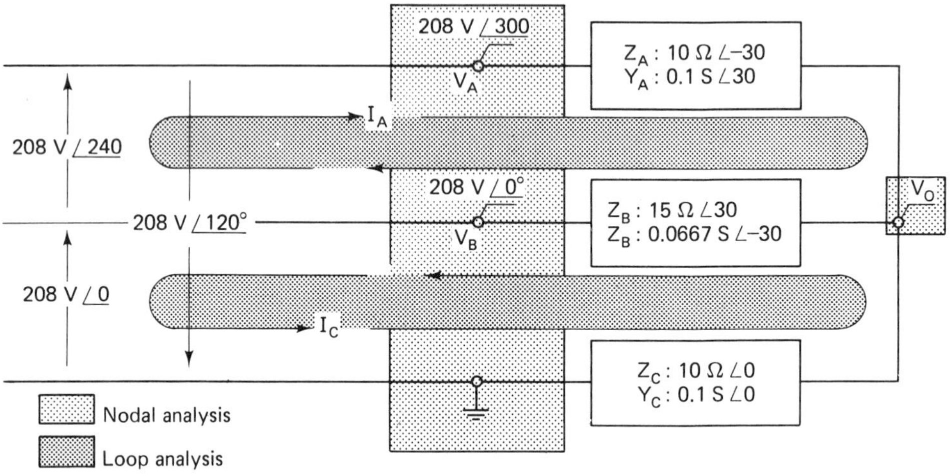

Figure A-10 Example A-3

$$I_A = I_{AB} - I_{CA} = \frac{V_{AB}}{Z_{AB}} - \frac{V_{CA}}{Z_{CA}}$$

$$= \frac{208\underline{/240}}{36.7\underline{/3.9}} - \frac{208\underline{/120}}{24.5\underline{/-26.1}}$$

$$= \mathbf{10.21\ A\underline{/292.4^\circ}}$$

$$V_{AO} = I_A Z_A = (10.21\underline{/292.4})(10\underline{/-30})$$

$$= \mathbf{102.1\ V\underline{/262.4^\circ}}$$

$$P_A = |V_{AO}| \times |I_A| \cos\theta_A = (102.1)(10.21)\cos(-30)$$

$$= \mathbf{0.903\ kW}$$

(b) The assumed ground and potentials V_A, V_B, and V_O appear on the sketch.

$$(V_O - V_A)Y_A + (V_O - V_B)Y_B + V_O Y_C = 0$$

$$V_O = \frac{V_A Y_A + V_B Y_B}{Y_A + Y_B + Y_C} = \frac{(208\underline{/300})(0.1\underline{/30}) + (208\underline{/0})(0.0667\underline{/-30})}{0.1\underline{/30} + 0.0667\underline{/-30} + 0.1\underline{/0}}$$

$$= 141.6\ V\underline{/326.1^\circ}$$

$$V_{BO} = V_B - V_O = 208\underline{/0} - 141.6\underline{/326.1}$$

$$= \mathbf{120.1\ V\underline{/41.1^\circ}}$$

$$I_B = \frac{V_{BO}}{Z_B} = \frac{120.1\underline{/41.1}}{15\underline{/30}}$$

$$= \mathbf{8.01\ A\underline{/11.1^\circ}}$$

$$P_B = (120.1)(8.01)\cos 30 = \mathbf{0.833\ kW}$$

(c) The selected loop currents are I_A and I_C (see the sketch).

$$I_A(Z_A + Z_B) + I_C Z_B = V_{AB}$$

$$I_A Z_B + I_C(Z_B + Z_C) = -V_{BC}$$

$$I_C = \frac{\begin{vmatrix} Z_A + Z_B & V_{AB} \\ Z_B & -V_{BC} \end{vmatrix}}{\begin{vmatrix} Z_A + Z_B & Z_B \\ Z_B & Z_B + Z_C \end{vmatrix}} = \frac{-V_{BC}(Z_A + Z_B) - V_{AB}Z_B}{Z_A Z_B + Z_B Z_C + Z_C Z_A}$$

$$I_C = \frac{208[(-1\underline{/0})(10\underline{/-30} + 15\underline{/30}) - (1\underline{/240})(15\underline{/30})]}{(10\underline{/-30})(15\underline{/30}) + (15\underline{/30})(10\underline{/0}) + (10\underline{/-30})(10\underline{/0})}$$

$$= \mathbf{14.16\ A\underline{/146.1^\circ}}$$

$$V_{CO} = I_C Z_C = (14.16\underline{/146.1})(10\underline{/0})$$

$$= \mathbf{141.6\ V\underline{/146.1^\circ}}$$

$$P_C = (141.6)(14.16)\cos 0 = \mathbf{2.004\ kW}$$

(d) $$P_t = P_A + P_B + P_C = 0.903 + 0.833 + 2.004$$

$$= \mathbf{3.74\ kW}$$

A-7. MEASUREMENT OF POWER IN UNBALANCED LOADS

The techniques for power measurement in balanced loads also apply to unbalanced. Three wattmeters, each measuring the power in one phase, can be used providing (1) the tie point of the wye impedances is accessible for a voltage coil connection or (2) each phase of a delta can be interrupted to insert a current coil.

The two-wattmeter method of Fig. A-7 is equally valid for measurement in unbalanced loading.

An unbalanced wye is illustrated in Fig. A-11. Since a delta can be replaced by an equivalent wye in which the line voltages, currents, and total power are identical to those of the original, proof that two wattmeters correctly measure total power in a wye by sensing line voltages and line currents also proves the truth of the method when applied to the delta.

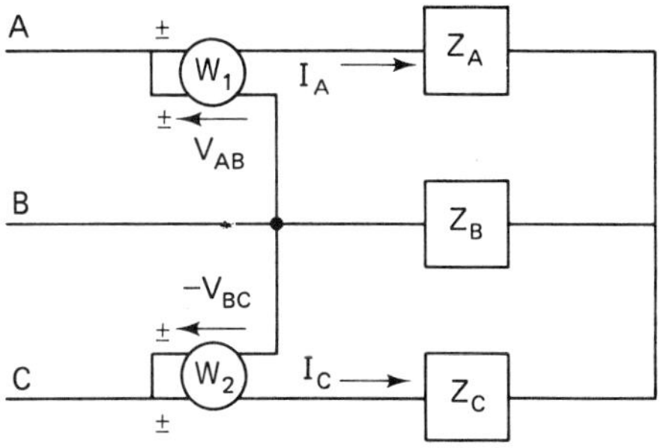

Figure A-11 Two-Wattmeter Power Measurement, Unbalanced

In the figure, the sum of the meter readings is

$$W_1 + W_2 = |V_{AB}||I_A| \cos \angle_A^{AB} + |V_{BC}||I_C| \cos \angle_C^{-BC}$$

In Fig. A-12, the vector V_{AB} is projected into the extension of I_A, establishing the segment $V_{AB} \cos \angle_A^{AB}$. $V_{AB} = V_{AO} - V_{BO}$; these vectors also appear on the diagram, projected into the I axis. The geometry shows that

$$|V_{AB}| \cos \angle_A^{AB} = |V_{AO}| \cos \angle_A^{AO} + |V_{BO}| \cos \angle_A^{-BO}$$

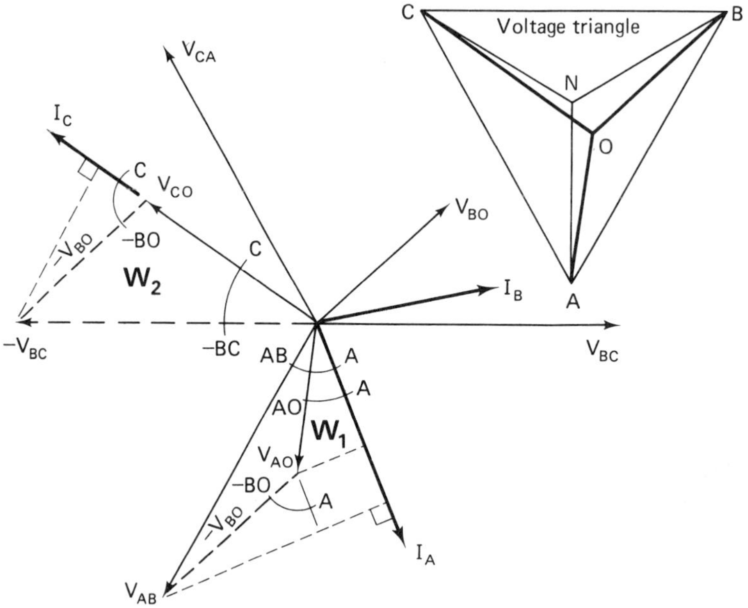

Figure A-12 Vector Diagram, Unbalanced Power Measurement

By a similar construction, it is shown that

$$|V_{BC}| \cos \angle_C^{-BC} = |V_{CO}| \cos \angle_C^{CO} + |V_{BO}| \cos \angle_C^{-BO}$$

Substituting these values in the power equation,

$$W_1 + W_2 = |V_{AO}||I_A| \cos \angle_A^{AO} + |V_{CO}||I_C| \cos \angle_C^{CO} + |V_{BO}|(I_A \cos \angle_A^{-BO} + I_C \cos \angle_C^{-BO})$$

The first two terms are the power in phases A and C, respectively. Recognizing that $I_A + I_C = -I_B$, the expression

$$(I_A \cos \angle_A^{-BO} + I_C \cos \angle_C^{-BO}) = -I_B \cos \angle_B^{-BO}$$

Therefore, the third term is the power in phase B and the sum of the wattmeters, $W_1 + W_2$, is the total power in the circuit.

B

Conversion Factors

[*US*]	× [*Multiplier*]	= [*SI*]
degree (angle) °	$1.745\ 329 \times 10^{-2}$	radian (rad)
degree Fahrenheit (F)	$t_C = (t_F - 32)/1.8$	degree Celsius (°C)
	$t_K = t_C + 273.15$	degree Kelvin (K)
foot (ft)	$3.048\ 000 \times 10^{-1}$	meter (m)
$foot^2$ (ft^2)	$9.290\ 304 \times 10^{-2}$	$meter^2$ (m^2)
$foot^3$ (ft^3)	$2.831\ 685 \times 10^{-2}$	$meter^3$ (m^3)
foot-pound (ft-lb)	1.355 818	joule (J)
foot-pound/second (ft-lb/sec)	1.355 818	watt (W)
horsepower (hp)	$7.460\ 000 \times 10^{2}$	watt (W)
inch (in)	$2.540\ 000 \times 10^{-2}$	meter (m)
$inch^2$ (in^2)	$6.451\ 600 \times 10^{-4}$	$meter^2$ (m^2)
$inch^3$ (in^3)	$1.638\ 706 \times 10^{-5}$	$meter^3$ (m^3)
$kiloline/inch^2$	$1.550\ 003 \times 10^{-2}$	tesla (T)
kilowatt-hour (kW-hr)	$3.600\ 000 \times 10^{6}$	joule (J)
line or maxwell	$1.000\ 000 \times 10^{-8}$	weber (Wb)
mile (mi)	$1.609\ 344 \times 10^{3}$	meter (m)
pound-foot (lb-ft)	1.355 818	newton-meter (N•m)
pound (force)	4.448 222	newton (N)
pound (mass) (lb)	$4.535\ 924 \times 10^{-1}$	kilogram (kg)

Note: The following units are identical is US and SI: farad (F), henry (H), siemen (S), volt (V), ohm (Ω), hertz (Hz), coulomb (C).

C

SI Prefixes

Multiple	*Term*	*Symbol*
10^{18}	exa	E
10^{15}	peta	P
10^{12}	tera	T
10^{9}	giga	G
10^{6}	mega	M
10^{3}	kilo	k
10^{2}	hecto*	h
10^{1}	deka*	da
10^{-1}	deci*	d
10^{-2}	centi*	c
10^{-3}	milli	m
10^{-6}	micro	μ
10^{-9}	nano	n
10^{-12}	pico	p
10^{-15}	femto	f
10^{-18}	atto	a

*Use of these prefixes should be avoided where possible. They have been approved only because they are in broad common usage with certain units—for example, cubic centimeter. Primarily, use should be confined to units of area and volume.

D

Mathematical Summary

1. FUNCTIONS OF AN ANGLE: FIG. D-1
2. SUBTENDED ARCS: FIG. D-2

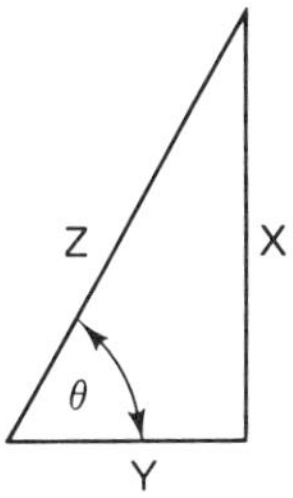

$$\sin\theta = \frac{X}{Z}$$

$$\cos\theta = \frac{Y}{Z}$$

$$\tan\theta = \frac{X}{Y} = \frac{\sin\theta}{\cos\theta}$$

Figure D-1 Functions of an Angle

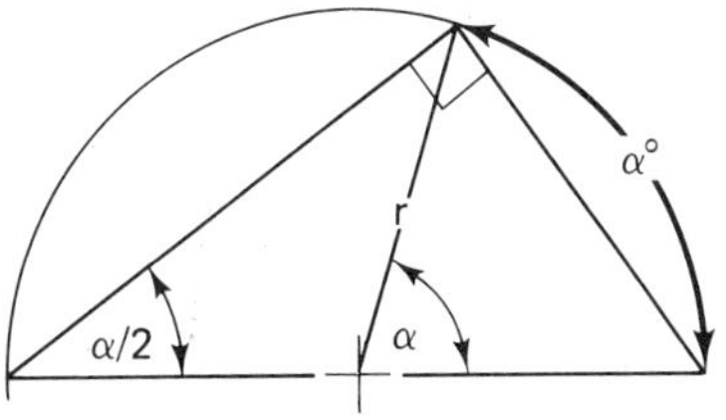

Figure D-2 Subtended Arcs

3. SINE AND COSINE CURVES: FIG. D-3
4. IDENTITIES:

$$\sin^2\alpha + \cos^2\alpha = 1$$

$$\sin^2\alpha = \tfrac{1}{2}(1 - \cos 2\alpha)$$

$$\cos^2\alpha = \tfrac{1}{2}(1 + \cos 2\alpha)$$

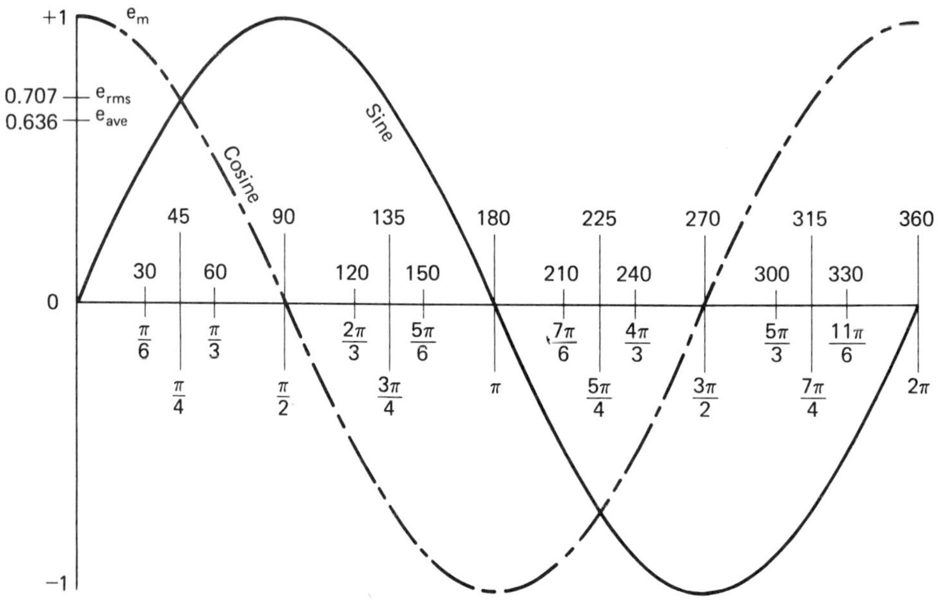

Angle	Sine	Cosine
$-\alpha$	$-\sin\alpha$	$\cos\alpha$
$90-\alpha$	$\cos\alpha$	$\sin\alpha$
$\alpha-90$	$-\cos\alpha$	$\sin\alpha$
$90+\alpha$	$\cos\alpha$	$-\sin\alpha$
$180-\alpha$	$\sin\alpha$	$-\cos\alpha$
$\alpha\pm 180$	$-\sin\alpha$	$-\cos\alpha$
$270-\alpha$	$-\cos\alpha$	$-\sin\alpha$
$270+\alpha$	$-\cos\alpha$	$\sin\alpha$
$\alpha\pm\beta$	$\sin\alpha\cos\beta\pm\cos\alpha\sin\beta$	$\cos\alpha\cos\beta\mp\sin\alpha\sin\beta$

Figure D-3 Sine and Cosine Curves

5. VECTOR OPERATIONS

a. Retangular ⟷ polar conversion:

$$R\underline{/\theta} = R\cos\theta + jR\sin\theta$$

$$x + jy = \sqrt{x^2+y^2}\ \underline{/\tan^{-1}\frac{y}{x}}$$

b. Addition (Fig. D-4):

$$A\ \underline{/\alpha} + B\ \underline{/\beta} = (A\cos\alpha + B\cos\beta) + j(A\sin\alpha + B\sin\beta)$$

c. Subtraction (Fig. D-4):

$$A\ \underline{/\alpha} - B\ \underline{/\beta} = (A\cos\alpha - B\cos\beta) + j(A\sin\alpha - B\sin\beta)$$

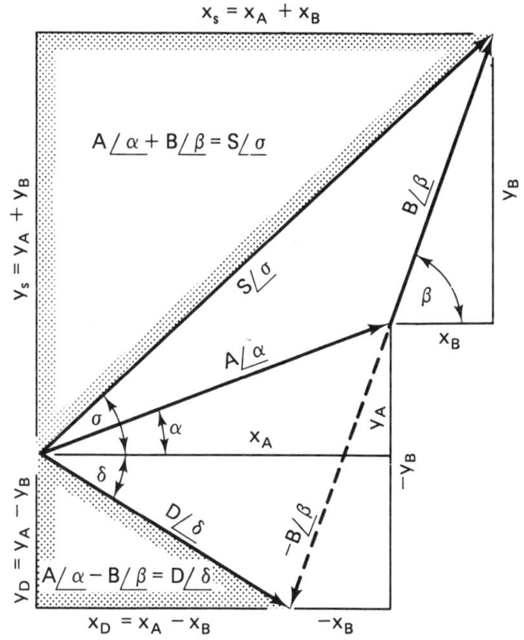

Figure D-4 Vector Addition and Subtraction

d. Multiplication (Fig. D-5):

$$A\,\underline{/\alpha} \times B\,\underline{/\beta} = (A \times B)\underline{/\alpha + \beta}$$

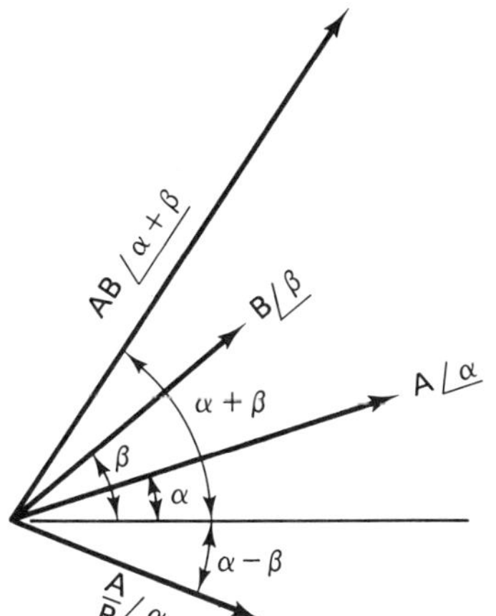

Figure D-5 Vector Multiplication and Division

e. Division (Fig. D-5):

$$\frac{A\,\underline{/\alpha}}{B\,\underline{/\beta}} = \frac{A}{B}\,\underline{/\alpha - \beta}$$

6. QUADRATIC EQUATION ($ax^2 + bx + c = 0$):

$$x = \frac{-b \pm \sqrt{b^2 - 4ac}}{2a}$$

7. DETERMINANT SOLUTION OF SIMULTANEOUS EQUATIONS:

a. Evaluation: $\begin{vmatrix} a_1 & b_1 \\ a_2 & b_2 \end{vmatrix} = (a_1b_2 - a_2b_1)$

$$\begin{vmatrix} a_1 & b_1 & c_1 \\ a_2 & b_2 & c_2 \\ a_3 & b_3 & c_3 \end{vmatrix} = (a_1b_2c_3 + b_1c_2a_3 + c_1a_2b_3) - (a_1c_2b_3 + b_1a_2c_3 + c_1b_2a_3)$$

b. $a_1x + b_1y = k_1$

$a_2x + b_2y = k_2$

$$x = \frac{\begin{vmatrix} k_1 & b_1 \\ k_2 & b_2 \end{vmatrix}}{\begin{vmatrix} a_1 & b_1 \\ a_2 & b_2 \end{vmatrix}} \qquad y = \frac{\begin{vmatrix} a_1 & k_1 \\ a_2 & k_2 \end{vmatrix}}{\begin{vmatrix} a_1 & b_1 \\ a_2 & b_2 \end{vmatrix}}$$

c. $a_1x + b_1y + c_1z = k_1$

$a_2x + b_2y + c_2z = k_2$

$a_3x + b_3y + c_3z = k_3$

$$D = \begin{vmatrix} a_1 & b_1 & c_1 \\ a_2 & b_2 & c_2 \\ a_3 & b_3 & c_3 \end{vmatrix}$$

$$x = \frac{\begin{vmatrix} k_1 & b_1 & c_1 \\ k_2 & b_2 & c_2 \\ k_3 & b_3 & c_3 \end{vmatrix}}{D}$$

$$y = \frac{\begin{vmatrix} a_1 & k_1 & c_1 \\ a_2 & k_2 & c_2 \\ a_3 & k_3 & c_3 \end{vmatrix}}{D}$$

$$z = \frac{\begin{vmatrix} a_1 & b_1 & k_1 \\ a_2 & b_2 & k_2 \\ a_3 & b_3 & k_3 \end{vmatrix}}{D}$$

E

Answers to Selected Problems

CHAPTER 1

1-3. (a) 1.829 m, (b) 801 N, (c) 4050 m^2, (d) 907 kg, (e) 9.78 m/s^2
1-5. 3.19×10^{-3}.
1-7. 2.9 A.
1-9. 5.66 A.
1-11. 354 A, 3/1
1-17. (a) 1.437 A, (b) 0.142 A.
1-21. (a) 70.1 hp, (b) 181.3 lb-ft.
1-23. (a) 10 N, (b) 2.5 N•m, (c) 6 kW, 5.4 kW, 0.6 kW

CHAPTER 2

2-1. (a) 2.22 $\omega\theta/\pi$, (b) 4.44 $\pi S/60$.
2-3. (a) 64.8 kl/in^2, 1.004 T, (b) 5.52 V, 20 Hz.
2-5. 4.67 V.
2.7. (a) 522 V, (b) 261 V, (c) 130.5 V.
2-9. 13%.
2-11. 0.175(17.5%).
2-13. 333 V.
2-15. (a) 4910 A, (b) 8.75×10^{-3}, (c) 910 A.

2-17. (a) 250 V, 21 kW, (b) 88.5%
2-19. 135 V, 8 V.
2-21. (a) (1) ±0.355 V, (2) No, (3) 138.8 V; (b) (1) 33.3, (2) 1.17 A; (c) (1) 2.58A, 23.4 Ω; (d) 120 V, −4.0%; (e) 77.4%.
2-27. (a) 288 W, (b) 500 W.
2-29. 0.036 Ω.
2-31. 95.8 lb-ft.
2-33. (a) 23.6 V.
2-37. (a) 50 A, (b) 6,200A.
2-39. (a) 39.7 hp, 119.1 lb-ft, 88.1% (b) 2044 r/min, 16.8%.
2-41. (a) 71.6 N•m, 12.83 kW; (b) 53.0 lb-ft, 12.1 hp.
2-43. (a) 0.703 lb-ft, 0.293 lb-ft; (b) 351 lb-ft.
2-45. 210.9 V, 18.98 kW.
2-47. 8 A, 615 A, 623 A, 12.21×10^3 A.
2-49. (a) 4.8 V, 72 W; (b) 0.382 N•m, 72 W, 0.282 lb-ft, 0.0965 hp.
2-51. 1725 r/min.
2-53. (a) 25.7 A, (b) 73.6% decrease.
2-57. (Shunt) (a) 123.1 V, (b) 22.5 Ω, (c) 124.8 V, 3.2 A, (d) (1) 1498 r/min, (2) 1774 r/min, (e) 16.2 lb-ft, (f) 1.4%, (g) 83.5%, (h) 5.4 hp, (i) 33.75 lb-ft

CHAPTER 3

3-1. $V_L = V_f k_f k_a / R_f R_{ac} - I_L R_{ac}$.
3-3. $G_1G_2G_3/(1 + G_2H)$.
3-5. Zero.
3-7. (a) $40 \times 10^3 / (R_L + 0.2 + 4 \times 10^3)$, (b) 2.0126 V, (c) 20.07 A, 20.12 A.
3-9. (a) 210 lb-ft, (b) 5.86 hp.

CHAPTER 4

4-1. 2,129 turns, 387 turns.
4-3. (a) 417 A, 1000 kW; (b) 417 A, 900 kW; (c) 417A, 850 kW.
4-5. (a) 1.5 V, (b) 160 turns, (c) 5.63×10^{-3} Wb. (d) 4.69×10^{-3} m^2, (e) 20 kVA.
4-7. (a) 36.2 A, 120.2 A; (b) 29.0 A, 96.2 A.
4-9. (a) 1.4%, (b) 3.0%, (c) −0.07%.
4-11. 15 V, 7.5 V.
4-13. 98.7%.
4-17. (a) 12.6Ω, 18.5Ω; (b) 1.2%; (c) 2 W, 615 W; (d) 245 W (all loading); (e) 98.1%
4-19. 5 kW.
4-24. 13.2/69 kV, 75.8/14.5 A.
4-27. 725 A, 1450 A.
4-31. (a) B, (b) 165.7 A, 72 kVA.
4-33. (a) B, (b) 344 A, 71.6 kVA

CHAPTER 5

5-1. 900 r/min.
5-2. 8, 3.
5-3. 1440 turns/phase.
5-4. 20°.
5-5. 7 (kp = 0.940) or 8 (kp = 0.985).
5-6. 0.960.
5-7. 0.946.
5-8. 2400 V.
5-9. 4160 V.
5-10. 1873kVA.
5-13. 11kV, 50 Hz, 83.3 MVA.
5-19. (a) 52.2%, 112 A; (b) 34.1%, 90 A, (c) 53.6%, 115 A.
5-21. (a) 10:1, (b) 12:1.
5-23. (a) 21.9 × 10^3 lb-ft (b) 20.2°, (c) 31.9°, (d) 317 A.
5-26. (a) 4086 hp, 94.3%, 23.9 × 10^3 lb-ft; (b) 2400 kVAr.
5-29. (a) 0.974, (b) 630 kVA, 1200 kW.

CHAPTER 6

6-1. (a) 900 r/min, (b) 50 r/min, (c) 0.0556.
6-3. (a) 0.6 Hz, (b) 2.5 Hz, (c) 60 Hz.
6-5. (Answers for rated conditions of 1160 r/min, 117 A): (a) 0.0283, (b) 0.899, (c) 3.29 kW, (d) 80.31 kW (e) 2.28 kW, (f) 78.04 kW, (g) 74.70 kW, (h) 100.1 hp, (i) 451 lb-ft, (j) 89.4%.
6-9. (a) 700 A, 535 lb-ft; (b) 560 A, 362 lb-ft.
6-11. 448 A, 560 A, 357 kVA.
6-13. 1200, 720, 450 r/min

CHAPTER 7

7-1. (a) 38.4 oz-in, (b) 875 r/min, 33.3 mph.
7-3. (a) 181 μf, (b) 0.167V$\angle -40$, 0.068V$\angle 50$, 0.181V$\angle -17.7$.
7-8. (a) +75°, (b) − 1590.

CHAPTER 8

8-1. (*a* only) 101 V, 118.5 V, 61.3%
8-2. (*a* only) 243 V, 247 V, 18.2%.
8-3. (1000 W only) Rheostat—83.3%, Rectifier—96.9%, α = 0.387 rad
8-7. (a) (1) 3ϕ—0.688 rad, 1ϕ—0.433 rad (2) 3ϕ—0.688 rad, 1ϕ—1.474 rad; (3) 3ϕ—1.167 rad, 1ϕ—0.433 rad; (b) 338 W, 332 W, 5.91 kW.

Index

A

AC motor (*see* Induction motor, Synchronous motor, Single-phase motors)
Alternator (*see* Synchronous generator)
Amplidyne, 106
Analytical methods, 25-28
Angular velocity, 26
ANSI, 28
Autotransformer, 150

B

Base speed, 89
Block diagram, 120
Breakdown torque, 244
Brushes, 67

C

Chopper, 313
Circuit analysis, ac, 299-306
- average, rms, and ripple voltages, 303
- steady-state, 299
- transient-state, 301

Circuit analysis, polyphase, 323-336
- CBA system, 325
- currents:
  - balanced delta, 329
  - balanced wye, 327
  - unbalanced, 332
- power:
  - balanced delta, 330
  - balanced wye, 329
  - unbalanced, 335
- power measurement:
  - balanced load, 331
  - unbalanced load, 335

Coercive force, 19
Commutation (*see* DC machine)
Control motor, ac, 292
Control transformer, 287
Conversion accuracy, 4
Conversion factors, 338
Converter (*see* Solid-state power and control)
Core loss, 20
Cycloconverter, 317

D

DC generator, 34-66
- armature current, 40

DC generator *(cont.)*:
- armature reaction, 50
- commutating pole, 53
- compensating winding, 54
- construction (*see* DC machine)
- developed power, 40
- effect of speed on E_g, 49
- field windings, 41
- forms, 41-49
  - compound, 46
  - degree of compounding, 56
  - shunt:
    - self-excited, 43
    - separately excited, 41
- generated voltage, 39
- homopolar, 107
- line voltage characteristics, 55
- operational principles, 34
- parallel paths, 39
- paralleling, 61-66
  - shunt, 62
  - compound, 64
- saturation curve, 44
- three-wire, 103
- torque, 60
- tuned, 107

DC machine, 34-119
- commutation, 76
- commutator segment, 68
- construction, 66
- windings, 67-76
  - frog-leg, 76
  - lap, 68
  - wave, 74

DC motor, 78-102
- construction, 66
- electrical braking, 92-94
  - dynamic, 93
  - plugging, 93
  - regenerative, 94
- jogging, 100
- power, developed and output, 79
- speed, 79
- speed control, 89-92
  - armature, 89
  - field, 90
  - Ward–Leonard, 101
- speed, no-load, 80
- speed regulation, 81
- speed-torque characteristics, 81

DC motor *(cont.)*:
starters and controllers, 94-102
above-and-below-base speed control, 96
automatic starter, 97
jogging, 100
three-point starter, 95
starting, 85
starting current, 85
starting torque, 88
torque, 79
Differential receiver, 287
Differential transmitter, 285
Diode (*see* Solid-state power and control)
Dual converter, 315
Dynamotor, 104

E

Eddy currents, 20
Efficiency, 43
Electromagnet, 7
Electromagnetic force, 21
Electromagnetic torque, 23
Electromagnetism, 6
Electromotive force:
induced, 16
Lenz's law, 17
Energy, 26-28
Energy balance, 27

F

Feedback regulation and control, 120-134
block algebra, 120
position control, 128-133
damping, 131
response, 131
stability, 131
stall torque, 130
regulation, 123-128
speed control, 128
transducer, 123
transfer function, 121
Flux, definition, 6
Flux density, 6
Force, electromagnetic, 21
Frequency converter, 316

G

Generator (*see* DC generator, Synchronous generator)

H

Horsepower, 27
Hysteresis, 18-19
Hysteresis motor, 278

I

IEEE, 28
Induction motor, polyphase, 230-264
control circuits, 255
design letter-class, 244
double-cage rotor, 243
efficiency, 237
electrical braking:
dynamic, 255
plugging, 253
regenerative, 255
equivalent circuit, 232
Thévenin rotor, 236
operational principles, 230
performance analysis, 237
power generation, 253
ratings and application, 261
rotor, 235
rotor power input (RPI), 237
rotor resistance, effect, 241
rotor speed, 232
service factor, 262
speed control, 248-253
auxiliary methods, 252
frequency, 252
pole-changing, 250
slip, 249
slip, 232
for maximum power, 240
for maximum torque, 240
slip speed, 232
starting current, 244
starting methods, 245-248
part-winding, 247
reduced voltage, 245
wound rotor, 247
wye-delta, 247
starting torque, 242, 245

Induction motor *(cont.)*:
synchronous speed, 232
testing, 260-61
torque:
definitive, 244
developed, 239
maximum, 240
starting, 242, 245
Induction motor, single-phase (*see* Single-phase motors)
Induction voltage regulator, 168
Inverter, 314

L

Lenz's law, 17

M

Magnetic amplifier, 166
Magnetic characteristics of materials, 8-10, 11
atomic basis, 8
classification, 8
domain theory, 9
permeability, 11
retentivity, 10
Magnetic circuit, 10-16
field intensity, 12
Kirchhoff's laws, 12
magnetomotive force, 10
Ohm's law, 11
permeability, 11
reluctance, 11
Magnetic domain theory, 9
Magnetic field intensity (*H*), 12
Magnetic force:
on a current-carrying conductor, 21
tractive, 24
Magnetic saturation, 9
Magnetism, 4-8
flux, 6
flux density, 6
force line concept, 4
residual, 10
Magnetomotive force, 10
Mathematical summary, 339-342
Mechanical analysis, 26-27

N

NEC, 28
NEMA, 28
NFPA, 28

P

Permeability, 11
Polyphase circuits, 323-336
advantages, 182
Power, 26
Power factor correction, 223

R

Receiver (synchro), 283
Rectifier (*see* Solid-state power and control)
Reluctance, 11
Reluctance motor, 276
Reluctance torque, 202
Repulsion motor, 279
Residual magnetism, 10
Resolver, 289
Retentivity, 10
Revolving flux, 184
Rotating amplifier, 105

S

Saturable reactor, 165
Saturation curve, 12
Speed characteristics of loads, 92
Servo components, 281-293
ac control motor, 292
resolver, 289
synchro, 281-288
control transformer, 287
differential receiver, 287
differential transmitter, 285
receiver, 283
transmitter, 282
Servomechanism, 132
Shaded-pole motor, 274
SI prefix, 338
Single-phase motors, 265-281
hysteresis, 278
induction, 265-276
operating principles, 265
phase-splitting, 268

Single-phase motors *(cont.):*
phase-splitting types:
capacitor-start, 272
permanent capacitor, 273
split phase, 271
two-value capacitor, 274
shaded pole, 274
reluctance, 276
repulsion, 279
repulsion-start, induction-run, 280
repulsion-induction, 280
universal, 280
Solid-state power and control, 298-322
circuit analysis, 299-306
diode, 306
free-wheeling, 307
diode circuits, 306-310
full-wave rectifier, 308
half-wave rectifier, 307
three-phase double-way rectifier, 309
three-phase single-way rectifier, 309
form factor, 319
letter-class power supplies, 319
motor applications, 319
pulse number, 306
switching function, 298
thristor (SCR), 311-317
thyristor circuits
chopper, 313
cycloconverter, 317
dual converter, 315
frequency converter, 316
inverter, 314
single-phase rectifier, 311
three-phase rectifier, 313
universal motor control, 314
transistor, 317-18
volt-time area, 301
Standards, 28
Synchro (*see* Servo components)
Synchronous generator, 176-198 (*see also* Synchronous machine)
excitation systems, 188
generated voltage, 176-182
loading, 195
parallel operation:
on an infinite bus, 215
two generators, 216
Synchronous generator *(cont.):*
rating and application, 212
synchronization, 217
synchronizing procedure, 222
voltage regulation, 190
Synchronous machine, 176-229
air-gap line, 192
amortisseur, 203
armature reaction, 186
coil pitch, 178
construction, 177
distribution factor, 179
efficiency, 196
equivalent circuit, 186
frequency, 179
generated voltage, 179
hunting, 208
open-circuit voltage, 192
phase belt, 178
pitch factor, 179
pole pitch, 178
power, 194
ratings and application, 212
revolving flux, 184
short-circuit current, 192
slot span, 178
stray-load loss, 197
synchronous impedance, 192
synchronous reactance, 188
testing, 208-212
open circuit, 209
resistance measurement, 209
short circuit, 211
torque, 194
torque angle, 195
vector diagrams, 201
winding factor, 179
winding impedance, 188
Synchronous motor, 199-208 (*see also* Synchronous machine)
circuit analysis, 200-203
maximum torque, 201
operating principles, 199-200
polarized frequency relay, 204
power factor, 207
power factor correction, 224
ratings and application, 214
torques, 206
starting, 203

T

Tachometer, ac, 293
Thyristor (*see* Solid-state power and control)
Torque, 26
Torque, electromagnetic, 21
Tractive force, magnetic, 24
Transformer, 135-175
  all-day efficiency, 144
  autotransformer, 150
  construction, 162
  efficiency, 144
  equivalent circuit:
    realistic, 139
    simplified, 140
  exciting current, 139
  functional principles, 138
  impedance, 140
  leakage flux, 140
  mutual flux, 140
  paralleling, 161
  polarity, 149
  polyphase, 154-161
    delta-delta, 154
    delta-wye, 157
    open delta, 155
    Scott-T, 160
    wye-delta, 157
    wye-wye, 155
Transformer *(cont.)*:
  power losses, 144
  ratings, 164
  role in power distribution, 135
  tapped, 148
  testing, 146-148
  turns ratio, 139
  vector diagram, 142
  voltage ratio, 139
  voltage regulation, 143
  winding impedance, 140
Transmitter (synchro), 282

U

Units, US and SI, 2-4
  conversion factors, 337
Universal motor, 280
Universal motor control, 314

V

Voltage classes, industrial and commercial, 137
Voltage regulation, 57
Volt-time area, 301

W

Work, 26